Lecture Notes in Computer Science 12049

More information about this series at http://www.springer.com/series/7407

M. Sohel Rahman · Kunihiko Sadakane ·
Wing-Kin Sung (Eds.)

WALCOM: Algorithms and Computation

14th International Conference, WALCOM 2020
Singapore, Singapore, March 31 – April 2, 2020
Proceedings

 Springer

Editors
M. Sohel Rahman (iD)
Department of CSE
BUET
Dhaka, Bangladesh

Kunihiko Sadakane (iD)
Mathematical Informatics
University of Tokyo
Tokyo, Japan

Wing-Kin Sung (iD)
School of Computing
National University of Singapore
Singapore, Singapore

ISSN 0302-9743   ISSN 1611-3349 (electronic)
Lecture Notes in Computer Science
ISBN 978-3-030-39880-4   ISBN 978-3-030-39881-1 (eBook)
https://doi.org/10.1007/978-3-030-39881-1

LNCS Sublibrary: SL1 – Theoretical Computer Science and General Issues

This Springer imprint is published by the registered company Springer Nature Switzerland AG
The registered company address is: Gewerbestrasse 11, 6330 Cham, Switzerland

Preface

This proceedings volume contains papers presented at WALCOM 2020, the 14th International Conference and Workshops on Algorithms and Computation, held during March 31 – April 2, 2020, at the Institute for Mathematical Science, National University of Singapore (NUS), Singapore.

WALCOM 2020 was organized in cooperation with the Institute of Mathematical Sciences (NUS), School of Computing (NUS), Special Interest Group for ALgorithms (SIGAL) of the Information Processing Society of Japan (IPSJ), and Technical Committees on Theoretical Foundations of Computing (COMP) of the Institute of Electronics, Information and Communication Engineers (IEICE).

The range of topics within the scope of WALCOM includes but is not limited to: Approximation Algorithms, Algorithmic Graph Theory and Combinatorics, Combinatorial Algorithms, Combinatorial Optimization, Computational Biology, Computational Complexity, Computational Geometry, Discrete Geometry, Data Structures, Experimental Algorithm Methodologies, Graph Algorithms, Graph Drawing, Parallel and Distributed Algorithms, Parameterized Algorithms, Parameterized Complexity, Network Optimization, Online Algorithms, Randomized Algorithms, and String Algorithms.

WALCOM is an annual conference series that provides an international forum for researchers working in the areas of algorithms and computation. The first edition (WALCOM 2007) was held on February 12, 2007, in Dhaka, Bangladesh, and was organized by the Bangladesh Academy of Sciences. WALCOM 2008, WALCOM 2010, WALCOM 2012, WALCOM 2015, and WALCOM 2018 were held in Bangladesh; whereas WALCOM 2009, WALCOM 2011, WALCOM 2013, WALCOM 2014, and WALCOM 2019 were held in India. WALCOM 2016 was held in Nepal and WALCOM 2017 was held in Taiwan.

WALCOM is led by a strong Steering Committee, whose members are Kyung-Yong Chwa (KAIST, South Korea), Costas S. Iliopoulos (KCL, UK), M. Kaykobad (BUET, Bangladesh), Petra Mutzel (TU Dortmund, Germany), Shin-ichi Nakano (Gunma University, Japan), Subhas Chandra Nandy (Indian Statistical Institute, Kolkata, India), Takao Nishizeki (Tohoku University, Japan), C. Pandu Rangan (IIT Madras, India), and Md. Saidur Rahman (BUET, Bangladesh).

The WALCOM Program Committees comprise computer scientists of international repute from different parts of the globe. Notably, the Program Committee of WALCOM 2020 comprised 34 eminent researchers from Australia, Bangladesh, Canada, Finland, France, Greece, Hong Kong, India, Israel, Italy, Japan, the Netherlands, Poland, Singapore, South Africa, South Korea, Taiwan, the UK, and the USA.

The technical program was finalized by selecting the highest quality papers from among 66 submitted papers. We had a rigorous review process where all papers had at least three reviews while most of the papers had four. Followed by an in-depth discussion by the Program Committee, this year we could only accept 23 full papers and

4 short papers. We selected "Shortest Covers of All Cyclic Shifts of a String" to be the best paper and "Approximability of the Independent Feedback Vertex Set Problem for Bipartite Graphs" to be the best student paper. Both paper awards were presented at the conference.

In addition to the 27 contributed talks, the scientific program of WALCOM 2020 included invited talks by 3 eminent researchers, namely, Prof. Osamu Watanabe (Tokyo Tech, Japan), Prof. Louxin Zhang (NUS, Singapore), and Prof. Md. Saidur Rahman (BUET, Bangladesh). We are extremely grateful to our invited speakers for their excellent talks at the conference. We thank all the authors who submitted their works for consideration to WALCOM 2020. We deeply appreciate the contribution of all Program Committee members and external reviewers for handling the submissions in a timely manner despite their extremely busy schedules. We must acknowledge the EasyChair conference management system again for providing us with their celebrated platform for conference administration. We are grateful to Springer for publishing the proceedings of WALCOM 2020, making the event a grand success.

April 2020   M. Sohel Rahman
                                                                  Kunihiko Sadakane
                                                                    Wing-Kin Sung

Organization

Program Committee

Hee-Kap Ahn	POSTECH, South Korea
Md. Shamsuzzoha Bayzid	BUET, Bangladesh
Guillaume Blin	Université de Bordeaux, LaBRI, UMR, CNRS, France
Hans Bodlaender	Utrecht University, The Netherlands
Gautam K. Das	IIT Guwahati, India
Jackie Daykin	Aberystwyth University, UK
Mark de Berg	Eindhoven University of Technology, The Netherlands
Naveen Garg	IIT Delhi, India
Wing-Kai Hon	National Tsing Hua University, Taiwan
Seok-Hee Hong	The University of Sydney, Australia
Jesper Jansson	The Hong Kong Polytechnic University, Hong Kong, China
Ralf Klasing	CNRS, University of Bordeaux, France
Gad M. Landau	University of Haifa, Israel, and NYU, USA
Inbok Lee	Korea Aerospace University, South Korea
Giuseppe Liotta	University of Perugia, Italy
Takaaki Mizuki	Tohoku University, Japan
Debajyoti Mondal	University of Saskatchewan, Canada
Krishnendu Mukhopadhyaya	Indian Statistical Institute, India
Shin-Ichi Nakano	Gunma University, Japan
Solon Pissis	CWI, The Netherlands
Simon Puglisi	University of Helsinki, Finland
Tomasz Radzik	King's College London, UK
Atif Rahman	BUET, Bangladesh
M. Sohel Rahman	BUET, Bangladesh
Wojciech Rytter	University of Warsaw, Poland
Kunihiko Sadakane	The University of Tokyo, Japan
William F. Smyth	McMaster University, Canada
Paul Spirakis	University of Patras, Greece and University of Liverpool, UK
Wing-Kin Sung	National University of Singapore, Singapore
Etsuji Tomita	The University of Electro-Communications, Japan
Ryuhei Uehara	Japan Advanced Institute of Science and Technology, Japan
Bruce Watson	Stellenbosch University, South Africa
Sue Whitesides	University of Victoria, Canada
Hsu-Chun Yen	National Taiwan University, Taiwan

Additional Reviewers

Ahn, Taehoon
Akrida, Eleni
Alanko, Jarno
Angelidakis, Haris
Bampas, Evangelos
Bannai, Hideo
Barua, Rana
Basappa, Manjanna
Bernardini, Giulia
Bilò, Davide
Binucci, Carla
Bousquet, Nicolas
Choi, Jongmin
Conte, Alessio
Deligkas, Argyrios
Devismes, Stéphane
Di Giacomo, Emilio
Di Stefano, Gabriele
Diarrassouba, Ibrahima
Epstein, Leah
Fici, Gabriele
Foucaud, Florent
Fuentes, Jose
Förster, Henry
Ghazawi, Samah
Ghosh, Arijit
Ghosh, Sasthi
Gorain, Barun
Goto, Keisuke
Grilli, Luca
Gu, Mei-Mei
Hermelin, Danny
Hoffmann, Michael
Ito, Takehiro
Jallu, Ramesh Kumar
Kamali, Shahin
Kang, Byeongwuk
Kare, Anjeneya Swami
Kaur, Dilpreet
Kawahara, Jun
Kim, Hwi
Kim, Hyeonjun
Kim, Jin Wook

Kim, Mincheol
Kociumaka, Tomasz
Komm, Dennis
Kondratovsky, Eitan
Kowaluk, Miroslaw
Kumar, Nikhil
Kuroki, Yuko
Kurpicz, Florian
Lampis, Michael
Lanir, Joel
Lee, Seungjun
Manlove, David
Marcus, Shoshana
Markou, Euripides
Mehrabi, Saeed
Mhaskar, Neerja
Michail, Othon
Miltzow, Till
Mincu, Radu-Stefan
Mishra, Pawan
Montecchiani, Fabrizio
Mukhopadhyaya, Srabani
Mutzel, Petra
Na, Joong Chae
Nandy, Subhas
Oh, Eunjin
Ortali, Giacomo
Page, Daniel R.
Park, Heejin
Rajendraprasad, Deepak
Rao, S. V.
Saitoh, Toshiki
Seara, Carlos
Sen, Sagnik
Shatabda, Swakkhar
Sikora, Florian
Sim, Jeong Seop
Simpson, Jamie
Sinha Mahapatra, Priya Ranjan
Suzuki, Akira
Takenaga, Yasuhiko
Tappini, Alessandra
Theofilatos, Michail

Toda, Takahisa
Unger, Walter
Uno, Yushi
van der Wegen, Marieke
Wallheimer, Nathan
White, Colin
Woeginger, Gerhard J.

Xiao, Mingyu
Yamanaka, Katsuhisa
Yoshinaka, Ryo
Zamaraev, Viktor
Zehavi, Meirav
Zuba, Wiktor

Contents

Short Papers

Invited Talks

Drawing Planar Graphs

Md. Saidur Rahman[1(✉)] and Md. Rezaul Karim[2]

[1] Graph Drawing and Information Visualization Laboratory,
Department of Computer Science and Engineering, Bangladesh University
of Engineering and Technology (BUET), Dhaka 1000, Bangladesh
saidurrahman@cse.buet.ac.bd
[2] Department of Computer Science and Engineering, University of Dhaka,
Dhaka 1000, Bangladesh
rkarim@du.ac.bd
https://saidurrahman.buet.ac.bd
http://www.cse.du.ac.bd/profile/?faculty=MRK

Abstract. A graph is planar if it can be drawn or embedded in the plane so that no two edges intersect geometrically except at a vertex to which they are both incident. A plane graph is a planar graph with a fixed planar embedding in the plane. A drawing problem X for a plane graph G asks to determine whether G has a drawing D satisfying a set P of given properties and to find D if it exists. The corresponding problem for a planar graph G asks to determine whether G has a planar embedding Γ such that Γ has a drawing D satisfying the set P of properties and find D if it exists. If every embedding of G has a drawing D satisfying P, then the problem is trivial, i.e., the problem for plane graphs and that for planar graphs are the same. Otherwise, the problem for planar graphs becomes difficult even if an efficient solution of the problem for a plane graph exists since a planar graph may have an exponential number of planar embeddings. Various techniques are found in literature that are used to solve the drawing problems for planar graphs. In this paper we review three of the widely used techniques, namely, (i) reduction to planarity testing, (ii) incremental modification and (iii) SPQR-tree decomposition.

Keywords: Graph drawing · Plane graph · Planar graph · Planarity testing · SPQR-tree

1 Introduction

A graph is *planar* if it can be drawn or embedded in the plane so that no two edges intersect geometrically except at a vertex to which they are both incident. A planar graph can have more than one planar embeddings. In fact a planar graph may have an exponential number of planar embeddings. One can differentiate two planar embeddings of a planar graph by observing the clockwise or counterclockwise ordering of the edges incident to each vertex. A *plane graph* G is a planar graph with a fixed planar embedding in the plane.

© Springer Nature Switzerland AG 2020
M. S. Rahman et al. (Eds.): WALCOM 2020, LNCS 12049, pp. 3–14, 2020.
https://doi.org/10.1007/978-3-030-39881-1_1

A drawing problem X for a *plane graph* G asks to determine whether G has a drawing D satisfying a set P of given properties and to find D if it exists. For an example, in a *rectangular drawing problem of a plane graph G*, it is asked to determine whether G has a drawing on a two dimensional plane keeping the embedding fixed such that every edge is drawn as either a horizontal line segment or a vertical line segment and each face including the outer face of G is drawn as a rectangle, and find such a drawing if it exists. However, in a *rectangular drawing problem for a planar graph G*, it is asked to determine whether G has a planar embedding Γ which has a rectangular drawing, and find such a drawing of Γ if it exists. That is, a drawing problem for a *planar graph G* asks to determine whether G has a planar embedding Γ such that Γ has a drawing D satisfying the set P of properties and find D if it exists. Suppose we have an efficient algorithm for a drawing problem of plane graphs and not every planar embedding has such a drawing. A straightforward algorithm checking each of all the embeddings by that efficient algorithm for plane graphs does not lead to an efficient algorithm for the drawing problem of planar graphs since a planar graph may have exponential number of planar embeddings.

Several approaches are found in literature for solving drawing problems on planar graphs. The key idea is to find a desirable embedding which has the drawing. For finding such a suitable embedding, in many cases the problem is reduced to planarity testing if some condition for having the drawing depends on the outer face of the embedding. This approach was first used for finding "convex drawings" of planar graphs by Chiba *et al.* in [9]. A *convex drawing of a plane graph G* is a drawing of G where each vertex is drawn as a point, each edge is drawn as a straight line segment and every face is drawn as a convex polygon. We say a *planar graph G has a convex drawing* if any of its planar embedding has a convex drawing. Not every planar graph has a convex drawing. Tutte established a necessary and sufficient condition for a plane graph to have a convex drawing [42] which was later slightly generalized by Thomassen [40]. Chiba *et al.* [9] gave a constructive proof of the condition of Thomassen and modified the condition into a form suitable for the convex testing of a plane graph. Using the form they have shown that the convex testing of a planar graph G can be reduced to the planarity testing of a certain graph obtained from G. Reduction to planarity testing also used for rectangular drawings [33], "box-rectangular drawings" [19] and "orthogonally convex drawings" [18] of planar graphs. In case of convex drawings of a planar graph, reducing the problem to planarity testing, one can identify only one embedding (among an exponential number of embeddings) for which he needs to check the properties of convex drawing [9]. In case of rectangular drawing one needs to check at most four embeddings [33] and for "box-rectangular drawing" one needs to check at most 81 (a constant number) planar embeddings [18].

Another approach is found in literature where starting from an arbitrary embedding one can find a desirable embedding by changing the embedding incrementally step by step. Hossain and Rahman [23] showed that every planar graph has a planar embedding which has a monotone drawing defined below, and find such an embedding by this approach. A *straight-line drawing* of a planar graph G is a drawing of G in which each vertex is drawn as a point and each edge is drawn as a straight-line segment without any edge crossing. A path P in a straight-line drawing of a planar graph is *monotone* if there exists a line l such that the orthogonal projections of the vertices of P on l appear along l in the order induced by P. A straight-line drawing Γ of a planar graph G is a *monotone drawing* of G if Γ contains at least one monotone path between every pair of vertices [2,4,23,24]. Note that every plane graph has a straight-line drawing [10,36], but not every plane graph has a monotone drawing.

Use of the SPQR-tree data structure that compactly represents all planar embeddings of a biconnected planar graph is also a widely used approach to find a desirable embedding efficiently. SPQR-tree decomposition has been used for finding "bend-minimum orthogonal drawing" of restricted classes of planar graphs [8,14]. An *orthogonal drawing* of a plane graph G is a drawing of G where each vertex is drawn as a point and each edge is drawn as a sequence of horizontal and vertical line segments. A point at which an edge changes its direction is called a *bend*. A *bend-minimum orthogonal drawing of a plane graph* has the minimum number of bends among all orthogonal drawings of G keeping the embedding fixed [30,32]. A *bend-minimum orthogonal drawing of a planar graph* G has the minimum number of bends among all orthogonal drawings of G where we are allowed to change the embedding. Note that there are polynomial-time algorithms for bend-minimum orthogonal drawings of plane graphs [15,39] whereas the problem of finding bend-minimum orthogonal drawings of planar graphs is NP-hard [39].

The remainder of the paper is organized as follows. In Sect. 2 we review how the rectangular drawing problem is solved by reducing to a planarity testing problem from [33]. In Sect. 3 we present the incremental modification method from [23] for finding a suitable embedding for monotone drawing of a planar graph. We describe the SPQR-tree decomposition approach in Sect. 4. Finally we conclude with future directions of research in Sect. 5. We refer books [26,27] for terminologies not defined in this paper.

2 Reduction to Planarity Testing

In this section we present the result of Rahman *et al.* [33] that we can determine whether a planar graph G has a rectangular drawing by checking at most four embeddings by the linear algorithm in [31]. The four planar embeddings are identified by reducing the problem to a planarity testing problem.

We denote by Δ the *maximum degree* of G. If a plane graph G has a rectangular drawing, then $\Delta \leq 4$ and G must be biconnected and have four or more vertices of degree 2 on the outer face. We call a vertex on the outer face of a plane graph an *outer vertex*. Thomassen [41] obtained a necessary and sufficient condition for a plane graph of $\Delta \leq 3$ to have a rectangular drawing when a quadruplet of outer vertices of degree two are designated as corners of a rectangular drawing of the outer face.

A linear-time algorithms is given in [29] to obtain a rectangular drawing of such a plane graph based on Thomassen's condition. Later Rahman *et al.* [31] slightly generalized the condition of Thomassen by giving a necessary and sufficient condition for a plane graph of $\Delta \leq 3$ to have a rectangular drawing (for some appropriately chosen quadruplet), and developed a linear-time algorithm to choose such a quadruplet and find a rectangular drawing of a plane graph for the chosen corners if they exist.

To present the conditions in Rahman *et al.* [31] we need some definitions. For a cycle C in a plane graph G, we denote by $G(C)$ the plane subgraph of G inside C (including C). An edge which is incident to exactly one vertex of C and located outside of C is called a *leg* of C, and the vertex of C to which the leg is incident is called a *leg-vertex* of C. A cycle C in G is called a *k-legged cycle* of G if C has exactly k legs. A k-legged cycle C is *minimal* if $G(C)$ does not contain any other k-legged cycle of G. A cycle C is called *regular* if $G - G(C)$ contains a cycle. We say that cycles C and C' in a plane graph G are *independent* if $G(C)$ and $G(C')$ have no common vertex. A set $\mathcal{S}$ of cycles is *independent* if any pair of cycles in $\mathcal{S}$ are independent. We now present the conditions in the following theorem.

Theorem 1 [31]. *Assume that G is a 2-connected plane graph with $\Delta \leq 3$ and has four or more outer vertices of degree two. Then four of them can be designated as the corners so that G has a rectangular drawing with the designated corners if and only if G satisfies the following three conditions: (a) every 2-legged cycle in G contains at least two outer vertices of degree two; (b) every 3-legged cycle in G contains at least one outer vertex of degree two; and (c) if an independent set $\mathcal{S}$ of cycles in G consists of c_2 2-legged cycles and c_3 3-legged cycles, then $2c_2 + c_3 \leq 4$.*

Not every plane graph has a rectangular drawing. Figures 1(a) and (b) depict two different planar embeddings of the same planar graph. The embedding in Fig. 1(a) has a rectangular drawing as illustrated in Fig. 1(c), and hence the planar graph has a rectangular drawing. On the other hand, the embedding in Fig. 1(b) does not have a rectangular drawing, because there are 3-legged cycles indicated by dotted lines which have no outer vertex of degree two. A straightforward algorithm checking each of all the embeddings by the linear algorithm in [31] does not run in polynomial time. In the rest of this section we show from Rahman *et al.* [33] that it is sufficient to check at most four embeddings of G which can be found through a planarity testing of a graph obtained from the input graph.

If the planar graph G is a subdivision of a planar 3-connected cubic graph, then one can determine whether G has a rectangular drawing or not by investigating any planar embedding of G [33]. We thus assume that G is a planar 2-connected graph with $\Delta \leq 3$ but is not a subdivision of a planar 3-connected cubic graph.

Let Γ be an arbitrary planar embedding of G. If G has at most two vertices of degree three, then one can easily examine whether G has a rectangular

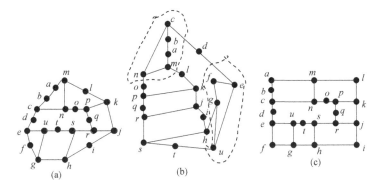

Fig. 1. (a)–(b) Two different planar embeddings of a same planar graph, (c) a rectangular drawing of the plane graph in (a).

drawing. We may thus assume that G has three or more vertices of degree three. Then Γ has a regular 2-legged cycle C. The pair of leg-vertices of C is a "critical separation pair" [26]. Let $(x_1, y_1), (x_2, y_2), \cdots, (x_l, y_l)$ be all "critical separation pairs" of G [26]. Clearly $l = O(n)$. If there is a planar embedding Γ' of G having a rectangular drawing, then the outer face $F_o(\Gamma')$ must contain all vertices $x_1, y_1, x_2, y_2, \cdots, x_l, y_l$. Construct a graph G^+ from G by adding a dummy vertex z and dummy edges (x_i, z) and (y_i, z) for all indices i, $1 \le i \le l$. Then G has a planar embedding whose outer face contains all vertices $x_1, y_1, x_2, y_2, \cdots, x_l, y_l$ if and only if G^+ is planar. (Figure 2(b) illustrates G^+ for G in Fig. 2(a).)

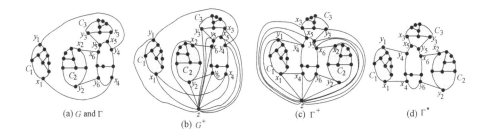

Fig. 2. G, Γ, G^+, Γ^+ and Γ^*.

We may thus assume that G^+ is planar. Let Γ^+ be an arbitrary planar embedding of G^+ such that z is embedded on the outer face, as illustrated in Fig. 2(c). We delete from Γ^+ the dummy vertex z and all dummy edges incident to z, and let Γ^* be the resulting planar embedding of G, in which $F_o(\Gamma^*)$ contains all vertices $x_1, y_1, x_2, y_2, \cdots, x_l, y_l$, as illustrated in Fig. 2(d). One can observe that every 2-legged cycle in Γ^* has the leg-vertices on $F_o(\Gamma^*)$.

Let p be the largest integer such that a number p of 2-legged cycles in Γ^* are independent with each other. Then $p \ge 2$ since Γ and hence Γ^* has a regular

2-legged cycle. In fact $p = 2$; if $p \geq 3$, then any planar embedding of G whose outer face contains all vertices $x_1, y_1, x_2, y_2, \cdots, x_l, y_l$ has three or more independent 2-legged cycles, and hence by Theorem 1(c) the embedding has no rectangular drawing, and consequently the planar graph G has no rectangular drawing.

Since $p = 2$, Γ^* has exactly two independent 2-legged cycles C_1 and C_2. We may assume without loss of generality that C_1 and C_2 are minimal 2-legged cycles, as illustrated in Fig. 3(a). By flipping $\Gamma^*(C_1)$ or $\Gamma^*(C_2)$ around the leg-vertices of C_1 or C_2, we have four different embeddings $\Gamma_1(= \Gamma^*)$, Γ_2, Γ_3 and Γ_4 such that each outer face $F_o(\Gamma_i)$, $1 \leq i \leq 4$, contains all vertices $x_1, y_1, x_2, y_2, \cdots, x_l, y_l$, as illustrated in Fig. 3. Since only the four embeddings $\Gamma_1, \Gamma_2, \Gamma_3$ and Γ_4 of G have all vertices $x_1, y_1, x_2, y_2, \cdots, x_l, y_l$ on the outer face, G has a rectangular drawing if and only if at least one of $\Gamma_1, \Gamma_2, \Gamma_3$ and Γ_4 has a rectangular drawing.

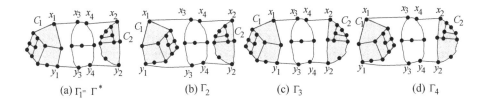

(a) $\Gamma_1 = \Gamma^*$ (b) Γ_2 (c) Γ_3 (d) Γ_4

Fig. 3. $\Gamma_1, \Gamma_2, \Gamma_3$ and Γ_4.

3   Incremental Modification

In this approach, starting from an arbitrary planar embedding, the desired planar embedding is obtained by modifying the embedding step by step. This approach is used for monotone straight-line drawings of planar graphs. We call a monotone drawing of a planar graph a *monotone grid drawing* if every vertex is drawn on a grid point.

It is known that not every plane graph (with a fixed embedding) admits a monotone drawing [2]. Hossain and Rahman [23] showed that every connected planar graph of n vertices has a planar embedding which has a monotone grid drawing on an $O(n) \times O(n^2)$ grid, and such a drawing can be computed in $O(n)$ time. An outline of their algorithm is as follows. First construct an embedding of G which has a "good spanning tree" T. Then find a monotone drawing of T by an algorithm in [2]. Finally draw each non-tree edge as a straight-line segment by shifting the drawing of some subtrees of T, if necessary.

Let G be a connected planar graph and let G_ϕ be a planar embedding of G. Let T be an ordered rooted spanning tree of G_ϕ such that the root r of T is an outer vertex of G_ϕ, and the ordering of the children of each vertex v in T is consistent with the ordering of the neighbors of v in G_ϕ. Let $P(r, v) = u_1(= r), u_2, \cdots, u_k(= v)$ be the path in T from the root r to a vertex $v \neq r$. The path $P(r, v)$ divides the children of u_i, $(1 \leq i < k)$, except u_{i+1}, into two groups; the left group L and the right group R. A child x of u_i is in the group L and denoted by u_i^L if the edge (u_i, x) appears before the edge (u_i, u_{i+1}) in clockwise ordering of the edges incident to u_i when the ordering is started from the edge (u_i, u_{i-1}), as illustrated in the Fig. 4(a). Similarly, a child x of u_i is in the group R and denoted by u_i^R if the edge (u_i, x) appears after the edge (u_i, u_{i+1}) in clockwise order of the edges incident to u_i when the ordering is started from the edge (u_i, u_{i-1}). We call T a *good spanning* tree of G_ϕ if every vertex v $(v \neq r)$ of G satisfies the following conditions (c1) and (c2) with respect to $P(r, v)$.

(c1) G does not have a non-tree edge (v, u_i), $i < k$; and

(c2) the edges of G incident to the vertex v excluding (u_{k-1}, v) can be partitioned into three disjoint (possibly empty) sets X_v, Y_v and Z_v satisfying the following conditions (a)–(c) (see Fig. 4(b)): (a) Each of X_v and Z_v is a set of consecutive non-tree edges and Y_v is a set of consecutive tree edges. (b) Edges of set X_v, Y_v and Z_v appear clockwise in this order from the edge (u_{k-1}, v). (c) For each edge $(v, v') \in X_v$, v' is contained in $T_{u_i^L}$, $i < k$, and for each edge $(v, v') \in Z_v$, v' is contained in $T_{u_i^R}$, $i < k$.

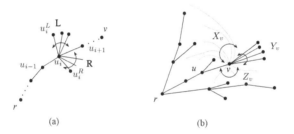

(a)   (b)

Fig. 4. (a) An illustration for $P(r, v)$, L and R groups, (b) an illustration for X_v, Y_v and Z_v sets of edges [23].

To find an embedding of G with a good spanning tree T, Hossain and Rahman [23] used an incremental modification approach which is described using an illustrative example in Fig. 5. We take an arbitrary planar embedding G_γ of G and start a breath-first-search (BFS) from an arbitrary outer vertex r of G_γ and regard r as the root of our desired spanning tree. In Fig. 5(a) the BFS is started from vertex a, and vertex b, c and d are visited from a in this order, as illustrated in Fig. 5(b). We next visit e from b, as illustrated in Fig. 5(c). When we visit a new vertex u, we check whether there is an edge (u, v) such that v is already visited and there is a (u, v)-split component or a u-component or a v-component inside the cycle induced by the edge (u, v) which does not contain the root r. The $\{e, d\}$-split component H_1 induced by the vertices $\{d, h, i, j, e\}$ is such a split component in Fig. 5(c) and the subgraph H_2 induced by vertices d, m, n is such a u-component for $u = d$ which are inside the cycle induced by the edge (e, d). We move the subgraphs H_1 and H_2 out of the cycle induced by the non-tree edge (e, d), as illustrated in Fig. 5(d). Since (b, e) is a tree edge and (e, d) is a non-tree edge, according to the definition of a good spanning tree, the edges (e, f) and (e, k) must be non-tree edges. Similarly, since (a, d) is a tree edge and (e, d) a non-tree edge, the edge (d, l) must be a non-tree edge. We mark (e, f), (e, k) and (d, l) as non-tree edges as shown in the Fig. 5(e). We then visit vertices f, l, m, n, g, h and j, as illustrated in Fig. 5(f). When we visit k, we find a k-component H induced by vertices $\{k, p, o\}$ and we move H out of the cycle induced by (e, k) as shown in Fig. 5(g). The BFS continues in this way and at the end of the BFS we find a planar embedding G_ϕ of G and a good spanning tree T as illustrated in Fig. 5(h), where the edges of the good spanning tree T are drawn as solid lines, and non-tree edges are drawn as dashed lines.

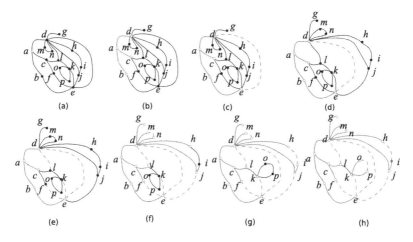

Fig. 5. Illustration for an outline of construction of a good spanning tree T. White vertices are visited vertices. Black vertices are not visited. Solid edges are tree edges. Dashed edges are non-tree edges [23].

Good spanning trees are also used for finding "2-visibility drawings" and "VPG drawings" of planar graphs [23].

4   SPQR-Tree Decomposition

An SPQR-tree, introduced by Di Battista and Tamassia [12], represents the decomposition of a biconnected graph with respect to its triconnected components. It is a data structure that compactly represents all planar embeddings of a biconnected planar graph.

Let $G = (V, E)$ be a biconnected planar graph. A pair $\{u, v\}$ of vertices of G is a *separation pair* if there exist two subgraphs $G_1 = (V_1, E_1)$ and $G_2 = (V_2, E_2)$ satisfying following two conditions: (a) $V = V_1 \cup V_2$, $V_1 \cap V_2 = \{u, v\}$; and (b) $E = E_1 \cup E_2$, $E_1 \cap E_2 = \phi$, $|E_1| \geq 2$, $|E_2| \geq 2$. A *split pair* of G is either a separation pair or a pair of adjacent vertices. Let $\{s, t\}$ be a split pair of G such that the edge (s, t) exists in G. The SPQR-tree $\mathcal{T}$ of G with respect to the reference edge $e = (s, t)$ describes a recursive decomposition of G induced by its split pairs. Tree $\mathcal{T}$ is a rooted ordered tree whose nodes are of four types: $S, P, Q,$ and R. Each node x of $\mathcal{T}$ corresponds to a subgraph of G, called its *pertinent graph* G_x. Each node x of $\mathcal{T}$ has an associated biconnected multigraph, called the *skeleton* of x, and denoted by *skeleton(x)*. The skeleton of a R-node corresponds to a triconnected graph, the skeleton of an S-node, corresponds to a simple cycle of length at least three, the skeleton of a P-node corresponds to a bundle of at least three parallel edges, and a Q-node represents a single edge of the graph and its skeleton consists of two parallel edges. Figure 6 illustrates an SPQR-tree decomposition of a graph.

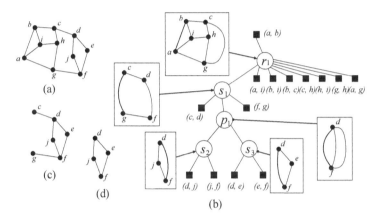

Fig. 6. (a) A biconnected graph G of maximum degree three, (b) SPQR-tree $\mathcal{T}$ of G with respect to reference edge (a, b) and skeleton of the P-, S-, and R-nodes, (c) the pertinent graph of S-node s_1 and (d) the pertinent graph of P-node p_1.

The reference edge is a Q node which is the natural root of an SPQR-tree. However an SPQR-tree can be re-rooted to any other nodes according to problem solving requirements. In general the problem is solved in bottom up approach on SPQR-tree, see the case of orthogonal drawings in [8,11]. Starting from leaf nodes the drawing problem at hand is solved on the pertinent graphs of each node by merging the solution of its children. Dynamic programming can also be used by keeping feasible solutions in a table at each node.

Finding a solution for a R-node is relatively easy [28] since the skeleton of a R-node is a triconnected graph which has a linear number of embeddings. However, merging the solution of a R-node with the solution of P-nodes or S-node becomes difficult. So this approach efficiently solve a drawing problem if it has no R-node, that is, the graph is a series-parallel graph [34,35].

5   Conclusions

In this paper we have reviewed three approaches for handling drawing problems of planar graphs. In graph drawing problems where some conditions for having the drawing depend on the outer face of the embedding, we may choose a smaller number of the candidate embeddings by reducing the problem to a planarity testing problem. It is interesting to investigate which of the other graph drawing problems can be reduced to a planarity testing problem.

There are several linear algorithms for planarity testing [6,17,21,25,37]. However, these algorithms are not conceptually simple. Finding a linear algorithm for planarity testing which is conceptually simple is still a need for the community. Finding planar embeddings of planar graphs with some desired properties is also an interesting direction of research. Sometimes the quality of a planar embedding of a planar graph can be measured in terms of the maximum distance of its vertices from the external face [3,5].

If the drawing of the graph depends on the properties of some embedded subgraph, we may try to solve it by incremental modification like the approach described in Sect. 3.

It looks that a problem can be solved if we can define a dynamic programming on SPQR-trees [1]. For graph drawing problems which are NP-hard, like bend-minimum orthogonal drawing [39] and upward planar drawings of digraphs [16], can we choose degree, height, treewidth etc. of SPQR-tree as parameters for developing parameterized algorithms?

We have not exhausted all the approaches for drawing planar graphs in this paper [22]. Combination of these approaches with some new approaches might be helpful. Very recently Didimo et. al gave a linear algorithm for orthogonal drawings of planar graphs of maximum degree 3 which uses SPQR-tree decomposition together with shape analysis, labeling and counting techniques etc. [13].

For the last few years graph drawing community has focused their interest in drawing non-planar graphs [7,20]. In some cases, an algorithm for a non-planar graph first somehow convert the non-planar graph to a planar graph, then find a drawing using an algorithm for a planar graph and finally modify the drawing for the original non-planar graph [38]. It is interesting to investigate which of the planar graph drawing approaches can be extended for the drawings of non-planar graphs.

Acknowledgement. We thank Debajyoti Mondal and Shin-ichi Nakano for their useful comments on the manuscript of this paper.

References

1. Angelini, P., et al.: Testing planarity of partially embedded graphs. In: Proceedings of the Twenty-First Annual ACM-SIAM Symposium on Discrete Algorithms, pp. 202–221. SIAM (2010)
2. Angelini, P., Colasante, E., Di Battista, G., Frati, F., Patrignani, M.: Monotone drawings of graphs. J. Graph Algorithms Appl. **16**(1), 5–35 (2012)

3. Angelini, P., Di Battista, G., Patrignani, M.: Finding a minimum-depth embedding of a planar graph in $O(n^4)$ time. Algorithmica **60**(4), 890–937 (2011)
4. Angelini, P., et al.: Monotone drawings of graphs with fixed embedding. Algorithmica **71**(2), 233–257 (2015)
5. Bienstock, D., Monma, C.L.: On the complexity of embedding planar graphs to minimize certain distance measures. Algorithmica **5**(1–4), 93–109 (1990)
6. Boyer, J.M., Cortese, P.F., Patrignani, M., Di Battista, G.: Stop minding your P's and Q's: implementing a fast and simple DFS-based planarity testing and embedding algorithm. In: Liotta, G. (ed.) GD 2003. LNCS, vol. 2912, pp. 25–36. Springer, Heidelberg (2004). https://doi.org/10.1007/978-3-540-24595-7_3
7. Brandenburg, F.J.: 1-visibility representations of 1-planar graphs. J. Graph Algorithms Appl. **18**(3), 421–438 (2014)
8. Chang, Y.J., Yen, H.C.: On bend-minimized orthogonal drawings of planar 3-graphs. In: Proceedings of 33rd International Symposium on Computational Geometry (SoCG 2017). Schloss Dagstuhl-Leibniz-Zentrum fuer Informatik (2017)
9. Chiba, N., Yamanouchi, T., Nishizeki, T.: Linear algorithms for convex drawings of planar graphs. Prog. Graph Theory **173**, 153–173 (1984)
10. De Fraysseix, H., Pach, J., Pollack, R.: How to draw a planar graph on a grid. Combinatorica **10**(1), 41–51 (1990)
11. Di Battista, G., Liotta, G., Vargiu, F.: Spirality and optimal orthogonal drawings. SIAM J. Comput. **27**(6), 1764–1811 (1998)
12. Di Battista, G., Tamassia, R.: On-line maintenance of triconnected components with SPQR-trees. Algorithmica **15**(4), 302–318 (1996)
13. Didimo, W., Liotta, G., Ortali, G., Patrignani, M.: Optimal orthogonal drawings of planar 3-graphs in linear time. arXiv preprint, arXiv:1910.11782 (2019)
14. Didimo, W., Liotta, G., Patrignani, M.: Bend-minimum orthogonal drawings in quadratic time. In: Biedl, T., Kerren, A. (eds.) GD 2018. LNCS, vol. 11282, pp. 481–494. Springer, Cham (2018). https://doi.org/10.1007/978-3-030-04414-5_34
15. Garg, A., Tamassia, R.: A new minimum cost flow algorithm with applications to graph drawing. In: North, S. (ed.) GD 1996. LNCS, vol. 1190, pp. 201–216. Springer, Heidelberg (1997). https://doi.org/10.1007/3-540-62495-3_49
16. Garg, A., Tamassia, R.: On the computational complexity of upward and rectilinear planarity testing. SIAM J. Comput. **31**(2), 601–625 (2001)
17. Haeupler, B., Tarjan, R.E.: Planarity algorithms via PQ-trees. Electron. Notes Discret. Math. **31**, 143–149 (2008)
18. Hasan, M.M., Rahman, M.S.: No-bend orthogonal drawings and no-bend orthogonally convex drawings of planar graphs (extended abstract). In: Du, D.-Z., Duan, Z., Tian, C. (eds.) COCOON 2019. LNCS, vol. 11653, pp. 254–265. Springer, Cham (2019). https://doi.org/10.1007/978-3-030-26176-4_21
19. Hasan, M.M., Rahman, M.S., Karim, M.R.: Box-rectangular drawings of planar graphs. J. Graph Algorithms Appl. **17**(6), 629–646 (2013)
20. Hong, S., Tokuyama, T.: Algorithmics for beyond planar graphs. In: NII Shonan Meeting Seminar, no. 27, pp. 51–63 (2016)
21. Hopcroft, J., Tarjan, R.: Efficient planarity testing. J. ACM (JACM) **21**(4), 549–568 (1974)
22. Hossain, M., Mondal, D., Rahman, M., Salma, S.: Universal line-sets for drawing planar 3-trees. J. Graph Algorithms Appl. **17**(2), 59–79 (2013)
23. Hossain, M.I., Rahman, M.S.: Good spanning trees in graph drawing. Theor. Comput. Sci. **607**, 149–165 (2015)
24. Hossain, M.I., Rahman, M.S.: Straight-line monotone grid drawings of series-parallel graphs. Discrete Math. Algorithms Appl. **7**(02), 1550007 (2015)

25. Mehlhorn, K., Mutzel, P.: On the embedding phase of the hopcroft and tarjan planarity testing algorithm. Algorithmica **16**(2), 233–242 (1996)
26. Nishizeki, T., Rahman, M.S.: Planar Graph Drawing, vol. 12. World Scientific Publishing Company, Singapore (2004)
27. Rahman, M.S.: Basic Graph Theory. Springer, Cham (2017). https://doi.org/10.1007/978-3-319-49475-3
28. Rahman, M.S., Egi, N., Nishizeki, T.: No-bend orthogonal drawings of subdivisions of planar triconnected cubic graphs. IEICE Trans. Inform. Syst. **88**(1), 23–30 (2005)
29. Rahman, M.S., Nakano, S., Nishizeki, T.: Rectangular grid drawings of plane graphs. Comput. Geom. **10**(3), 203–220 (1998)
30. Rahman, M.S., Nakano, S., Nishizeki, T.: A linear algorithm for bend-optimal orthogonal drawings of triconnected cubic plane graphs. J. Graph Algorithms Appl. **3**, 31–62 (1999)
31. Rahman, M.S., Nakano, S., Nishizeki, T.: Rectangular drawings of plane graphs without designated corners. Comput. Geom. **21**(3), 121–138 (2002)
32. Rahman, M.S., Nishizeki, T.: Bend-minimum orthogonal drawings of plane 3-graphs. In: Goos, G., Hartmanis, J., van Leeuwen, J., Kučera, L. (eds.) WG 2002. LNCS, vol. 2573, pp. 367–378. Springer, Heidelberg (2002). https://doi.org/10.1007/3-540-36379-3_32
33. Rahman, M.S., Nishizeki, T., Ghosh, S.: Rectangular drawings of planar graphs. J. Algorithms **50**(1), 62–78 (2004)
34. Samee, M.A.H., Alam, M.J., Adnan, M.A., Rahman, M.S.: Minimum segment drawings of series-parallel graphs with the maximum degree three. In: Tollis, I.G., Patrignani, M. (eds.) GD 2008. LNCS, vol. 5417, pp. 408–419. Springer, Heidelberg (2009). https://doi.org/10.1007/978-3-642-00219-9_40
35. Samee, M.A.H., Rahman, M.S.: Upward planar drawings of series-parallel digraphs with maximum degree three. In: Proceedings of WALCOM 2007, pp. 28–45. Bangladesh Academy of Sciences (2007)
36. Schnyder, W.: Embedding planar graphs on the grid. In: Proceedings of the First Annual ACM-SIAM Symposium on Discrete Algorithms, pp. 138–148. Society for Industrial and Applied Mathematics (1990)
37. Shih, W.K., Hsu, W.L.: A new planarity test. Theor. Comput. Sci. **223**(1), 179–192 (1999)
38. Sultana, S., Rahman, M.S., Roy, A., Tairin, S.: Bar 1-visibility drawings of 1-planar graphs. In: Gupta, P., Zaroliagis, C. (eds.) ICAA 2014. LNCS, vol. 8321, pp. 62–76. Springer, Cham (2014). https://doi.org/10.1007/978-3-319-04126-1_6
39. Tamassia, R.: On embedding a graph in the grid with the minimum number of bends. SIAM J. Comput. **16**(3), 421–444 (1987)
40. Thomassen, C.: Planarity and duality of finite and infinite graphs. J. Comb. Theory Ser. B **29**(2), 244–271 (1980)
41. Thomassen, C.: Plane representations of graphs. In: Bondy, J.A., Murty, U.S.R. (eds.) Progress in Graph Theory. Academic Press, New York (1984)
42. Tutte, W.T.: Convex representations of graphs. Proc. London Math. Soc. **3**(1), 304–320 (1960)

Space Efficient Separator Algorithms
for Planar Graphs

Osamu Watanabe$^{(\boxtimes)}$ (ID)

Tokyo Institute of Technology, Tokyo 152-8550, Japan
`watanabe@c.titech.ac.jp`

Abstract. The *Separator Theorem* states that any planar graph G with n vertices has a *separator* of size $O(\sqrt{n})$, that is, a set S of $O(\sqrt{n})$ vertices of G such that by removing S, G is split into disconnected subgraphs of almost equal size, say, each having more than $n/3$ vertices. A separator algorithm is an algorithm that computes a separator for a given planar graph. We consider two algorithms that have been developed recently by the author and his colleagues [4,6] for designing a "sublinear-space" and polynomial-time separator algorithm.

Keywords: Separator Theorem · Space-efficient algorithms

1 Summary

In this talk I explain two algorithms that have been developed by the author and his colleagues [4,6] to prove the following theorem. (We have not optimized the constant α and prove the theorem with $\alpha = 1/31$.)

Theorem 1. *For some constant $\alpha > 0$, there exists an algorithm that takes an undirected planar graph with n vertices as input and outputs its α-separator of size $O(\sqrt{n})$ in polynomial-time and $\widetilde{O}(\sqrt{n})$-space.* ☐

The *Separator Theorem* states that any planar graph G with n vertices has a *separator* of size $O(\sqrt{n})$, that is, a set S of $O(\sqrt{n})$ vertices of G such that by removing S, G is split into disconnected subgraphs of almost equal size, say, each having more than $n/3$ vertices. In fact, in their seminal work that first proved the Separator Theorem, Lipton and Tarjan [11] gave an efficient separator algorithm, an algorithm for computing an $O(\sqrt{n})$-size separator for planar graphs.

Since the work of Lipton and Tarjan, several versions of separator algorithms have been proposed, and they have been applied to design various algorithms for planar graphs. The purpose of our work is to give yet another type of separator algorithm stated in the above theorem. Its motivation stems from its relation to the graph reachability problem.

A major part of this work was done during the project "Exploring the Limits of Computation (ELC)" supported by MEXT KAKENHI Grant No. 24106008.

M. S. Rahman et al. (Eds.): WALCOM 2020, LNCS 12049, pp. 15–21, 2020.
https://doi.org/10.1007/978-3-030-39881-1_2

The *graph reachability* or the *st-connectivity* problem asks, for a given *directed* graph G and two of its vertices s and t, whether there is a path in G from s to t or not. This problem is one of the fundamental computational problems in both algorithm design and in complexity theory. In particular, the $O(\log n)$-space *non*-computability of the graph reachability problem is a typical example of the L $\neq$ NL conjecture, one of the fundamental open problems in computational complexity theory. Interestingly, for its undirected version, that is, for the *undirected* graph reachability problem, Reingold [12] gave a remarkable $O(\log n)$-space algorithm. Note that, by using standard algorithmic techniques such as BFS or DFS we can design an algorithm that runs in almost linear-space and in almost linear-time for solving reachability over directed graphs. Savitch gave an algorithm that solves reachability using $O((\log n)^2)$-space. However Savitch's algorithm takes super polynomial-time. While the BFS/DFS algorithm is time-efficient, Savitch's algorithm is space-efficient. Thus it is natural to ask whether there exists an algorithm, for the directed graph reachability problem, that runs in sublinear-space and yet polynomial-time. For this question, the best known answer is the algorithm given by Barnes et al. [7] that runs in $O(n/2^{\sqrt{\log n}})$-space and polynomial-time. Unfortunately, though the bound $O(n/2^{\sqrt{\log n}})$ is "sublinear", it is quite close to being linear. If we view $O(n^{1-\varepsilon})$ (for some $\varepsilon > 0$), as "sublinear", it has been open whether such a sublinear-space and polynomial-time algorithm exists for the reachability problem.

Recently, there have been some advancements on this sublinear-space and polynomial-time computability for restricted graph classes. (Since all algorithms stated below run in polynomial-time, we omit mentioning "polynomial-time" in the following.) The first break through was given by Asano and Doerr [2], who showed an algorithm that solves the reachability problem on directed grid graphs in $\widetilde{O}(n^{1/2+\epsilon})$-space. Inspired by this work, Imai, Nakagawa, Pavan, Vinodchandran, and Watanabe [9] gave an $\widetilde{O}(n^{1/2+\epsilon})$-space algorithm for the reachability problem on directed planar graphs. Later Asano et al. [3] gave an improved $\widetilde{O}(\sqrt{n})$-space algorithm. For directed grid graphs, Ashida and Nakano [5] showed an $\widetilde{O}(n^{1/3})$-space algorithm. Note that in all of the above algorithms the input graphs are planar and they all critically rely on the existence of $O(\sqrt{n})$-size separator. The last algorithm of Ashida and Nakano is interesting in this sense because its space bound is much less than the separator size. These algorithms except for the first one use a separator algorithm proposed by Imai *et al.* in [9] that produces a $O(\sqrt{n})$-size separator in $\widetilde{O}(\sqrt{n})$-space. This separator algorithm is based on an algorithm of Gazit and Miller [8] that is designed for computing a separator efficiently on a parallel computation model. Since space-bounded computation and parallel computation share common features, one can expect that the algorithm of Gazit and Miller can be modified naturally to obtain an algorithm that runs in polynomial-time and uses $\widetilde{O}(\sqrt{n})$-space. Unfortunately, it is not easy to translate Gazit and Miller's algorithm to obtain a $\widetilde{O}(\sqrt{n})$-space algorithm for computing a planar separator. The work of [9] provided a sketch of a $\widetilde{O}(\sqrt{n})$-space, polynomial-time algorithm for computing a planar separator, however, several nontrivial details have been left unexplained in that work.

Right after the presentation of [9], the author and his colleagues have been trying to give a complete and detailed explanation to the algorithm claimed in [9], and finally, we have been able to publish a report [6] in which we give an algorithm stated as Theorem 1 in detail. In fact, we needed to give another separator algorithm [4] as a basis of this algorithm. By our deitaled analysis, we have realized that the computational relations between the separator problem and some other generic computational problems. The purpose of this talk is to explain these two algorithms from this view point.

2   Preliminaries for Our Discussion

We explain basic notions and notation on graphs, in particular, planar graphs. We also explain our computational model and some algorithms that are used as key tools in our discussion.

Throughout this paper, by a graph we usually mean an undirected graph. For any graph G, we formally use $V(G)$ and $E(G)$ to denote the set of vertices of G and the set of edges of G respectively. On the other hand, we mainly discuss with some fixed graph G, and in this case, we simply use V and E to denote $V(G)$ and $E(G)$. Also unless otherwise stated, we use n to denote the number of vertices of G, which is a typical size parameter for disucssing the complexity of algorithms.

For any vertex $v \in V$, let $N_G(v)$ denote the set of its *neighbors*, namely, vertices adjacent to v in G. For any $U \subseteq V$ (resp., $D \subseteq E$) we use $G[U]$ (resp., $G[D]$) to denote a subgraph of G *induced* by U (resp., by D).

We assume that an input graph is *simple*, that is, each pair of vertices has at most one edge. Though a non-simple graph may be created by our transformations (in particular, in our base dual-graph), all our arguments should be valid by treating multiple edges (between the same pair of vertices) as different edges. By a *path* of $G = (V, E)$, we mean a "simple" path, that is, a sequence adjacent edges of E where no vertex is visited more than once. Similarly, a *cycle* of $G = (V, E)$ is a sequence adjacent edges of E that comes back to the first vertex where no vertex except for the first one is visited more than once. We simply represent a path and a cycle by a sequence $(v_1, \ldots, v_m)$ of vertices of V such that every $\{v_i, v_{i+1}\}$ (and also $\{v_m, v_1\}$ for a cycle) is an edge of E. Although the same sequence could be used for representing both a path and a cycle, the difference should be clear from the context. We may also use this sequence representation for indicating the "direction" of a path or a cycle.

We avoid rigorous but tedious topological treatments of planar graphs, and following the previous work in the literature (see, e.g., [10]), we prepare a framework for discussing plane graphs in a combinatorial way.

Definition 1. *A graph G is* planar *if it has a* planar embedding, *a way to arrange the vertices of $N_G(v)$ on the plane (i.e., $\mathbb{R}^2$) for all $v \in V$ and a drawing of the edges so that no pair of edges intersect each other except at their endpoints. A planar embedding is specified by a* combinatorial embedding, *that is, a set π $:= \{\pi_v : v \in V\}$, where each π_v is an enumeration of vertices of $N_G(v)$ in the clockwise order around v under the embedding.*

[Remark]. *Throughout this paper, we use a* plane graph *to mean a planar graph that is embedded in the plane following one of such combinatorial embeddings.* □

Consider any plane graph G embedded in the plane under a combinatorial embedding π. One of the keys for discussing plane graphs is a way to define the faces of G. Intuitively, the graph G (embedded in the plane) separates the plane into "subplanes," each of which is a "face" of G. Here we formally define the notion of "face" in a combinatorial way.

First define the notion of "left-traverse" of G. Consider any edge $\{v_1, v_2\}$ of G, and fix its direction as, e.g., (v_1, v_2), which we regard as the first directed edge e_1 (of the left-traverse). Consider the enumeration π_{v_2} of adjacent vertices of v_2, and let v_3 be the next vertex of v_1 in the enumeration, that is, v_3 is clockwise the first vertex adjacent to v_2 after v_1. Then define $e_2 = (v_2, v_3)$ as the second directed edge. Continue this process of identifying directed edges (based on edges of G) until we come back to the first directed edge e_1. We call this process *left-traverse process*, and a sequence of vertices visited during the process *left-traverse* of G started from (v_1, v_2). Although a left-traverse is a sequence of vertices, this can be regraded as the sequence of the directed edges in the order that they are identified by the left-traverse process.

Consider any plane graph G with at least two faces. Then any left-traverse $t = (v_1, \ldots, v_m)$ of G separates the plane where G is embedded into at least two "subplanes." From the choice of directed edges, it is clear that there is no vertex of G in the "subplane" left of the directed edges of t; that is, a "face" is identified by this left-traverse. In fact, it has been known that for any plane graph, all its faces are identified by some left-traverse as stated in Proposition 1 below. In this paper, we regard this proposition as our definition of the notion of face.

Proposition 1. *Consider any plane graph G. Then for any left-traverse of G, there is a unique subplane, i.e., face, located left of the traverse w.r.t. its direction. On the other hand, for any face of G, there is a left-traverse of G that defines it as a subplane located left of the left-traverse w.r.t. its direction.*

[Remark]. *Since each left-traverse is determined by a starting directed edge, there are more than one essentially the same left-traverses. Thus, the correspondence between faces and left-traverses is not one-to-one but one-to-many.* □

Note that a left-traverse is not a cycle in general. But if G is 2-connected, then any left-traverse of G is a cycle. (A graph is *2-connected* if every pair $\{u, v\}$ of its vertices are connected by two paths not sharing vertices except for $\{u, v\}$, or equivalently, it has no *cut vertex*.)

Proposition 2. *Consider any plane graph G that is also 2-connected. Then any left-traverse is a cycle. That is, every face of G has a boundary consisting of one cycle, which we call a* face boundary cycle. □

We mainly consider 2-connected graphs. Thus, we may identify a face with its boundary cycle; this gives us a way to represent a face.

Definition 2. *For any 2-connected plane graph G, for any face, its* face bound-ary representation *is its face boundary cycle c, or more precisely, a sequence of vertices of c in the order so that the face is located left of c w.r.t. the order. The* size *of a face is the number of vertices of its face boundary cycle.* □

We say that two faces are *incident* (resp., *edge-incident*) if their boundaries share some vertex (resp., some edge). In general, we say that two graph objects are *incident* if they share some vertex. For example, we say that two paths are incident if they share some vertex (and possibly more). Here is our way to describe all faces of a given graph G.

Definition 3. *A* complete face information *of a graph G is the lists of (i) all face boundary representations of G, (ii) all incident pairs of faces of G, and (iii) all edge-incident pairs of faces of G.* □

We introduce a generalized separator notion, which is mainly used in this paper. First we define the standard separator notion formally.

Definition 4. *For any graph G, and for any parameter $\alpha \in (0,1)$, an α-*separator *of G is a set S of vertices of G such that the removal of S creates two disconnected subgraphs each of which has at least αn vertices.* □

A generalized separator notion is defined for weighted graphs. Here a weighted graph is a plane graph G for which we additionally give a weight to each face of G. Formally, we may represent a weighted graph $G = (V, E)$ by V, E, its combinatorial planar embedding, and its complete *weighted* face infor-mation where each face representation is also given its weight. By normalizing weights, we may assume that the total weight is always 1.

Definition 5. *For any 2-connected weighted graph, and for any $\rho \in (0,1)$, a* cycle weighted ρ-separator *of the graph is a cycle C that creates two subplanes each of which has weight $\geq \rho$.*

[Remark]. *The weight of a subplane is the sum of the weights of all faces in the subplane.* □

Computational Model

For discussing sublinear-space algorithms, we use the standard multi-tape Turing machine model. A multi-tape Turing machine consists of a read-only input tape, a write-only output tape, and a constant number of work tapes. The space complexity of this Turing machine is measured by the total number of cells that can be used as its work tapes. As a unit for space complexity, we often use "word", by which we mean $c_0 \log n$ cells, where c_0 is a constant large enough such that index of a vertex (i.e., a number between 1 and n) can be stored in one word. We use $\tilde{O}(s(n))$ to mean $O(s(n))$ words. (Recall that n is the number of vertices of an input graph, which is the main complexity parameter throughout this paper.)

For the sake of explanation, we will follow a standard convention and describe a sublinear-space algorithm by a sequence of constant number of sublinear-space subroutines $A_1, \ldots, A_k$ such that each A_i computes, from its given input, some output that is passed to A_{i+1} as input. Note that some of these outputs cannot be stored in a sublinear-size work tape; nevertheless, there is a standard way to design a sublinear-space algorithm based on these subroutines. The key idea is to compute intermediate inputs every time when they are necessary. For example, while running A_i, when it is necessary to see the jth bit of the input to A_i, simply execute A_{i-1} (from the beginning) until it yields the desired jth bit on its work tape, and then resume the computation of A_i using this obtained bit. It is easy to see that this computation can be executed in sublinear-space. Furthermore, while a large amount of extra computation time is needed, we can show that the total running time can be polynomially bounded if all subroutines run in polynomial-time.

Basic Algorithms

We recall some basic algorithms given by the previous work that are used several times in our algorithms.

The first one is the algorithm of Lipton and Tarjan [11], which can be stated as follows.

Proposition 3. *For any planar graph with n vertices, there exists a $1/3$-separator of size $2\sqrt{2n}$. Furthermore, there exists a polynomial-time algorithm that finds one of such separators by a breadth-first search. The algorithm can be modified so that when a given (breadth-first search) tree of G is given as an additional input, then the algorithm runs in polynomial-time w.r.t. n and in $\widetilde{O}(k_{\mathrm{bfs}})$-space where k_{bfs} is the depth of the BFS tree.*

We also recall the log-space algorithm of Reingold [12] for the *undirected graph reachability problem*. More specifically, we have an $O(\log n)$-space algorithm that determines, for a given graph G and a pair of vertices u and v of G, whether u is reachable to v in $O(\log n)$-space. This algorithm is used several places in the following; we call it *Reinglod's algorithm* for the undirected graph reachability test.

We then use Reingold's algorithm with the algorithmic method of Allender and Mahajan [1] to design an $O(\log n)$-space algorithm that tests, for a given graph G, whether it is planar and (if it is so) computes one of its combinatorial embeddings. We refer this algorithm as the *planarity test algorithm* of Allender and Mahajan.

For a given planar graph G, once one of its combinatorial embeddings π is computed, we can compute from π a triangulation of G w.r.t. π in $O(\log n)$-space. (Recall that we say that a plane graph is *triangulated* (w.r.t. π) if addition of any edge results in a nonplanar graph. For a plane graph, its *triangulation* means to add edges to the plane graph until it gets triangulated w.r.t. π.)

Acknowledgements. A major part of this work was done during the project "Exploring the Limits of Computation (ELC)." During the ELC project, many researchers have joined our discussion which helped us very much for finding some errors and improving the presentation of our idea of separator algorithms. In particular, I thank to Dr. Mohit Garg and Dr. Sebastian Kuhnert for their technical contributions. I would like to express my deep appliciations to the coauthors of the sequence of papers on this and related topics; among them, I especially thank to my students, Dr. Ryo Ashida, Dr. Tatsuya Imai, and Dr. Kotaro Nakagawa, and wish brilliant futures to them.

References

1. Allender, E., Mahajan, M.: The complexity of planarity testing. Inf. Comput. **189**(1), 117–134 (2004)
2. Asano, T., Doerr, B.: Memory-constrained algorithms for shortest path problem. In: Proceedings of the 23rd CCCG (2011)
3. Asano, T., Kirkpatrick, D., Nakagawa, K., Watanabe, O.: $\widetilde{O}(\sqrt{n})$-Space and polynomial-time algorithm for planar directed graph reachability. In: Csuhaj-Varjú, E., Dietzfelbinger, M., Ésik, Z. (eds.) MFCS 2014. LNCS, vol. 8635, pp. 45–56. Springer, Heidelberg (2014). https://doi.org/10.1007/978-3-662-44465-8_5
4. Ashida, R., Kuhnert, S., Watanabe, O.: A space-efficient separator algorithm for planar graphs. IEICE Trans. Fundam. **E102−A**(9), 1007–1016 (2019). A version with a simplified proof is available from Tokyo Tech Research Repository, search "T2R2 Osamu Watanabe."
5. Ashida, R., Nakagawa, K.: $\widetilde{O}(n^{1/3})$-space algorithm for the grid graph reachability problem. In: Proceedings of the 34th SoCG, pp. 5:1–5:13 (2018)
6. Ashida, R., Imai, T., Nakagawa, K., Pavan, A., Vinodchandran, N.V., Watanabe, O.: A sublinear-space and polynomial-time separator algorithm for planar graphs. Electron. Colloquium Comput. Complex. (ECCC) **26**(91) (2019)
7. Barnes, G., Buss, J.F., Ruzzo, W.L., Schieber, B.: A sublinear space, polynomial time algorithm for directed s-t connectivity. In: Proceedings Structure in Complexity Theory Conference, pp. 27–33. IEEE (1992)
8. Gazit, H., Miller, G.L.: A parallel algorithm for finding a separator in planer graphs. In: Proceedings of the 28th FOCS, pp. 238–248 (1987)
9. Imai, T., Nakagawa, K., Pavan, A., Vinodchandran, N.V., Watanabe, O.: An $O(n^{\frac{1}{2}+\epsilon})$-space and polynomial-time algorithm for directed planar reachability. In: Proceedings of the 28th CCC, pp. 277–286 (2013)
10. Miller, G.L.: Finding small simple cycle separators for 2-connected planar graphs. J. Comput. Syst. Sci. **32**(3), 265–279 (1986)
11. Lipton, R.J., Tarjan, R.E.: A separator theorem for planar graphs. SIAM J. Appl. Math. **36**(2), 177–189 (1979)
12. Reingold, O.: Undirected connectivity in log-space. J. ACM **55**(4), 17 (2008)

Recent Progresses in the Combinatorial and Algorithmic Study of Rooted Phylogenetic Networks

Louxin Zhang$^{(\boxtimes)}$

Department of Mathematics, National University of Singapore,
Singapore 119076, Singapore
`matzlx@nus.edu.sg`

Abstract. Galled trees are studied as a recombination model in theoretical population genetics. Tree-child networks, reticulation-visible networks and tree-based networks can be considered as the generalizations of galled trees through relaxing a structural condition. Although these networks are simple, their topological structures have yet to be fully understood. Here, recent progresses in the tree and cluster containment problems and network counting problems are summarized.

Phylogenetic networks have been used to model horizontal gene transfers [2], recombinations and other reticulate evolutionary event in the past two decades [9]. A rooted phylogenetic network (RPN) on a set of taxa X (e.g, species, genes, or individuals in a population) is a directed acyclic digraph (DAG) with a unique start node called the root, in which all the leaves that are the nodes of indegree 1 and outdegree 0 represent X and the root represents the least common ancestor of the taxa. In a phylogenetic network, all the nodes other than the leaves and the root are divided into tree nodes and reticulate nodes. In this short survey, tree nodes include the root and all the nodes of indegree 1 and outdegree greater than 1, whereas reticulate nodes include all the nodes of indegree greater than 1 and outdegree 1. Notice that phylogenetic trees are simply RPNs with no reticulate nodes. A phylogenetic network is binary if the sum of indegree and outdegree is 1 for its leaves, 2 for its root and 3 for all the other nodes.

The topological properties of RPNs are more complicated and reconstructing phylogenetic networks are more challenging than phylogenetic trees [9,12,13] and thus different mathematical issues arise in the study of RPNs. One of the issues is the relationships between phylogenetic trees and RPNs. Are RPNs uniquely determined by the set of clusters or phylogenetic trees displayed in them? Are RPNs uniquely determined by the distances between their nodes? How to determine whether or not a tree is displayed in a phylogenetic network? How to determine whether or not a cluster is displayed in a phylogenetic network?

Another issue is the topological structures of RPNs. Different classes of RPNs have been introduced, including galled trees [10,14], tree-child networks (TCNs) [1], normal networks [16], reticulation-visible networks [12] and tree-based networks [3,18]. Many algorithmic problems that are NP-complete in general become

© Springer Nature Switzerland AG 2020
M. S. Rahman et al. (Eds.): WALCOM 2020, LNCS 12049, pp. 22–27, 2020.
https://doi.org/10.1007/978-3-030-39881-1_3

tractable on RPNs in these classes, including the tree containment problem and the cluster containment problem [19]. Recently, we have studied the hierarchical structures of RPNs in these special classes. The relatively simple structures of galled networks and tree-child networks allows us to count and enumerate these networks.

In this talk, recent progresses in the study of the problems mentioned above will be discussed.

1 Basic Definitions

In this survey, we focus on binary rooted phylogenetic networks. For such a network N, we use $\rho(N), \mathcal{R}(N), \mathcal{T}(N)$ and $\mathcal{L}(N)$ to denote the root, the set of reticulate nodes, the set of tree nodes and the set of leaves of N, respectively.

Let X be a set of taxa and N be a binary rooted phylogenetic network on X. For two distinct nodes u and v in N, v is a *descendant* of u if there is a directed path from $\rho(N)$ to v that contains u. Since N is acyclic, the descendant relation is asymmetric and transitive. The set of all the leaves that are descendants of u is said to be the (hard) *cluster* of u in N.

Since each reticulate node is of indegree 2 and outdegree 1, deleting exactly one incoming edge for every reticulate node leads to a spanning directed tree T' with the root $\rho(N)$ in N. Every leaf of N is a leaf of T'. However, T' may contain other leaves. In other words, some internal nodes of N become the leaves of T'.

For a subset $L \subseteq \mathcal{L}(N)$ and a spanning tree T' obtained by the removal of one incoming edge for every reticulate node from N, removing all the nodes from T' that are not in the path from $\rho(N)$ to the leaves in L produces a tree on L, denoted by T'_L. We call T'_L the restriction of T' on L.

For a phylogenetic tree T on $X' \subseteq X$, T is said to be contained in N if there is a spanning tree T' in N such that $T'_{X'}$ is a subdivision of T.

A subset $C \subseteq X$ is said to be a *soft cluster* of N if C is the cluster of a node in some phylogenetic tree contained in N.

For a set S of nodes in N, $N - S$ denotes the subnetwork obtained by the removal of all the nodes of S from N together with all the edges incident to these nodes.

2 Component Graphs

Let N be a binary rooted phylogenetic network on X. The connected components of $N - \mathcal{R}(N)$ are said to be the *tree-node components* of N; the connected components of $N - [\mathcal{T}(N) \cup \mathcal{L}(N)]$ are said to be the *reticulate-node components* of N.

Let N have m tree-node components and p reticulate-node components. The *component graph* of N is the graph $G(N)$ with $m + p$ nodes $u_1, u_2, \cdots, u_m$ and $v_1, v_2, \cdots, v_p$ such that each u_i represents a tree-node component C_i and each v_j represents a reticulate node component C_j and an edge (u_i, v_j) exists if and only if N contains a directed edge from a node in C_i to a node in C_j. An illustration of the concept of component graph is given in Fig. 1. Note that the component graph is a rooted acyclic graph that might contain multiple parallel edges between two nodes for a phylogenetic network.

The concept of component graph is useful for studying phylogenetic networks in which each reticulate node component contains exactly a reticulate node.

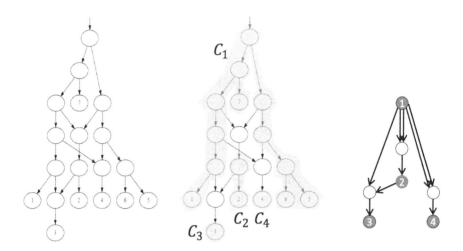

Fig. 1. A rooted phylogenetic network on $X = \{1, 2, \cdots, 7\}$ (left) that has four tree-node components (shaded) and three reticulate node components that are each singleton (right) and its component graph (right).

3   Network Classes and Inclusion Relations

Let N be a rooted phylogenetic network on X. N is a *tree-child network* if each internal node has at least a child that is a tree-node.

N is a *normal network* if it is tree-child and neither of parents is an ancestor of the other for any reticulate node.

N is a *galled network* if the two parents of every reticulate node are contained in the same tree node component.

For two nodes u and v in N, u is a *dominator* of v if every path from $\rho(N)$ to v contains u. N is a *reticulation-visible* if for every reticulate node r there exists a leaf ℓ_u such that r is a dominator of ℓ_u.

N is a *tree-based network* if the removal of an incoming edge for every reticulate node always produces a spanning tree that does not contain a leaf that is not in X. It can be determine whether a rooted phylogenetic network is tree-based in a linear time [3,18].

The following inclusion relations can be easily verified (see [21]).

Theorem 1. *Let* $\mathcal{NN}_k(X), \mathcal{TC}_k(X), \mathcal{GN}_k(X), \mathcal{RV}_k(X), \mathcal{TB}_k(X), \mathcal{A}_k(X)$ *denote the sets of binary normal, tree-child, galled, reticulation-visible, tree-based and arbitrary networks that have k reticulation nodes on X.*

(1.) *For any X and integer $k > 0$,* $\left\{ \begin{matrix} \mathcal{NN}_k(X) \subseteq \mathcal{TC}_k(X) \\ \mathcal{GN}_k(X) \end{matrix} \right\} \subseteq \mathcal{RV}_k(X) \subseteq$
$\mathcal{TB}_k(X) \subseteq \mathcal{A}_k(X).$

(2.) *For any X,* $\left\{ \begin{matrix} \mathcal{NN}_2(X) \subset \mathcal{TC}_2(X) \\ \mathcal{GN}_2(X) \end{matrix} \right\} \subset \mathcal{RV}_2(X) \subset \mathcal{TB}_2(X) = \mathcal{A}_2(X).$

(3.) *For any X such that $|X| > 1$, $\mathcal{TC}_1(X) = \mathcal{GN}_1(X) = \mathcal{RV}_1(X) = \mathcal{TB}_1(X) = \mathcal{A}_1(X)$.*

The following connections between different network classes are also interesting.

Theorem 2. *(1.)* [6,11] *The component graph of a galled network is a tree.*
(2.) [8] *The component graph of a reticulation-visible network is a tree-child network.*

4   Generating and Counting Tree-Child Networks

In [20], a fast procedure was presented for generating all tree-child networks. It is a generalization of the following well-known procedure for generation of phylogenetic trees:

For any X such that $|X| = k$ and $t \in X$, all the $\frac{(2k-2)!}{2^{k-1}(n-1)!}$ phylogenetic trees on X can be generated by attaching the leaf t into each edge of the phylogenetic trees on $X - \{t\}$.

Using the above generation approach and Theorem 1.3, the following counting results were obtained in [21].

Theorem 3. *Let X be a set of n taxa. Then, the numbers of binary normal networks and arbitrary networks with a reticulate node are respectively:*

$$|\mathcal{NN}_1(X)| = \frac{(n+2)(2n)!}{2^n n!} - 3 \cdot 2^{n-1} n!.$$

$$|\mathcal{A}_1(X)| = |\mathcal{TC}_1(X)| = \frac{n(2n)!}{2^n n!} - 2^{n-1} n!.$$

Theorem 4. *Let X be a set of n taxa. The numbers of binary tree-child and galled networks with two reticulations are respectively:*

$$|\mathcal{TC}_2(X)| = \frac{n!}{2^n} \sum_{j=1}^{n-2} \binom{2j}{j} \binom{2n-2j}{n-j} \frac{j(2j+1)(2n-j-1)}{2n-2j-1} + n(n-1)2^{n-3} n! - \frac{(2n-1)!n}{3 \cdot 2^{n-1}(n-2)!},$$

$$|\mathcal{GN}_2(X)| = \frac{n!}{2^{n-1}} \sum_{j=0}^{n-2} \binom{2j}{j} \binom{2n-2j}{n-j} \frac{(j+1)^2(2j+3)}{(n-j)(2n-2j-1)} + n(n-1)2^{n-3} n! - \frac{(2n-1)!n}{3 \cdot 2^{n-1}(n-2)!}.$$

A phylogenetic network is an *one-component* network if the unique child of every reticulate node is a network leaf. For a network class $\mathcal{C}$, we use $\mathcal{C}_{1,k}$ to denote the set of all one-component networks with k reticulations in $\mathcal{C}$.

Theorem 5. *Let X be a set of n taxa. The number of binary one-component tree-child networks with k reticulations on X is*

$$\mathcal{TC}_{1,k}(X) = \binom{n}{k} \frac{(2n-2)!}{2^{n-1}(n-k-1)!}$$

Theorem 6. *Let $g_{n,k}$ be the number of one-component galled networks with k reticulate nodes on n taxa. Then,*

$$g_{n,k+1} = \frac{(n-k)}{(k+1)}(n+k-1)\left(g_{n,k} + g_{n,k-1}\right)$$

$$+ \frac{n!(n-k)}{2(k+1)} \sum_{j=1}^{k} \binom{2j}{j} \frac{(n+1-j)g_{n-j,k-j} - (n+1-k)g_{n+1-j,k-j}}{2^j(n+1-j)!}.$$

Finally, Cardona proved the following beautiful fact recently.

Theorem 7. *There are twice as many tree-child networks with $n-1$ reticulations on X as tree-child networks with $n-2$ reticulations on X, where $n = |X|$.*

5   The Cluster and Tree Containment Problems

The cluster containment problem is to determine whether or not N contains X' given a rooted phylogenetic network N and a cluster $X' \subseteq X$.

The tree containment problem is to determine whether or not N contains T given a phylogenetic network N and a phylogenetic tree T on the same set of taxa.

These two decision problems are NP-complete. Therefore, there are unlikely polynomial time algorithms for solving the problems. Recently, we developed fast algorithms for solving the cluster and tree containment problems [7,17]. Additionally, linear-time algorithms exist for the two problems on reticulation-visible networks or nearly stable phylogenetic networks [4].

Theorem 8. *(1.) A linear-time algorithm exists for solving the tree containment problem [5,15].*
(2.) A linear-time algorithm exists for solving the cluster containment problem [4].

Acknowledgments. The author thanks B. DasGupta, P. Gambette, A. Gunawan, A. Labarre, S. Vialette, BX Liu and HW Yan for their collaboration in studying the combinatorial and algorithmic aspects of phylogenetic networks. This work was supported by Singapore-France Institute Merlion project and Singapore's Ministry of Education Academic Research Fund Tier-1 [grant R-146-000-238-114].

References

1. Cardona, G., Rossello, F., Valiente, G.: Comparison of tree-child phylogenetic networks. IEEE/ACM Trans. Comput. Biol. Bioinform. **6**(4), 552–569 (2009)

2. Doolittle, W.F.: Phylogenetic classification and the universal tree. Science **284**, 2124–2128 (1999)
3. Francis, A., Steel, M.: Which phylogenetic networks are merely trees with additional arcs? Syst. Biol. **64**(5), 768–777 (2015)
4. Gambette, P., Gunawan, A.D., Labarre, A., Vialette, S., Zhang, L.: Solving the tree containment problem in linear time for nearly stable phylogenetic networks. Discret. Appl. Math. **246**, 62–79 (2018)
5. Gunawan, A.D.M.: Solving the tree containment problem for reticulation-visible networks in linear time. In: Jansson, J., Martín-Vide, C., Vega-Rodríguez, M.A. (eds.) AlCoB 2018. LNCS, vol. 10849, pp. 24–36. Springer, Cham (2018). https://doi.org/10.1007/978-3-319-91938-6_3
6. Gunawan, A.D., DasGupta, B., Zhang, L.: A decomposition theorem and two algorithms for reticulation-visible networks. Inform. Comput. **252**, 161–175 (2017)
7. Gunawan, A.D., Lu, B., Zhang, L.: A program for verification of phylogenetic network models. Bioinformatics **32**, i503–i510 (2016)
8. Gunawan, A.D., Yan, H., Zhang, L.: Compression of phylogenetic networks and algorithm for the tree containment problem. J. Comput. Biol. **26**, 285–294 (2019)
9. Gusfield, D.: ReCombinatorics: The Algorithmics of Ancestral Recombination Graphs and Explicit Phylogenetic Networks. MIT Press, Boston (2014)
10. Gusfield, D., Eddhu, S., Langley, C.: The fine structure of galls in phylogenetic networks. INFORMS J. Comput. **16**(4), 459–469 (2004)
11. Huson, D.H., Klöpper, T.H.: Beyond galled trees - decomposition and computation of galled networks. In: Speed, T., Huang, H. (eds.) RECOMB 2007. LNCS, vol. 4453, pp. 211–225. Springer, Heidelberg (2007). https://doi.org/10.1007/978-3-540-71681-5_15
12. Huson, D.H., Rupp, R., Scornavacca, C.: Phylogenetic Networks: Concepts Algorithms and Applications. Cambridge University Press, Cambridge (2010)
13. Steel, M.: Phylogeny: Discrete and Random Processes in Evolution. SIAM, Philadelphia (2016)
14. Wang, L., Zhang, K., Zhang, L.: Perfect phylogenetic networks with recombination. J. Comput. Biol. **8**(1), 69–78 (2001)
15. Weller, M.: Linear-time tree containment in phylogenetic networks. In: Blanchette, M., Ouangraoua, A. (eds.) RECOMB-CG 2018. LNCS, vol. 11183, pp. 309–323. Springer, Cham (2018). https://doi.org/10.1007/978-3-030-00834-5_18
16. Willson, S.J.: Unique determination of some homoplasies at hybridization events. Bull. Math. Biol. **69**, 1709–1725 (2007)
17. Yan, H., Gunawan, A.D., Zhang, L.: S-cluster++: a fast program for solving the cluster containment problem for phylogenetic networks. Bioinformatics **34**(17), i680–i686 (2018)
18. Zhang, L.: On tree-based phylogenetic networks. J. Comput. Biol. **23**(7), 553–565 (2016)
19. Zhang, L.: Clusters, trees, and phylogenetic network classes. In: Warnow, T. (ed.) Bioinformatics and Phylogenetics. CB, vol. 29, pp. 277–315. Springer, Cham (2019). https://doi.org/10.1007/978-3-030-10837-3_12
20. Zhang, L.: Generating normal networks via leaf insertion and nearest neighbor interchange. BMC Bioinform. 20, article no. 642 (2019)
21. Zhang, L.: Counting tree-child networks and their subclasses. arXiv preprint arXiv:1908.01917 (2019)

Long Papers

Optimum Algorithm for the Mutual Visibility Problem

Subhash Bhagat$^{(\boxtimes)}$

ACM Unit, Indian Statistical Institute, Kolkata, India
subhash.bhagat.math@gmail.com

Abstract. We consider a distributed system of $n \geq 3$ opaque robots deployed in the Euclidean plane. If three robots lie on a line, the middle robot obstructs the visions of the two other robots. The *mutual visibility* problem requires the robots to form a configuration in which no three robots are collinear i.e., all the robots in the system are mutually visible. Robots work without any centralized control. We considers the FSTATE computational model in which each robot is endowed with an additional constant amount of persistent memory to retain some information of their previous states [3]. This information is not available to the other robots in the system. Except from this persistent memory, the robots are oblivious i.e., they do not carry forward any other information from their previous computational cycles. The robots do not have any explicit message passing capabilities. Under these weak settings, we present a deterministic distributed algorithm to solve the *mutual visibility* problem for a set of synchronous robots using only 1 bit of persistent memory. The proposed algorithm solves the mutual visibility problem in 2 rounds and guarantees collision-free movements for the robots. The algorithm is optimum in terms of round complexity, the amount of memory for the FSTATE computational model and number of movements for the robots.

Keywords: Swarm robots · Mutual visibility problem · Synchronous · Persistent memory.

1 Introduction

A *swarm* of robots is a distributed multi-robot system consisting of autonomous, homogeneous, small mobile robots. The robots work cooperatively to achieve some global task. The robots are represented as points on the Euclidean plane in which they can move freely. The robots are indistinguishable by their appearances or nature of actions. They have identical capabilities and run the same distributed algorithm. They do not have any global coordinate system. However, each robot has its own local coordinate system. The directions and orientations of the local coordinate axes of a robot may vary from the others. They do not have any explicit communication skill. The computational cycle of each robot has

Research supported in part by DST INSPIRE Faculty research grant DST/INS-PIRE/04/2015/002801 from Department of Science & Technology, India.

M. S. Rahman et al. (Eds.): WALCOM 2020, LNCS 12049, pp. 31–42, 2020.
https://doi.org/10.1007/978-3-030-39881-1_4

three phases *Look-Compute-Move*. Each robot executes the computational cycle repeatedly. In the *Look* phase, a robot takes the snapshot of its surroundings to obtain the positions of the other robots in the system. In the *Compute* phase, a robot uses the information gathered in the *Look* phase to compute a destination point. Finally, in the *Move* phase, it moves to its computed destination point. The activations of the robots are controlled by a synchronous scheduler (FSYNC model) which activates all the robots in the system in each round.

The robots may be endowed with *visible lights*. The lights can assume a constant number of predefined colors to represent different states of the robots. These visible lights can be used in three different ways by the robots: (i) the robots use these lights to retain and communicate some constant amount of information about their states, the lights are visible to all the robots in the system (we refer, this model as FALL model) or (ii) the light of a robot is visible to itself only and not to the others (FSTATE model) or (iii) the light of a robot is visible to all other robots in the system but not to itself, it is used to communicate with other robots in the system and the robot does not know the color of its light (FCOMM model) [3]. The lights can be used for communication or for internal memory or for both. This is one of the ways to implement the persistent memory. In this work, a robot uses persistent memory only to remember its states (FSTATE model) and this piece of information is not communicated to the other robots.

1.1   Earlier Works

The study of the *mutual visibility* problem was initiated by Di Luna et al. [1]. They presented a solution to the problem for a set of semi-synchronous, oblivious robots when they know n, the total number of robots in the system, and their algorithm runs in $\Omega(n^2)$ rounds for synchronous robots. Bhagat et al. presented a solution to the *mutual visibility* problem for oblivious asynchronous robots under the assumption that the robots have an agreement in one coordinate axis and knowledge of n and it takes $O(n)$ rounds for synchronous robots [4].

All the efficient algorithms in the literature for the *mutual visibility* problem for the robots with lights have considered the FALL model. Di Luna et al. were the first to present a solution to the problem under the FALL model [15]. They solved the problem for the semi-synchronous robots with 3 colors and for asynchronous robots with 3 colors under one axis agreement. Their algorithm takes $O(n)$ rounds for synchronous robots. Later, Sharma et al. provided a solution to the problem which uses only 2 colors for semi-synchronous robots and 2 colors for asynchronous robots under one-axis agreement [13]. Sharma et al. modified the algorithm presented in [1] to improve the round complexity of the algorithm for synchronous robots [14]. Their algorithm uses 3 colors for the lights and runs in $O(n\log(n))$ rounds. Vaidyanathan et al. proposed a distributed algorithm for synchronous robots using 12 colors which runs in $O(\log(n))$ rounds for $n \geq 4$ robots and assumes common *chirality* i.e., common notion of clockwise direction [2]. Sharma et al. presented algorithms which can solve the problem in 13 rounds with 12 colors for semi-synchronous robots and in 9 rounds with 9 colors for synchronous robots [12]. Sharma et al. proposed a solution to the problem which runs in $O(\log(n))$ rounds for asynchronous robots using 25 colors [11]. Sharma et al. further improved the solution which runs in $O(1)$ rounds for a set of asynchronous robots using 47 colors [6]. Bhagat and Mukhopadhyaya solved the problem for a set of asynchronous robots using 7 colors [7]. In their solution, each robot moves exactly once and it takes $O(n)$ for synchronous robots. In [8], Sharma et al. proposed a solution to the problem for a set of synchronous robots without lights. Their algorithm uses the knowledge of n and runs in $O(n)$ rounds. Recently, Bhagat and Mukhopadhyay proposed an algorithm for the problem under the FSTATE model for semi-synchronous robots using one internal bit memory [16]. In [17], Bhagat et al. proposed an algorithm to solve the problem under the FSTATE for asynchronous robots. Their algorithm assumes knowledge of n and one bit internal memory. The mutual visibility problem has also been considered under different fault models [5, 9, 10].

1.2   Our Contribution

This paper presents study of the *mutual visibility* problem for a set of synchronous robots in the Euclidean plane. A distributed algorithm has been proposed to solve the problem for a set of robots under the FSTATE model which does not have the communication overhead of the FCOMM model. The persistent memory is used only to remember information about previous states and the information is not communicated to the other robots. The proposed algorithm does not assume any other extra assumptions like agreement on the coordinate axes or chirality, knowledge of n. Our proposed solution to the *mutual visibility* problem has the following advantages;

- While all the existing solutions of the mutual visibility problem for the robots with persistent memory have considered the FALL model (both communication and internal memory purposes), our approach assumes the FSTATE model. Thus, our solution does not require explicit message passing capabilities for the robots.
- The proposed algorithm does not use the value of n which makes the algorithm easily applicable to any scalable system. This is an improvement over the algorithm proposed in [8] which uses knowledge of n to solve the problem in $O(n)$ rounds for synchronous robots.
- The proposed algorithm uses only 1 bit of persistent memory and achieves mutual visibility within 2 rounds. While the round complexity is optimum in general, the amount of memory is also optimum for the FSTATE computational model. This is an improvement over the best known efficient algorithm presented in [12] which takes 9 colors for the lights of the robots (both for communication and internal memory purposes) and runs in 9 rounds for synchronous robots.
- In our algorithm, each robot moves at most once. This implies an energy efficient solution to the *mutual visibility* problem.
- The solution also provides collision free movements for the robots.

– Our algorithm also solves the *convex hull formation* problem which requires the robots to obtain different vertices of a convex hull. One of the applications of this problem includes formation and surveillance of a protected region.

2   Model and Notations

This paper considers a set of n homogeneous, autonomous, synchronous robots in the Euclidean plane. The robots are opaque. They operate under the FSTATE model. The persistent memory of each robots holds two values: *off* and *on*. Let $s_i(t)$ denote this value for the robot r_i at time t. Initially, all robots have value *off*. These values are persistent and do not change automatically. Except for this persistent memory, the robots are oblivious i.e., they do not remember any other data from their previous computational cycles. We assume that a robot cannot be stopped by an adversary before reaching its destination point. Initially all the robots occupy distinct locations and they are stationary.

– **configurations of the robots:** Let $\mathcal{R} = \{r_1, r_2, \ldots, r_n\}$ denote the set of n robots and $r_i(t)$ denote the position of robot r_i at time $t \in \mathbb{N}$. If there is no ambiguity, we use r_i to denote both the robot and the location occupied by it. A configuration of the robot positions, $\mathcal{R}(t) = \{r_1(t), r_2(t), \ldots, r_n(t)\}$, is the set of distinct positions occupied by the robots at time t. Let $\widetilde{C}_L$ be the collection of all configurations in which all the points in $\mathcal{R}(t)$ lie on a straight line and $\widetilde{C}_{NL}$ be the set of all configurations in which there exist at least three non-collinear points in $\mathcal{R}(t)$.
– The smallest enclosing circle of the points in $\mathcal{R}(t)$ is denoted by *SEC*. Let $\mathcal{O}$ denote the center of *SEC*.
– **Vision of a robot:** If three robots r_i, r_j and r_k are collinear with r_j lying in between r_i and r_k, then r_i and r_k are not visible to each other. We define the vision, $\mathcal{V}_i(t)$, of robot r_i at time t to be the set of robot positions visible to r_i (excluding $r_i(t)$). The *visibility polygon* of r_i at time t, denoted by $\mathcal{VP}_i(t)$, is defined as follows: sort the points in $\mathcal{V}_i(t)$ angularly in anti clockwise direction w.r.t. r_i, starting from any robot position in $\mathcal{V}_i(t)$. Then connect them in that order to generate the polygon $\mathcal{VP}_i(t)$. For a robot $r_i \in \mathcal{R}$, let $\Gamma_i(t)$ be the set of angles defined as follows:
$\Gamma_i(t) = \{\angle r_j r_i r_k : r_j \text{ and } r_k \text{ are two adjacent vertices of } r_i \text{ on } \mathcal{VP}_i(t) \}$
– A straight line $\mathcal{L}$ is called a *line of collinearity* if it contains more than two distinct robot positions. Robots occupying distinct positions on $\mathcal{L}$ are termed *collinear* robots.
– By $\overline{pq}$, we denote the closed line segment joining two points p and q, including the end points p and q. Let (p, q) denote the open line segment joining the points p and q, excluding the two end points p and q. Let $|\overline{pq}|$ denote the length of $\overline{pq}$.
– Let $\mathcal{CH}(t)$ denote the convex hull of the robot positions in $\mathcal{R}(t)$. We partition the robot positions in $\mathcal{R}(t)$ into two sub-classes:

- **Vertex robots:** A robot which lie on a vertex of $\mathcal{CH}(t)$, is called a *vertex robot*. Robot r_i lies on a vertex of $\mathcal{CH}(t)$ if any one of the following is true: (i) if $\mathcal{R}(t) \in \widetilde{C}_{NL}$ and the maximum angle in $\Gamma_i(t)$ is greater than π (ii) if $\mathcal{R}(t) \in \widetilde{C}_L$ and r_i does not lie in between two other distinct robot positions.
- **Interior robots:** A robot which lies inside $\mathcal{CH}(t)$ or in between two distinct robot positions on a side of $\mathcal{CH}(t)$, is called an *interior robot*. Robot r_i is a interior robot if any one of the following is true: (i) if $\mathcal{R}(t) \in \widetilde{C}_{NL}$ and the maximum angle in $\Gamma_i(t)$ is less than or equal to π or (ii) if $\mathcal{R}(t) \in \widetilde{C}_L$ and r_i lies in between two robot positions.

- $d_{ij}^k(t)$: Let $\mathcal{L}_{ij}(t)$ denote the straight line joining r_i and r_j. The perpendicular distance of the line $\mathcal{L}_{ij}(t)$ from the point r_k is denoted by $d_{ij}^k(t)$.
- $D_i(t)$: $D_i(t)$ is the minimum distance of any two robot positions in $\{r_i(t)\} \cup \mathcal{V}_i(t)$.

Observation 1. *Let $\triangle ABC$ be a triangle such that $\angle ABC = 90^\circ$. Consider a point D on the side BC. Suppose the bisectors of the angles $\angle ACB$ and $\angle ADB$ intersect the side AB at the points P_c and P_d. Then, the point P_d lies on the open segment (B, P_c).*

3   Mutual Visibility Under the *FSYNC* model

This section presents a deterministic distributed algorithm to solve the *mutual visibility* problem for a set of fully synchronous robots. The following lemma shows that any deterministic distributed algorithm for the *mutual visibility* problem will take at least 2 rounds.

Lemma 1. *Consider an initial robot configuration in which all the robots lie on exactly one line and they are equally spaced on this line. Starting from this configuration, any deterministic distributed algorithm will take at least 2 rounds to solve the mutual visibility problem.*

3.1   Algorithm MutualVisibilityFsync()

Consider an initial robot configuration $\mathcal{R}(t_0)$. The outline of our algorithm is as follows; to obtain mutual visibility, our algorithm places all the robots on the vertices of a convex hull. Since robots are opaque, they may not have the complete view of all the robot positions in $\mathcal{R}(t_0)$ and hence they cannot compute $\mathcal{CH}(t_0)$. We divide our algorithm into two phases.

The phase 1 is the *vertex adjustment* phase in which all the robots having positions on the vertices of $\mathcal{CH}(t_0)$ (i.e., the vertex robots) move in such a way that (i) they become visible to all the interior robots and (ii) they remain as the vertex robots on $\mathcal{CH}(t)$. The phase 2 is *interior adjustment phase* in which all robots lying within or on the edges of $\mathcal{CH}(t)$, for some $t > t_0$, move to become vertex robots on a convex hull $\mathcal{CH}(t^*)$, for some $t* > t$. Details of the algorithm are as follows.

3.1.1   Different Actions of the Robots

A robot r_i acts according to the following:

1. **Vertex adjustment phase:** If r_i is a vertex robot and $s_i = off$, then it changes the value of s_i to on, computes a destination point and moves to it. If r_i is an interior robot and $s_i = off$, then it changes the value of s_i to on and does not move.
2. **Interior adjustment phase:** If r_i is a vertex robot with $s_i = on$, it does nothing. If r_i is an interior robot and $s_i = on$, then it computes a destination point and move straight to it without changing the value of s_i.

Note that a robot r_i identifies the phase of the algorithm by checking the value of s_i. Robots use algorithm *ComputeDestination()* to compute their destination points.

3.2   Algorithm ComputeDestination()

This section describes algorithm *ComputeDestination()* which is used by the robots to compute destination points. Algorithm *ComputeDestination()* should satisfy the following conditions: (i) for two different robots it should return two distinct destination points and (ii) the paths of movements of two distinct robots should not intersect each other in between and (iii) the destination points of all the robots lie on the vertices of a convex hull. Condition (i) guarantees that two different robots occupy two distinct positions in the final configuration, condition (ii) provides collision free movements for the robots and condition (iii) guarantees mutual visibility among all the robots.

(A) **Vertex adjustment phase:** Consider a vertex robot r_i on $\mathcal{CH}(t_0)$. Robot r_i computes its destination point in the following way:
  – **The direction of movement:** First suppose that $\mathcal{R}(t_0) \in \widetilde{C}_{NL}$. Let r_j and r_k be the two adjacent robot positions of r_i on $\mathcal{CH}(t_0)$ (note that r_j and r_k may be two interior robot positions lying on the edges of $\mathcal{CH}(t_0)$). Let ray_{ij} and ray_{ik} be the two rays originated at r_i and they lie along the straight lines $\mathcal{L}_{ij}$ and $\mathcal{L}_{ik}$ respectively such that they do not contain r_j and r_k. Let W_i be the wedge defined by the rays ray_{ij} and ray_{ik} which does not contain any robot position. Consider a robot position $r_m \in V_i(t_0)$ and let θ_m denote the smallest angle made by $\mathcal{L}_{im}$ with the boundary of W_i. Consider the set $\Gamma_i^*(t_0)$ of all such θ_m such that θ_m completely lies within W_i. Let β_i be the maximum angle in Γ_i^* (ties are broken arbitrarily) and $Bisec_i$ denote the bisector of β_i. The direction of movement of r_i is along $Bisec_i$.
  Suppose $\mathcal{R}(t_0) \in \widetilde{C}_L$. In this case, $|V_i(t_0)| = 1$. Let r_j be the visible robot to r_i. Let $\mathcal{L}^*$ be the perpendicular line to the line of collinearity $\hat{\mathcal{L}}$ at the point r_j. The robot r_i arbitrarily chooses a direction along $\mathcal{L}^*$ and let $\mathcal{L}^+$ denote the ray along this direction. Let v_i be a point on $\mathcal{L}^+$ such that $\angle r_j r_i v_i = \frac{\pi}{4}$. Let ray_i be the ray originated at r_i and passing through v_i.

The direction of movement of r_i is along $DIR_i(t_0)$ which is defined as follows:

$$DIR_i(t_0) = \begin{cases} Bisec_i & \text{if } \mathcal{R}(t_0) \in \tilde{C}_{NL} \\ ray_i & \text{if } \mathcal{R}(t_0) \in \tilde{C}_L \end{cases}$$

- **The amount of displacement:**
  Let $d_i = \min\{d_{ij}^k(t_0), d_{ik}^j(t_0), d_{jk}^i(t_0) : \forall r_j, r_k \in V_i(t_0)\}$. Let σ_i denote the amount of displacement of r_i and it is defined as follows,

$$\sigma_i = \begin{cases} \frac{1}{34}\min\{d_i, D_i(t_0)\} & \text{if } \mathcal{R}(t_0) \in \tilde{C}_{NL} \\ |\overline{r_i v_i}| & \text{if } \mathcal{R}(t_0) \in \tilde{C}_L \end{cases}$$

  The fraction in the computation of $\sigma_i(t_0)$ is chosen in such a way that three initially non-collinear robots do not become collinear after their movements. Other suitable values will also work.
- **The destination point:** Let z_i be the point on $DIR_i(t_0)$ at distance σ_i from r_i. The destination point of r_i is z_i.

(B) **Interior adjustment phase:** In this phase, all interior robots move to become vertices of a convex hull $\mathcal{CH}(t^*)$. The interior robots move in such a way that the existing vertex robots remain the same. First, we consider the scenario in which $\mathcal{R}(t)$ is reached from $\mathcal{R}(t_0) \in \tilde{C}_L$. An interior robot r_i can easily identify this by checking (i) $s_i = on$ and (ii) $|V_i| = 4$ (two of them are from $V_i(t_0)$ and two other were the vertex robots in $\mathcal{R}(t_0)$). Let r_a and r_b be the two distinct robot positions in $V_i(t)$ such that they lie on perpendicular lines to the line segment $\overline{r_c r_d}$ where $r_c, r_d \in V_i(t)\backslash\{r_a, r_b\}$. Let $\mathcal{C}_{ab}$ be the circle having $\overline{r_a r_b}$ as diameter. Let $\mathcal{L}_i$ be the perpendicular to $\overline{r_c r_d}$ at r_i and it intersects $\mathcal{C}_{ab}$ at the points u_i and v_i. Robot r_i arbitrarily chooses any one of u_i, v_i as its destination point.

Otherwise, suppose $\mathcal{R}(t)$ is reached from $\mathcal{R}(t_0) \in \tilde{C}_{NL}$. Suppose r_i is a robot not lying on a vertex of $\mathcal{CH}(t)$. In this case, r_i computes a destination point to become a vertex of a larger convex hull containing $\mathcal{CH}$. Note that, in this phase, all the robots lying on the vertices of $\mathcal{CH}(t)$ are visible to the interior robots and thus the interior robots can easily compute the same SEC. To describe the method, we define the following notations:

Consider the smallest enclosing circle SEC of $\mathcal{R}(t)$. For a robot $r_i \in \mathcal{R}$, not lying at $\mathcal{O}$, let rad_i denote the ray originated from $\mathcal{O}$ and passing through r_i. We call such half-lines *radial rays*. Radial ray rad_i may contain other robot positions. Let $\mathcal{H}_{rad}$ denote the set of all such distinct radial rays passing through the robot positions in $\mathcal{R}(t)$ not lying at $\mathcal{O}$. Our algorithm places the interior robots on some circular arcs. These circular arcs have vertices of $\mathcal{CH}(t)$ as their end points and any set of points lying on these circular arcs are in convex position. We call these arcs as *safe arcs*. To compute *safe arcs*, we use the notion of *safe regions* used in [2].

Safe region: Consider an edge e_{xy} of $\mathcal{CH}(t)$ formed by vertex robot positions r_x and r_y (Fig. 1). Let r_k be the other neighbour of r_x on the boundary of $\mathcal{CH}(t)$. Note that both the robots r_y and r_k are vertex robots and visible

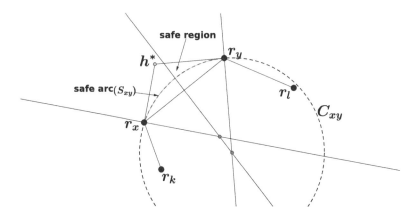

Fig. 1. An illustration of computation in interior adjustment phase: safe arc

to r_x (even if initially they were not visible to r_x due to some robot positions on the edges of $\mathcal{CH}(t_0)$, after Phase 1 they become visible to r_x). Similarly, consider the other neighbour, say r_l, of r_y on the boundary of $\mathcal{CH}(t)$. Consider α_x and α_y, the interior angles of $\mathcal{CH}$ at r_x and r_y respectively. Now, we define the notion of *safe region* used in [2]. The *safe region* with respect to e_{xy} is the interior of the triangle $\triangle r_x h r_y$ where h lies outside of $\mathcal{CH}$, $\angle r_y r_x h = \frac{\pi - \alpha_x}{4}$, and $\angle r_x r_y h = \frac{\pi - \alpha_y}{4}$. The safe region can be computed using the positions r_x, r_y, r_k and r_l.

Safe arc: For an edge e_{xy} of $\mathcal{CH}(t)$, a *safe arc* is a circular arc between r_x and r_y lying completely within the intersection area of SEC and the *safe region* w.r.t. e_{xy}. We constrain all the *safe arcs* to lie within SEC and this with the Observation 1 helps us to prove that during movements, no two paths of two distinct robots cross each other (this is to avoid collisions). In order to compute such a *safe arc*, we define a point h^* as follows: if the *safe region* $\triangle r_x h r_y$, lies completely within SEC, then $h^* = h$. Otherwise, h^* is the intersection point between SEC and T_{xy} which is closest to e_{xy}, where T_{xy} is the perpendicular bisector of the edge e_{xy}. The *safe arc* lies within the triangle $\triangle r_x h^* r_y$. Since a *safe arc* is circular and lies within a *safe region*, all robots lying on it are in convex positions and thus mutually visibility. Let S_{xy} denote the *safe arc* w.r.t. e_{xy} and C_{xy} be the corresponding circle of which S_{xy} is an arc. A safe arc w.r.t. an edge e_{xy} of $\mathcal{CH}(t)$ is computed in the following way: let L_x and L_y be the lines perpendicular to the line segments $\overline{h^* r_x}$ and $\overline{h^* r_y}$ at the points r_x and r_y respectively (Fig. 1). Let L_x and L_y intersect T_{xy} at the points v_x and v_y. These two intersection points are equally distanced from r_x and r_y. Draw circles having v_x and v_y as centers and passing through r_x and r_y. It can be easily proven that at least one of them have arc between r_x and r_y which completely lies within the $\triangle r_x h^* r_y$. Then C_{xy} is the circle which has an arc lying within the $\triangle r_x h^* r_y$ and closer to e_{xy}. Arc S_{xy} is the portion of C_{xy} lying within $\triangle r_x h^* r_y$.

Destination point: An interior robot r_i computes its destination point in the following way:

- If r_i lies on $\mathcal{O}$, it chooses two arbitrary rays rad_j and rad_k which are adjacent to each other (i.e., there is no other rad_m lying in between them). Let $\mathcal{L}_i$ be the bisector of the angle made by rad_j and rad_k at r_i. We consider $\mathcal{L}_i$ as rad_i in this case. Let $\mathcal{L}_i$ intersect S_{ab} at the point p_i, for some vertex robot positions r_a and r_b on $\mathcal{CH}$. Robot r_i marks p_i as its destination point.
- Suppose $r_i \neq \mathcal{O}$. If there is no other robot position on the line segment $\overline{r_i p_i}$, then the intersection point between $\overline{r_i p_i}$ and S_{ab} is the destination point for r_i, for some vertex robot positions r_a and r_b. Otherwise, robot r_i computes its destination in the following way:
  - Let r_c and r_d be the two robot positions, none of them lying on $\mathcal{O}$, such that the wedge defined by rad_c and rad_d contains exactly one radial ray rad_i. Let W_c and W_d denote the wedges defined by rad_i with rad_c and rad_b respectively.
  - Without loss of generality, suppose $\angle r_i \mathcal{O} r_c > \angle r_i \mathcal{O} r_d$ (if tie, broken arbitrarily). The wedge W_c is divided into four equal parts and the part closest to rad_i is considered. We denote this part by W_1. Let $\mathcal{L}_{w_1} \neq rad_i$ be the other half-line which defines W_1.
  - Let $\mathcal{L}_i^*$ be the line perpendicular to rad_i at the point p_i (this line is a tangent to SEC at p_i) and u_i denote the intersection point between $\mathcal{L}_i^*$ and $\mathcal{L}_{w_1}$.
  - Let V_i be the bisector of the angle $\angle p_i r_i u_i$. Let V_i intersect S_{ab} at the point q_i, for some vertex robot positions r_a and r_b.
  - robot r_i marks q_i as its destination and moves to this point along the line segment $\overline{r_i q_i}$.

A robot terminates the execution of algorithm $MutualVisibility()$ when it finds itself on a hull vertex with $s_i = on$.

3.3   Correctness of MutualVisibilityFsync()

In this section we prove that algorithm $MutualVisibilityFsync()$ solves the mutual visibility problem in finite time.

Lemma 2. *Let r_i, r_j and r_k be three arbitrary robots, which are not initially collinear. They do not become collinear after the vertex adjustment phase.*

Proof. The lemma is obvious if the three robots do not move in the first round. Suppose at least one of them moves. We prove the lemma by showing that none of the three values $d_{ij}^k(t_0)$, $d_{ik}^j(t_0)$ and $d_{jk}^i(t_0)$ becomes zero. Without loss of generality, we prove that $d_{ij}^k(t_0)$ never vanishes. Robots move in synchronous round and in each movement a robot can reduce the value of $d_{ij}^k(t_0)$ by maximum an amount of $\frac{1}{34} d_{ij}^k(t_0)$. In a single round, the maximum reduction in the value of $d_{ij}^k(t_0)$, due to the movements of all the three robots, is bounded above by

$\frac{3}{3^4}d_{ij}^k(t_0)$. Thus, in a single round, the three robots can optimally reduce the value of $d_{ij}^k(t_0)$ to $(1 - \frac{3}{3^4})d_{ij}^k(t_0)$. This implies that the value of $d_{ij}^k(t_0)$ does not become zero, after the first round.

Lemma 3. *If a new line of collinearity $\mathcal{L}_{new}$ is created after the vertex adjustment phase, then $\mathcal{L}_{new}$ contains exactly three robots and all these robots were initially lying on the same line of collinearity.*

Proof. In the first round of the algorithm, only the robots on the hull vertices of $\mathcal{CH}(t_0)$ move. Since no three non-collinear robots become collinear, all the robots on $\mathcal{L}_{new}$ were initially lying on the same line of collinearity. The robots lying interior of $\mathcal{CH}(t_0)$ and on the edges of $\mathcal{CH}(t_0)$ in between two other robots, do not move in the first round. This imples that from each line of collinearity at most two robots move and these two robots lie on the two corner robot positions on that line. These two robots may have directions of movements lying on two different sides of the line of collinearity. Thus, the line, joining their new position after the first round, may pass through a robot position which was initially collinear with them. Hence, the lemma is true.

Lemma 4. *Suppose $\mathcal{R}(t_0) \in \widetilde{C}_{NL}$. Then, after the vertex adjustment phase, all the vertex robots on $\mathcal{CH}(t)$ are visible to all the interior robots.*

Lemma 5. *Algorithm ComputeDestination() returns two distinct destination point for two different robots. Furthermore, paths to the destination points of two different robots do not intersect each other in between.*

Proof. Consider two distinct robots r_i and r_j. Following are the different scenarios:

- r_i **and** r_j **move in the vertex adjustment phase:** In this scenario, r_i and r_j are vertex robots. If initially all the robots are collinear i.e., $\mathcal{R}(t_0) \in \widetilde{C}_L$, then r_i and r_j move along the parallel lines and hence the lemma is true in this case. Otherwise, the destination points of these two robots lie within the distance $\frac{1}{3}dist(r_i, r_j)$ from each of the points r_i and r_j. This implies the lemma in this scenario.
- r_i **and** r_j **move in the interior adjustment phase:** If the initial configuration $\mathcal{R}(t_0)$ was in $\widetilde{C}_L$, then robots move along parallel lines (these lines are perpendicular to the initial line of collinearity of the robot positions). This implies the lemma in this case. Otherwise, first suppose that one of them, say r_i, lies on $\mathcal{O}$. Let the destination point of r_j lies in the wedge W_c for some neighbour robot r_c of r_j. Since r_i lies on $\mathcal{O}$, it moves along a bisector of an angle in $\Gamma_i(t)$. This bisector passes either passes through the middle of W_c or it does not pass through W_c. The lemma is obvious in the later case. Consider the former case. The destination point of r_j lies in the sub-wedge W_1 of W_c where $W_1 = \frac{1}{4}W_c$ and W_1 is adjacent to rad_j. Thus, the destination points of r_i and r_j lie in two completely disjoint wedges. This implies the lemma in this case. Now suppose that none of them lies at $\mathcal{O}$. Consider the case when rad_i and rad_j are two different lines. Let W_{ij} be the wedge defined by rad_i

and rad_j. If the destination points of these two robots lie in two different wedges adjacent to them, then we are done. Suppose the destination points of the both robots lie in W_{ij}. In this case, the destination points of r_i and r_j lie in two wedges completely separated by the line passing through the middle of W_{ij}. Thus the lemma true in this case. Finally, consider the case in which rad_i coincides which rad_j i.e., r_i and r_j are collinear with $\mathcal{O}$. If their destination point lies in two different wedges adjacent to rad_i, then we are done. Otherwise, W_{ij} contains the destination points of r_i and r_j. Without loss of generality, suppose that r_j lies in between r_i and p_i (the intersection point between rad_i and SEC). Consider the triangle $\triangle p_i r_i u_i$ (u_i is the intersection point between the tangent to SEC at p_i and the other line defining the $\frac{1}{4}^{th}$ sector of W_{ij}, closest to rad_i). Since $\angle r_i p_i u_i = \frac{\pi}{2}$, by observation 1, the destination point of r_j lies within the line segment $p_i h_i$, where h_i is the intersection point between the bisector of $\angle p_i r_i u_i$ and the line segment $\overline{p_i u_i}$. This implies that r_i and r_j have distinct destination points. Since the segment $\overline{p_i u_i}$ lies outside the circle SEC and robots move in straight lines, the paths of r_i and r_j to their respective destination points do not intersect each other in between.

Corollary 1. *During the whole execution of algorithm MutualVisibilityFsync(), no two robots collide.*

Lemma 6. *After the interior adjustment phase, all the interior robots in $\mathcal{CH}(t_0)$ become vertex robots on $\mathcal{CH}(t^*)$, $t^* > t_0$.*

Proof. The lemma follows from the lemma 5 and the fact that all the safe arcs are circular and lie in the safe regions.

Lemma 7. *Algorithm MutualVisibilityFsync() takes exactly two rounds to place all the robots on the boundary of $\mathcal{CH}(t^*)$, $t^* > t_0$. Furthermore, during the whole execution of the algorithm, each robot moves at most once.*

Theorem 1. *Algorithm MutualVisibilityFsync() solves the mutual visibility problem for a set of fully-synchronous robots under FSTATE model using 1 bit persistent memory and within 2 rounds.*

References

1. Di Luna, G.A., Flocchini, P., Poloni, F., Santoro, N., Viglietta, G.: The mutual visibility problem for oblivious robots. In: Proceedings 26th Canadian Conference on Computational Geometry (CCCG 2014) (2014)
2. Vaidyanathan, R., Busch, C., Trahan, J.L., Sharma, G., Rai, S.: Logarithmic-time complete visibility for robots with lights. In: Proceedings Parallel and Distributed Processing Symposium (IPDPS), pp. 375–384 (2015)
3. Flocchini, P., Santoro, N., Viglietta, G., Yamashita, M.: Rendezvous of two robots with constant memory. In: Moscibroda, T., Rescigno, A.A. (eds.) SIROCCO 2013. LNCS, vol. 8179, pp. 189–200. Springer, Cham (2013). https://doi.org/10.1007/978-3-319-03578-9_16

4. Bhagat, S., Chaudhuri, S.G., Mukhopadhyaya, K.: Formation of general position by asynchronous mobile robots under one-axis agreement. In: Kaykobad, M., Petreschi, R. (eds.) WALCOM 2016. LNCS, vol. 9627, pp. 80–91. Springer, Cham (2016). https://doi.org/10.1007/978-3-319-30139-6_7

5. Aljohani, A., Sharma, G.: Complete visibility for mobile robots with lights tolerating faults. Int. J. Netw. Comput. **8**(1), 32–52 (2018)

6. Sharma, G., Vaidyanathan, R., Trahan, J.L.: Constant-time complete visibility for asynchronous robots with lights. In: Spirakis, P., Tsigas, P. (eds.) SSS 2017. LNCS, vol. 10616, pp. 265–281. Springer, Cham (2017). https://doi.org/10.1007/978-3-319-69084-1_18

7. Bhagat, S., Mukhopadhyaya, K.: Optimum algorithm for mutual visibility among asynchronous robots with lights. In: Spirakis, P., Tsigas, P. (eds.) SSS 2017. LNCS, vol. 10616, pp. 341–355. Springer, Cham (2017). https://doi.org/10.1007/978-3-319-69084-1_24

8. Sharma, G., Busch, C., Mukhopadhyay, S.: Complete visibility for oblivious robots in $\mathcal{O}(N)$ time. In: Podelski, A., Taïani, F. (eds.) NETYS 2018. LNCS, vol. 11028, pp. 67–84. Springer, Cham (2019). https://doi.org/10.1007/978-3-030-05529-5_5

9. Aljohani, A., Poudel, P., Sharma, G.: Fault-tolerant complete visibility for asynchronous robots with lights under one-axis agreement. In: Rahman, M.S., Sung, W.-K., Uehara, R. (eds.) WALCOM 2018. LNCS, vol. 10755, pp. 169–182. Springer, Cham (2018). https://doi.org/10.1007/978-3-319-75172-6_15

10. Sharma, G.: Mutual visibility for robots with lights tolerating light faults. In: Proceedings IEEE International Parallel and Distributed Processing Symposium Workshops (IPDPSW), pp. 829–836 (2018)

11. Sharma, G., Vaidyanathan, R., Trahan, J.L., Busch, C., Rai, S.: O(log N)-time complete visibility for asynchronous robots with lights. In: Proceedings Parallel and Distributed Processing Symposium (IPDPS), pp. 513–522 (2017)

12. Sharma, G., Vaidyanathan, R., Trahan, J.L., Busch, C., Rai, S.: Complete visibility for robots with lights in $O(1)$ Time. In: Bonakdarpour, B., Petit, F. (eds.) SSS 2016. LNCS, vol. 10083, pp. 327–345. Springer, Cham (2016). https://doi.org/10.1007/978-3-319-49259-9_26

13. Sharma, G., Busch, C., Mukhopadhyay, S.: Mutual visibility with an optimal number of colors. In: Bose, P., Gąsieniec, L.A., Römer, K., Wattenhofer, R. (eds.) ALGOSENSORS 2015. LNCS, vol. 9536, pp. 196–210. Springer, Cham (2015). https://doi.org/10.1007/978-3-319-28472-9_15

14. Sharma, G., Busch, C., Mukhopadhyay, S.: Bounds on mutual visibility algorithms. In: Proceedings 27th Canadian Conference on Computational Geometry (CCCG 2015) (2015)

15. Di Luna, G.A., Flocchini, P., Gan Chaudhuri, S., Poloni, F., Santoro, N., Viglietta, G.: Mutual visibility by luminous robots without collisions. Inf. Comput. **254**, 392–418 (2017)

16. Bhagat, S., Mukhopadhyaya, K.: Mutual visibility by robots with persistent memory. In: Chen, Y., Deng, X., Lu, M. (eds.) FAW 2019. LNCS, vol. 11458, pp. 144–155. Springer, Cham (2019). https://doi.org/10.1007/978-3-030-18126-0_13

17. Bhagat, S., Gan Chaudhuri, S., Mukhopadhyaya, K.: Mutual visibility for asynchronous robots. In: Censor-Hillel, K., Flammini, M. (eds.) SIROCCO 2019. LNCS, vol. 11639, pp. 336–339. Springer, Cham (2019). https://doi.org/10.1007/978-3-030-24922-9_24

Routing in Histograms

Man-Kwun Chiu[1], Jonas Cleve[1] (ID), Katharina Klost[1], Matias Korman[2],
Wolfgang Mulzer[1] (ID), André van Renssen[3] (ID), Marcel Roeloffzen[4],
and Max Willert[1]([✉])

[1] Institut für Informatik, Freie Universität Berlin, 14195 Berlin, Germany
`chiumk@zedat.fu-berlin.de`,
{`jonascleve,kathklost,mulzerm,willerma`}`@inf.fu-berlin.de`
[2] Department of Computer Science, Tufts University, Medford, MA, USA
`matias.korman@tufts.edu`
[3] School of Computer Science, University of Sydney, Sydney, Australia
`andre.vanrenssen@sydney.edu.au`
[4] Department of Mathematics and Computer Science, TU Eindhoven,
Eindhoven, The Netherlands
`m.j.m.roeloffzen@tue.nl`

Abstract. Let P be an x-monotone orthogonal polygon with n vertices.
We call P a *simple histogram* if its upper boundary is a single edge; and
a *double histogram* if it has a horizontal chord from the left boundary to
the right boundary. Two points p and q in P are *co-visible* if and only if
the (axis-parallel) rectangle spanned by p and q completely lies in P. In
the *r-visibility graph* $G(P)$ of P, we connect two vertices of P with an
edge if and only if they are co-visible. We consider *routing with preprocessing* in $G(P)$. We may preprocess P to obtain a *label* and a *routing
table* for each vertex of P. Then, we must be able to route a packet
between any two vertices s and t of P, where each step may use only
the label of the target node t, the routing table and the neighborhood of
the current node, and the packet header. The routing problem has been
studied extensively for general graphs, where truly compact and efficient
routing schemes with polylogartihmic routing tables have turned out to
be impossible. Thus, special graph classes are of interest.

We present a routing scheme for double histograms that sends any
data packet along a path of length at most twice the (unweighted) shortest path distance between the endpoints. The labels, routing tables, and
headers need $O(\log n)$ bits. For the simple histograms, we obtain a routing scheme with optimal routing paths, $O(\log n)$-bit labels, one-bit routing tables, and no headers.

M.-K. Chiu, J. Cleve and W. Mulzer partially supported by ERC STG 757609. J.
Cleve partially supported by DFG grant MU 3501/1-2. M. Korman partially sup-
ported by MEXT KAKENHI No. 17K12635 and the NSF award CCF-1422311. A. van
Renssen partially supported by JST ERATO Grant Number JPMJER1201, Japan.

M. S. Rahman et al. (Eds.): WALCOM 2020, LNCS 12049, pp. 43–54, 2020.
https://doi.org/10.1007/978-3-030-39881-1_5

1   Introduction

The *routing problem* is a classic question in distributed graph algorithms [6,10]. We would like to preprocess a graph G for the following task: given a data packet located at a *source* vertex s, route the packet to a *target* vertex t, identified by its *label*. We strive for three main properties: *locality* (to find the next step of the packet, the scheme should use only information at the current vertex or in the packet header), *efficiency* (the packet should choose a path that is similar to a shortest path between s and t), and *compactness* (the space for labels, routing tables, and packet headers should be as small as possible). The ratio between the length of the *routing path* and a shortest path is called *stretch factor*.

Obviously, we could store at each vertex v of G the complete shortest path tree of v. This routing scheme is local and perfectly efficient: we can send the packet along a shortest path. However, the scheme lacks compactness. Thus, the general challenge is to balance the (potentially) conflicting goals of compactness and efficiency, while maintaining locality.

There are many routing schemes for general graphs (e.g., [11] and the references therein). For example, the scheme by Roditty and Tov [11] stores a poly-logarithmic number of bits in the packet header, and it routes a packet from s to t on a path of length $O(k\Delta + m^{1/k})$, where $k > 2$ is any fixed integer, Δ is the shortest path distance between s and t, and m is the number of edges. The routing tables use $mn^{O(1/\sqrt{\log n})}$ total space, where n is the number of vertices. In the late 1980's, Peleg and Upfal [10] proved that in general graphs, any routing scheme with constant stretch factor must store $\Omega(n^c)$ bits per vertex, for some constant $c > 0$. This provides ample motivation to focus on special graph classes to obtain better routing schemes. For instance, trees admit routing schemes that always follow the shortest path and that store $O(\log n)$ bits at each node [5,13]. Moreover, in planar graphs, for any fixed $\varepsilon > 0$, there is a routing scheme with a poly-logarithmic number of bits in each routing table that always finds a path that is within a factor of $1 + \varepsilon$ from optimal [12]. Similar results are also available for unit disk graphs [9,14].

Another approach is *geometric routing*: the graph lies in a geometric space, and the routing algorithm must find the next vertex for the packet based on the coordinates of the source and the target vertex, the current vertex, and its neighborhood; e.g., [3] and the references therein. In contrast to compact routing schemes, there are no routing tables, and the routing is purely based on the local geometry (and possibly the packet header). For example, the routing algorithm for triangulations by Bose and Morin [4] uses the line segment between the source and the target for its routing decisions. In a recent result, Bose *et al.* [3] show that if vertices do not store any routing tables, no geometric routing scheme can achieve stretch factor $o(\sqrt{n})$. This holds irrespective of the header size.

Here, we combine the approaches from routing in abstract graphs and from geometric routing. For this, we consider a particularly interesting and practically relevant class of geometric graphs, namely visibility graphs of polygons. Banyassady *et al.* [1] presented a routing scheme for polygonal domains with n vertices and h holes that uses $O(\log n)$ bits for the label, $O((\varepsilon^{-1} + h) \log n)$ bits for the routing tables, and achieves a stretch of $1 + \varepsilon$, for any fixed $\varepsilon > 0$. However, their approach is efficient only if the edges of the visibility graph are weighted with their Euclidean lengths. Banyassady *et al.* ask whether there is an efficient routing scheme for visibility graphs with unit weights (the *hop-distance*). This setting seems to be more relevant in practice, and similar results have already been obtained in unit disk graphs for routing schemes [9,14] and for spanners [2,8].

To address this problem, we use *routing tables* at the vertices to represent information about the structure of the graph (as in abstract routing), but we also assume that the labels of all adjacent vertices are directly visible at a node in a *link table* (as in geometric routing). This aligns well with the practical situation, as a node must be aware of its physical neighbors and their labels for meaningful communication to be possible. The size of this link table does not count for the compactness, as it depends on the graph and cannot be influenced during preprocessing. We focus on r-visibility graphs of orthogonal simple and double histograms. At first, this may seem a strong restriction. However, even this case turns out to be quite challenging and reveals the whole richness of the compact routing problem in unweighted, geometrically defined graphs: on the one hand, the problem is still highly nontrivial, while on the other hand, much better results than in general graphs are possible. Histograms constitute a natural starting point, as they are often crucial building blocks in visibility problems. Moreover, r-visibility is a popular concept in orthogonal polygons that enjoys many useful structural properties; see, e.g., [7] and the references therein for more background on histograms and r-visibility.

A simple histogram is a monotone orthogonal polygon whose upper boundary consists of a single edge; a double histogram is a monotone orthogonal polygon that has a horizontal chord that touches the boundary of P only at the left and the right boundary. Let P be a (simple or double) histogram with n vertices. Two vertices v and w in P are connected in the visibility graph $G(P)$ by an unweighted edge if and only if the axis-parallel rectangle spanned by v and w is contained in the (closed) region P. We say that v and w are *co-visible*. We present the first efficient and compact routing schemes for polygonal domains under the hop-distance. In particular, in simple histograms, we can route along a shortest path with no headers, $O(\log n)$-bit labels, and $O(1)$-bit routing tables. In double histograms, we achieve stretch factor 2 and need labels, routing tables, and headers of $O(\log n)$ bits. The precise results are in Theorems 3.4 and 4.12. For space reasons, all proofs are deferred to the full version (available at http://arxiv.org/abs/1902.06599).

2 Preliminaries

Let $G = (V, E)$ be a *simple, undirected, unweighted, connected graph*. The (closed) *neighborhood* $N(v)$ of a vertex $v \in V$ is the set containing v and its adjacent nodes. Let $v, w \in V$. A sequence $\pi : \langle v = p_0, p_1, \ldots, p_k = w \rangle$ of vertices with $p_{i-1} p_i \in E$, for $i = 1, \ldots, k$, is called a *path* of length k between v and w.

The length of π is denoted $|\pi|$. We define $d(v,w) = \min_\pi |\pi|$ as the length of a shortest path between v and w, where π goes over all paths between v and w. The distance $d(v,w)$ between two vertices $v, w \in V$ is called the *hop distance* of v and w. Next, we define a *routing scheme*. The algorithm that decides the next step of the packet is modeled by a *routing function*. During preprocessing, every node is assigned a (binary) *label* that identifies it in the network. The routing function uses local information at the current node, the label of the target node, and the *header* stored in the packet. The local information of a node v has two parts: (i) the *link table*, a list of the labels of $N(v)$, and (ii) the *routing table*, a bitstring chosen during preprocessing to represent relevant topological properties of G. Formally, a routing scheme of a graph G consists of (a) a label $\text{lab}(v) \in \{0,1\}^+$ for each node $v \in V$; (b) a routing table $\rho(v) \in \{0,1\}^*$ for each node $v \in V$; and (c) a routing function $f\colon \left(\{0,1\}^*\right)^4 \to \left(\{0,1\}^*\right)^2$. The routing function takes the link table and routing table of a current node $s \in V$, the label $\text{lab}(t)$ of a target t, and a header $h \in \{0,1\}^*$. From these, it determines the label $\text{lab}(v)$ of a node adjacent to s and a new header h'. The local information in the packet is updated to h', and it is sent to v. The routing scheme is *correct* if: for any two sites $s, t \in V$, consider the sequence $(\ell_0, h_0) = (\text{lab}(s), \varepsilon)$ and $(\text{lab}(p_{i+1}), h_{i+1}) = f\Big(\text{lab}(N(p_i)), \rho(p_i), \text{lab}(t), h_i\Big)$, for $i \geq 0$. Then, there is a $k = k(s,t) \geq 0$ with $p_k = t$ and $p_i \neq t$, for $i = 0, \ldots, k-1$. We say the routing scheme *reaches* t in k steps, and $\pi : \langle p_0, \ldots, p_k \rangle$ is the *routing path* from s to t. The *routing distance* is denoted $d_\rho(s,t) = |\pi|$. Let $\mathcal{R}$ be a correct routing scheme for a graph class $\mathcal{G}$, i.e., $\mathcal{R}$ is a correct routing scheme for every graph in $\mathcal{G}$. There are several measures for the quality of $\mathcal{R}$. For one, the various pieces of information used for the routing should be small. This is measured by the *maximum label size* $\text{Lab}(n)$, the *maximum routing table size* $\text{Tab}(n)$, and the *maximum header size* $H(n)$, over all graphs in $\mathcal{G}$ of a certain size. They are defined as $\text{Lab}(n) = \max_{|V|=n} \max_{v \in V} |\text{lab}(v)|$, $\text{Tab}(n) = \max_{|V|=n} \max_{v \in V} |\rho(v)|$, and $H(n) = \max_{|V|=n} \max_{s \neq t \in V} \max_{i=0,\ldots,k(s,t)} |h_i|$. Furthermore, the *stretch* $\zeta(n)$ relates the length of the routing path to the shortest path: $\zeta(n) = \max_{|V|=n} \max_{s \neq t \in V} d_\rho(s,t)/d(s,t)$.

Let P be a *simple orthogonal (axis-aligned) polygon* with n vertices $V(P)$ so that no three vertices in $V(P)$ are on the same vertical or horizontal line. The vertices are indexed counterclockwise from 0 to $n-1$; the lexicographically largest vertex has index $n-1$. For $v \in V(P)$, we write v_x and v_y for the x- and y-coordinate, and v_{id} for the index. We consider r-*visibility*: $p, q \in P$ *see each other* (are *co-visible*) if and only if the axis-aligned rectangle spanned by p and q is inside (the closed set) P. The *visibility graph* $G(P) = \big(V(P), E(P)\big)$ of P has an edge between two vertices $v, w \in V(P)$ if and only if v and w are co-visible.

A *histogram* is an x-monotone orthogonal polygon where the upper boundary consists of exactly one horizontal edge, the *base edge*. By our convention, the endpoints of the base edge have index 0 (left) and $n-1$ (right). They are called the *base vertices*. A *double histogram* is an x-monotone orthogonal polygon P with a *base line*, a horizontal line segment whose relative interior lies in the interior of P and whose left and right endpoints are on the left and right boundary edge of P. We assume that the base line lies on the x-axis. Two vertices v, w in P lie *on the same side* if both are below or above the base line, i.e., if $v_y w_y > 0$. Every histogram is also a double histogram. From now on, we let P denote a (double) histogram.

Next, we classify the vertices of P. A vertex v in P is incident to exactly one horizontal edge h. We call v a *left* vertex if it is the left endpoint of h; otherwise, v is a *right* vertex. Furthermore, v is *convex* if the interior angle at v is $\pi/2$; otherwise, v is *reflex*. Accordingly, every vertex of P is either ℓ-*convex*, r-*convex*, ℓ-*reflex*, or r-*reflex*.

To understand the shortest paths in P, we associate with each $v \in V(P)$ three landmark points in P (not necessarily vertices); see Fig. 1. The *corresponding vertex* of v, $cv(v)$, is the unique vertex with the same horizontal edge as v. To obtain the *left point* $\ell(v)$ of v, we shoot a leftward horizontal ray r from v. Let e be the vertical edge where r first hits the boundary of P. If e is the left boundary of P; then if P is a simple histogram, we let $\ell(v)$ be the left base vertex; and otherwise $\ell(v)$ is the point where r hits e. If e is not the left boundary of P, we let $\ell(v)$ be the endpoint of e closer to the base line. The *right point* $r(v)$ of v is defined analogously, by shooting the horizontal ray to the right.

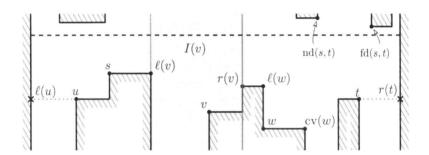

Fig. 1. Left and right points, the corresponding vertex, and the near and far dominators. The interval $I(v)$ of v is the *set of vertices* between $\ell(v)$ and $r(v)$. The dashed line is the base line.

Let p and q be two points in P. We say that p is *(strictly) to the left of* q, if $p_x \le q_x$ (or $p_x < q_x$). The term *(strictly) to the right of* is defined analogously. The *interval* $[p, q]$ of p and q is the set of vertices in P between p and q, i.e., $[p, q] = \{v \in V(P) \mid p_x \le v_x \le q_x\}$. By general position, this corresponds to index intervals in simple histograms. More precisely, if P is a simple histogram and p is either an r-reflex vertex or the left base vertex and q is either ℓ-reflex or the right base vertex, then $[p, q] = \{v \in V(P) \mid p_{id} \le v_{id} \le q_{id}\}$. The *interval of a vertex* v, $I(v)$, is the interval of the left and right point of v, $I(v) = [\ell(v), r(v)]$. Every vertex visible from v is in $I(v)$, i.e., $N(v) \subseteq I(v)$. This interval will be crucial in our routing schemes and gives a very useful characterization of visibility in double histograms.

Let s and t be two vertices with $t \in I(s) \setminus N(s)$. We define two more land-marks for s and t. Assume that t lies strictly to the right of s, the other case is symmetric. The *near dominator* $\mathrm{nd}(s,t)$ of t with respect to s is the rightmost vertex in $N(s)$ to the left of t. If there is more than one such vertex, $\mathrm{nd}(s,t)$ is the vertex closest to the base line. Since t is not visible from s, the near dominator always exists. The *far dominator* $\mathrm{fd}(s,t)$ of t with respect to s is the leftmost vertex in $N(s)$ to the right of t. If there is no such vertex, we set $\mathrm{fd}(s,t) = r(s)$, the projection of s on the right boundary. The interval $I(s,t) = [\mathrm{nd}(s,t), \mathrm{fd}(s,t)]$ has all vertices between the near and far dominator; see Fig. 1.

3   Simple Histograms

Let P be a simple histogram with n vertices. The idea for our routing scheme is as follows: as long as the target vertex t is not in the interval $I(s)$ of the current vertex s, i.e., as long as there is a higher vertex that blocks visibility between s and t, we have to leave the interval of s as fast as possible. Once $t \in I(s)$, we have to find the interval containing t. To do this, we must analyze in detail how shortest paths between vertices in P behave.

Paths in a Simple Histogram. We analyze the (shortest) paths in a simple histogram. The following lemma identifies certain "bottleneck" vertices that appear on any path; see Fig. 2.

Lemma 3.1. *Let $v, w \in V(P)$ be co-visible vertices such that v is either r-reflex or the left base vertex and w is either ℓ-reflex or the right base vertex. Let s and t be two vertices with $s \in [v, w]$ and $t \notin [v, w]$. Then, any path between s and t includes v or w.*

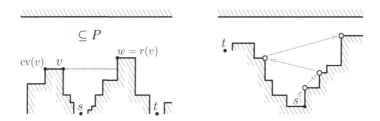

Fig. 2. Left: Any path from s to t includes v or w, since the blue rectangle contains only v and w as vertices. Right: A shortest path from s to t using the highest vertex. (Color figure online)

An immediate consequence of Lemma 3.1 is that if $t \notin I(s)$, then any path from s to t uses $\ell(s)$ or $r(s)$. The next lemma shows that if $t \notin I(s)$, there is a shortest path from s to t that uses the higher vertex of $\ell(s)$ and $r(s)$, see Fig. 2.

Lemma 3.2. *Let s and t be two vertices with $t \notin I(s)$. If $\ell(s)_y > r(s)_y$ (resp., $\ell(s)_y < r(s)_y$), then there is a shortest path from s to t using $\ell(s)$ $(r(s))$.*

The next lemma considers the case where t is in $I(s)$. Then, the near and far dominator are the potential vertices that lie on a shortest path from s to t (see also Fig. 3).

Lemma 3.3. *Let s and t be two vertices with $t \in I(s) \setminus N(s)$. Then, $\mathrm{nd}(s,t)$ is reflex and either $\mathrm{fd}(s,t) = \ell(\mathrm{nd}(s,t))$ or $\mathrm{fd}(s,t) = r(\mathrm{nd}(s,t))$.*

The Routing Scheme. We now describe our routing scheme and prove that it gives a shortest path. Let $v \in V(P)$. If v is convex and not a base vertex, it is labeled with its id, i.e., $\mathrm{lab}(v) = v_{\mathrm{id}}$. Otherwise, suppose that v is an r-reflex vertex or the left base vertex. The *breakpoint of v*, $\mathrm{br}(v)$, is defined as the left endpoint of the horizontal edge with the highest y-coordinate to the right of and below v that is visible from v; analogous definitions apply to ℓ-reflex vertices and the right base vertex. The label of v consists of the ids of v and its breakpoint, i.e., $\mathrm{lab}(v) = (v_{\mathrm{id}}, \mathrm{br}(v)_{\mathrm{id}})$. Therefore, $\mathrm{Lab}(n) = 2 \cdot \lceil \log n \rceil$. The routing table of v stores one bit, indicating whether $\ell(v)_y > r(v)_y$. Hence, $\mathrm{Tab}(n) = 1$.

We are given the current vertex s and the label $\mathrm{lab}(t)$ of the target vertex t. The routing function does not need a header, i.e., $H(n) = 0$. If t is visible from s, i.e., if $\mathrm{lab}(t) \in \mathrm{lab}(N(s))$, we directly go from s to t on a shortest path. Thus, assume that t is not visible from s. First, we check if $t \in I(s)$. This is done as follows: we determine the smallest and largest id in the link table $\mathrm{lab}(N(s))$ of s. The corresponding vertices are $\ell(s)$ and $r(s)$. Then, we can check if $t_{\mathrm{id}} \in [\ell(s)_{\mathrm{id}}, r(s)_{\mathrm{id}}]$, which is the case if and only if $t \in I(s)$. Now, there are two cases, illustrated in Fig. 3. First, suppose $t \notin I(s)$. If the bit in the routing table of s indicates that $\ell(s)$ is higher than $r(s)$, we take the hop to $\ell(s)$; otherwise, we hop to $r(s)$. By Lemma 3.2, this hop lies on a shortest path from s to t.

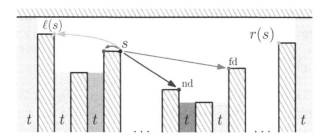

Fig. 3. The cases where the vertex t lies and the corresponding vertices where the data packet is sent to. If $t \in [\ell(s), s]$ we have $\mathrm{nd}(s,t) = \mathrm{cv}(s)$ and $\mathrm{fd}(s,t) = \ell(s)$.

Second, suppose that $t \in I(s) \setminus N(s)$. This case is slightly more involved. We use the link table $\mathrm{lab}(N(s))$ of s and the label $\mathrm{lab}(t)$ of t to determine $\mathrm{fd}(s,t)$ and $\mathrm{nd}(s,t)$. Again, we can do this by comparing the ids. Lemma 3.3 states that either $\mathrm{fd}(s,t) = \ell(\mathrm{nd}(s,t))$ or $\mathrm{fd}(s,t) = r(\mathrm{nd}(s,t))$. We discuss the case that $\mathrm{fd}(s,t) = r(\mathrm{nd}(s,t))$, the other case is symmetric. By Lemma 3.1, any shortest path from s to t includes $\mathrm{fd}(s,t)$ or $\mathrm{nd}(s,t)$. Moreover, due to Lemma 3.3,

$\text{nd}(s,t)$ is reflex, and we can use its label to access $b_{\text{id}} = \text{br}(\text{nd}(s,t))_{\text{id}}$. The vertex b splits $I(s,t) = [\text{nd}(s,t), \text{fd}(s,t)]$ into two disjoint subintervals $[\text{nd}(s,t), b]$ and $[\text{cv}(b), \text{fd}(s,t)]$. Also, b and $\text{cv}(b)$ are not visible from s, as they are located strictly between the far and the near dominator. Based on b_{id}, we can now decide on the next hop.

If $t \in [\text{nd}(s,t), b]$, we take the hop to $\text{nd}(s,t)$. If $t = b$, our packet uses a shortest path of length 2. Thus, assume that t lies between $\text{nd}(s,t)$ and b. This is only possible if b is ℓ-reflex, and we can apply Lemma 3.1 to see that any shortest path from s to t includes $\text{nd}(s,t)$ or b. But since $d(s,b) = 2$, our data packet routes along a shortest path.

If $t \in [\text{cv}(b), \text{fd}(s,t)]$, we take the hop to $\text{fd}(s,t)$. If $t = \text{cv}(b)$, our packet uses a shortest path of length 2. Thus, assume that t lies between $\text{cv}(b)$ and $\text{fd}(s,t)$. This is only possible if $\text{cv}(b)$ is r-reflex, so we can apply Lemma 3.1 to see that any shortest path from s to t uses $\text{fd}(s,t)$ or $\text{cv}(b)$. Since $d(s, \text{cv}(b)) = 2$, our packet routes along a shortest path. Thus:

Theorem 3.4. *Let P be a simple histogram with n vertices. There is a routing scheme for $G(P)$ with 1-bit routing tables, no header, and label size $2 \cdot \lceil \log n \rceil$, such that we can route between any two vertices on a shortest path.*

4   Double Histograms

Let P be a double histogram with n vertices. Similar to the simple histogram case, we first focus on the structure of shortest paths in P. Again, if the target vertex t is not in the interval $I(s)$ of the current vertex s, we should widen the interval (i.e., extend it to include more vertices) as fast as possible. However, in contrast to simple histograms, we can now change sides arbitrarily often. Nevertheless, we can guarantee that in each step, the interval comes closer to t. Once we have reached the case that t is in the interval of the current vertex, we have to find the sub-interval that contains t. Unlike in simple histograms, this case is now simpler to describe.

Paths in a Double Histogram. To understand shortest paths in double histograms, we distinguish three cases, depending on where t lies relative to s. First, if t is close, i.e., if $t \in I(s)$, we focus on the near and far dominators. Second, if $t \notin I(s)$ but there is a vertex v visible from s with $t \in I(v)$, then we can find a vertex on a shortest path from s to t. Third, if there is no visible vertex v from s such that $t \in I(v)$, we can apply our intuition from simple histograms: go as fast as possible towards the base line.

Let s, t be two vertices with $t \in I(s) \setminus N(s)$. In contrast to simple histograms, $\text{fd}(s,t)$ now might not be a vertex. Furthermore, $\text{fd}(s,t)$ and $\text{nd}(s,t)$ might be on different sides of the base line. In this case, Lemma 3.3 no longer holds. However, the next lemma establishes a visibility relation between them.

Lemma 4.1. *Let $s, t \in V(P)$ with $t \in I(s) \setminus N(s)$. Then, $\text{nd}(s,t)$ and $\text{fd}(s,t)$ are co-visible.*

The proof of the next lemma uses Lemma 4.1 to find a shortest path vertex.

Lemma 4.2. *One of $nd(s,t)$ or $fd(s,t)$ is on a shortest path from s to t. If $fd(s,t)$ is not a vertex, then $nd(s,t)$ is on a shortest path from s to t.*

Next, we consider the case where $fd(s,t)$ is a vertex but not on a shortest path from s to t. Then, $fd(s,t)$ cannot see t, and we define $fd^2(s,t) = fd(fd(s,t),t)$. By Lemma 4.1, $nd(s,t)$ and $fd(s,t)$ are co-visible, so $fd^2(s,t)$ has to be in the interval $[nd(s,t),t]$, and therefore it is a vertex. The following lemma states that $fd^2(s,t)$ is strictly closer to t than s.

Lemma 4.3. *If $fd(s,t)$ is a vertex but not on a shortest path from s to t, then we have $d(fd^2(s,t),t) = d(s,t) - 1$.*

Let s,t be two vertices so that $t \notin I(s)$ but there is a vertex $v \in N(s)$ with $t \in I(v)$. For clarity of presentation, we will always assume that s is below the base line. The crux of this case is this: there might be many vertices visible from s that have t in their interval. However, we can find a best vertex as follows: once t is in the interval of a vertex, the goal is to shrink the interval (i.e., reduce it to include fewer vertices) as fast as possible. Therefore, we must find a vertex $v \in N(s)$ whose left or right interval boundary is closest to t among all vertices in $N(s)$. This leads to the following inductive definition of two sequences $a^i(s)$ and $b^i(s)$ of vertices in $N(s)$. For $i = 0$, we let $a^0(s) = b^0(s) = s$. For $i > 0$, if the set $A^i(s) = \{v \in N(s) \mid \ell(v)_x < \ell(a^{i-1}(s))_x\}$ is nonempty, we define $a^i(s) = \operatorname{argmin}\{v_x \mid v \in A^i(s)\}$; and $a^i(s) = a^{i-1}(s)$, otherwise. If the set $B^i(s) = \{v \in N(s) \mid r(v)_x > r(b^{i-1}(s))_x\}$ is nonempty, we define $b^i(s) = \operatorname{argmax}\{v_x \mid v \in B^i(s)\}$; and $b^i(s) = b^{i-1}(s)$, otherwise. We force unambiguity by choosing the vertex closer to the base line. Let $a^*(s)$ be the vertex with $a^*(s) = a^i(s) = a^{i-1}(s)$, for an $i > 0$, and $b^*(s)$ the vertex with $b^*(s) = b^i(s) = b^{i-1}(s)$, for an $i > 0$. If the context is clear, we write a^i instead of $a^i(s)$ and b^i instead of $b^i(s)$.

Let us try to understand this definition. For $i \geq 0$, we write ℓ^i for $\ell(a^i)$; and we write ℓ^* for $\ell(a^*)$. Then, we have $a^0 = s$ and $\ell^0 = \ell(s)$. Now, if $\ell(s)$ is not a vertex, then $a^* = s$, because there is no vertex whose left point is strictly to the left of the left boundary of P. On the other hand, if ℓ^0 is a vertex in P, we have $a^1 = \ell^0 = \ell(s)$, and $[\ell^1, a^1]$ is an interval between points on the lower side of P. Then comes a (possibly empty) sequence of intervals $[\ell^2, a^2], [\ell^3, a^3], \ldots, [\ell^k, a^k]$ between points on the upper side of P; possibly followed by the interval $[\ell(r(s)), r(s)]$. There are four possibilities for a^*: it could be s, $\ell(s)$, a vertex a^i on the upper side of P, or $r(s)$. If $a^* \neq s$, then the intervals $[\ell^1, a^1] \subset [\ell^2, a^2] \subset \cdots \subset [\ell^*, a^*]$ are strictly increasing: ℓ^i is strictly to the left of ℓ^{i-1} and a^i is strictly to the right of a^{i-1}; see Fig. 4. Symmetric observations apply for the b^i; we write r^i for $r(b^i)$ and r^* for $r(b^*)$.

Lemma 4.4. *For $i \geq 1$, the vertices ℓ^{i-1}, a^i as well as r^{i-1}, b^i are co-visible.*

Finally, the next lemma tells us the following: if $t \in [\ell^*, r^*]$ we find a vertex $v \in N(s)$ with $t \in I(v)$. Its quite technical proof needs Lemma 4.4.

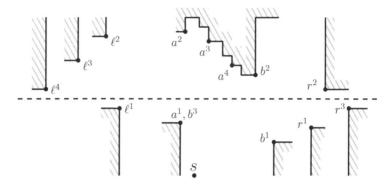

Fig. 4. The vertices a^i and b^i are illustrated. Observe that $\ell(s) = a^1 = b^3$ and $r(s) = b^1$.

Lemma 4.5. *If $t \in [\ell^i, \ell^{i-1}]$, for some $i \geq 1$, then a^i is on a shortest path from s to t. If $t \in [r^{i-1}, r^i]$, for some $i \geq 1$, then b^i is on a shortest path from s to t.*

Finally, we consider the case that there is no vertex $v \in N(s)$ with $t \in I(v)$, i.e., $t \notin [\ell^*, r^*]$. The intuition now is as follows: to widen the interval, we should go to a vertex that is visible from s, but closest to the base line. In simple histograms, there was only one such vertex, but in double histograms there might be a second one on the other side. These two vertices are the *dominators* *of s*. These two dominators might have their own dominators, and so on. This leads to the following inductive definition.

For $k \geq 0$, we define the k-th *bottom dominator* $\mathrm{bd}^k(s)$, the k-th *top dominator* $\mathrm{td}^k(s)$, and the k-th *interval* $I^k(s)$ of s. For any set $Q \subset V(P)$, we write Q^- (resp. Q^+) for all points in Q below (resp. above) the base line. We set $\mathrm{bd}^0(s) = \mathrm{td}^0(s) = s$ and $I^0(s) = \{s\}$. For $k > 0$, we set $I^k(s) = I(\mathrm{bd}^{k-1}(s)) \cup I(\mathrm{td}^{k-1}(s))$. If $I^k(s)^-$ is nonempty, we let $\mathrm{bd}^k(s)$ be the leftmost vertex inside $I^k(s)^-$ that minimizes the distance to the base line. If $I^k(s)^+$ is nonempty, we let $\mathrm{td}^k(s)$ be the leftmost vertex inside $I^k(s)^+$ that minimizes the distance to the base line. If one of the two sets is empty, the other one has to be nonempty, since $s \in I^k(s)$. In this case, we let $\mathrm{td}^k(s) = \mathrm{bd}^k(s)$. We write $\mathrm{bd}(s)$ for $\mathrm{bd}^1(s)$ and $\mathrm{td}(s)$ for $\mathrm{td}^1(s)$. Observe, that $I^1(s) = I(s)$ and $I^2(s) = [\ell^*, r^*]$. If $I(\mathrm{bd}^{k-1}(s)) = V(P)$, we have $\mathrm{bd}^k(s) = \mathrm{bd}^{k-1}(s)$. The same holds for the top dominator. We provide a few technical properties concerning the k-th interval as well as the k-th dominators.

Lemma 4.6. *For any $s \in V(P)$ and $k \geq 0$, we have $I^k(s) \subseteq I(\mathrm{bd}^k(s)) \cap I(\mathrm{td}^k(s))$ and $\mathrm{bd}^k(s)$, $\mathrm{td}^k(s)$ are co-visible.*

The following lemma seems rather specific, but will be needed later to deal with short paths.

Lemma 4.7. *For any $s \in V(P)$, we have $I^3(s) = I^2(\mathrm{bd}(s)) \cup I^2(\mathrm{td}(s))$.*

Intuitively, the meaning of $I^k(s)$ is as follows: let ℓ be the leftmost and r be the rightmost vertex with hop distance exactly k from s, then, $I^k(s) = [\ell, r]$. We do not really need this property, so we leave it as an exercise for the reader to find a proof for this. Instead, we prove the following weaker statement. For this, recall that due to its definition, $\text{bd}^k(s)$ might not be on the lower side of the histogram (and $\text{td}^k(s)$ might not be on the upper side).

Lemma 4.8. *Let $k \geq 0$ and let $s, t \in V(P)$ with $d(s, t) \leq k$. Then, $t \in I^k(s)$.*

Let $k \geq 0$ and $s \in V(P)$. For $i = 1, \ldots, k$, by Lemma 4.6, $\text{bd}^{i-1}(s), \text{td}^{i-1}(s) \in I(\text{bd}^i(s)) \cap I(\text{td}^i(s))$. Moreover, by definition, $\text{bd}^i(s), \text{td}^i(s) \in I(\text{bd}^{i-1}(s)) \cup I(\text{td}^{i-1}(s))$. As we show in the full version, now both $\text{bd}^i(s)$ and $\text{td}^i(s)$ can see at least one of $\text{bd}^{i-1}(s)$ or $\text{td}^{i-1}(s)$. Therefore, there is a path $\pi_b(s, k) : \langle s = p_0, \ldots, p_k = \text{bd}^k(s) \rangle$ from s to $\text{bd}^k(s)$ and a path $\pi_t(s, k) : \langle s = q_0, \ldots, q_k = \text{td}^k(s) \rangle$ from s to $\text{td}^k(s)$ with $p_i, q_i \in \{\text{bd}^i(s), \text{td}^i(s)\}$, for $i = 0, \ldots, k$. We call $\pi_b(s, k)$ and $\pi_t(s, k)$ the *canonical path* from s to $\text{bd}^k(s)$ and from s to $\text{td}^k(s)$, respectively. The following two lemmas show that for every $t \notin I^{k+1}(s)$ one of the canonical paths is the prefix of a shortest path from s to t. To show Lemma 4.10 we need Lemmas 4.8 and 4.9.

Lemma 4.9. *Let $k \geq 1$ and $s \in V(P)$. If $I(\text{bd}^{k-1}(s)) \neq V(P)$, we have that $d(s, \text{bd}^k(s)) = k$. If $I(\text{td}^{k-1}(s)) \neq V(P)$ we have $d(s, \text{td}^k(s)) = k$.*

Lemma 4.10. *Let s and t be vertices and $k \geq 1$ an integer such that $t \notin I^{k+1}(s)$. Then $\text{bd}^k(s)$ or $\text{td}^k(s)$ is on a shortest path from s to t.*

Routing Scheme. Let v be a vertex. The label of v consists of its x- and y-coordinate as well as the bounding x-coordinates of $I(v)$. We do not need v_{id} since (v_x, v_y) identifies the vertex in the network. Thus, $\text{Lab}(n) = 4 \cdot \lceil \log n \rceil$ since we can assume that $v_x, v_y \in \{0, \ldots, n-1\}$. In the routing table of v, we store the bounding x-coordinates of $I^2(\text{bd}(v))$ as well as the bounding x-coordinates of $I^2(\text{td}(v))$. Furthermore, we store $(\text{bd}^2(v)_x, \text{bd}^2(v)_y, bit)$ where bit indicates whether $\text{td}(v)$ or $\text{bd}(v)$ is on the path $\pi_b(v, 2)$. Thus, $\text{Tab}(n) = 6 \cdot \lceil \log n \rceil + 1$.

We are given a current vertex s together with its routing table and link table, the label of a target vertex t, and a header. If $t \in N(s)$, then $\text{lab}(t)$ is in the link table of s, and we send the data packet directly to t. If the header is non-empty, it will contain the coordinates of exactly one vertex visible from s. We clear the header and go to this respective vertex. The remaining discussion assumes that the header is empty and that $t \notin N(s)$. The routing function now distinguishes four cases depending on whether $t \in I(s)$, $t \in I^2(s)$ or $t \in I^3(s)$. We can check the first and the second condition locally, using the link table of s as well as the label of t (note that from the link table of s, we can deduce $a^*(s)$ and $b^*(s)$, and their interval boundaries). To check the third condition locally, we use Lemma 4.7 which shows that $I^3(s) = I^2(\text{bd}(s)) \cup I^2(\text{td}(s))$. Since we stored the bounding x-coordinates of these two intervals in the routing table of s, we can check $t \in I^3(s)$ easily.

Case 1 ($t \in I(s) \setminus N(s)$): if $\text{fd}(s, t)$ is a vertex, we can determine it by using the link table and the label of t. The packet is sent to $\text{fd}(s, t)$. If $\text{fd}(s, t)$ is not a vertex, we determine $\text{nd}(s, t)$ and send the packet there. The header remains empty.

Case 2 ($t \in I^2(s) \setminus I(s)$): there is an $i \geq 1$ with $t \in [\ell^i, \ell^{i-1}]$ or $t \in [r^{i-1}, r^i]$. We find i using the link table and $\mathrm{lab}(t)$. The packet is sent to a^i or b^i. The header remains empty.

Case 3 ($t \in I^3(s) \setminus I^2(s)$): if $t \in I^2(\mathrm{bd}(s))$, we send the packet to $\mathrm{bd}(s)$. Otherwise, $t \in I^2(\mathrm{td}(s))$, and we send the packet to $\mathrm{td}(s)$. In both cases, the header remains empty.

Case 4 ($t \notin I^3(s)$): the routing table has the entry $(\mathrm{bd}^2(s)_x, \mathrm{bd}^2(s)_y, b)$. We store $(\mathrm{bd}^2(s)_x, \mathrm{bd}^2(s)_y)$ in the header and send the packet to $\mathrm{bd}(s)$ or $\mathrm{td}(s)$, as indicated by b.

Obviously, $H(n) = 2 \cdot \lceil \log n \rceil$. It remains to analyze the stretch factor. For this, we use the following lemma:

Lemma 4.11. *Let $s, t \in V(P)$. After at most two steps of the routing scheme from s with target label $\mathrm{lab}(t)$, we reach a vertex v with $d(v, t) \leq d(s, t) - 1$.*

This immediately gives a stretch factor of 2 and our main theorem.

Theorem 4.12. *Let P be a double histogram with n vertices. There is a routing scheme for $G(P)$ with routing table, label and header size $O(\log n)$, such that we can route between any two vertices with stretch at most 2.*

References

1. Banyassady, B., et al.: Routing in polygonal domains. In: 28th ISAAC, pp. 10:1–10:13 (2017)
2. Biniaz, A.: Plane hop spanners for unit disk graphs: simpler and better. arXiv:1902.10051 (2019)
3. Bose, P., Fagerberg, R., van Renssen, A., Verdonschot, S.: Competitive local routing with constraints. JoCG **8**, 125–152 (2017)
4. Bose, P., Morin, P.: Competitive online routing in geometric graphs. TCS **324**, 273–288 (2004)
5. Fraigniaud, P., Gavoille, C.: Routing in trees. In: 28th ICALP, pp. 757–772 (2001)
6. Giordano, S., Stojmenovic, I.: Position based routing algorithms for ad hoc networks: a taxonomy. In: Cheng, X., Huang, X., Du, D.Z. (eds.) Ad Hoc Wireless Networking. NETA, vol. 14, pp. 103–136. Springer, Boston (2004). https://doi.org/10.1007/978-1-4613-0223-0_4
7. Hoffmann, F., Kriegel, K., Suri, S., Verbeek, K., Willert, M.: Tight bounds for conflict-free chromatic guarding of orthogonal art galleries. CGTA **73**, 24–34 (2018)
8. Jacquet, P., Viennot, L.: Remote-spanners: what to know beyond neighbors. In: 23rd IPDPS, pp. 1–10 (2009)
9. Kaplan, H., Mulzer, W., Roditty, L., Seiferth, P.: Routing in unit disk graphs. Algorithmica **80**, 830–848 (2018)
10. Peleg, D., Upfal, E.: A trade-off between space and efficiency for routing tables. JACM **36**, 510–530 (1989)
11. Roditty, L., Tov, R.: Close to linear space routing schemes. Distrib. Comput. **29**, 65–74 (2016)
12. Thorup, M.: Compact oracles for reachability and approximate distances in planar digraphs. JACM **51**, 993–1024 (2004)
13. Thorup, M., Zwick, U.: Compact routing schemes. In: 13th SPAA, pp. 1–10 (2001)
14. Yan, C., Xiang, Y., Dragan, F.F.: Compact and low delay routing labeling scheme for unit disk graphs. CGTA **45**, 305–325 (2012)

A Waste-Efficient Algorithm for Single-Droplet Sample Preparation on Microfluidic Chips

Miguel Coviello Gonzalez and Marek Chrobak[(⊠)]

University of California at Riverside, Riverside, USA
`marek@cs.ucr.edu`

Abstract. We address the problem of designing microfluidic chips for sample preparation, a crucial step in many experimental processes in chemical and biological sciences. One of the objectives of sample preparation is to dilute the sample fluid, called reactant, using another fluid called buffer, to produce desired volumes of fluid with prespecified reactant concentrations. In our model these fluids are manipulated in discrete volumes called droplets. The dilution process is represented by a *mixing graph* whose nodes represent 1–1 micro-mixers and edges represent channels for transporting fluids. We focus on designing such mixing graphs when the given sample (also referred to as the *target*) consists of a single-droplet, and the objective is to minimize total fluid waste. Our main contribution is an efficient algorithm called `RPRIS` that guarantees a better provable worst-case bound on waste and significantly outperforms state-of-the-art algorithms in experimental comparison.

1 Introduction

Microfluidic chips are miniature devices that manipulate tiny amounts of fluids on a small chip and can perform various laboratory functions such as dispensing, mixing, filtering and detection. They play an increasingly important role in today's science and technology, with applications in environmental or medical monitoring, protein or DNA analysis, drug discovery, physiological sample analysis, and cancer research.

These chips often contain modules whose function is to mix fluids. One application where fluid mixing plays a crucial role is sample preparation for some biological or chemical experiments. When preparing such samples, one of the objectives is to produce desired volumes of the fluid of interest, called *reactant*, diluted to some specified concentrations by mixing it with another fluid called *buffer*. As an example, an experimental study may require a sample that consists of $6\,\mu L$ of reactant with concentration 10%, $9\,\mu L$ of reactant with concentration 20%, and $3\,\mu L$ of reactant with concentration 40%. Samples of this form are often required in toxicology or pharmaceutical studies, among other applications.

Supported by NSF grant CCF-1536026.

There are different models for fluid mixing in the literature and multiple technologies for manufacturing fluid-mixing microfluidic chips. (See the survey in [2] or the recent book [1] for more information on different models and algorithmic issues related to fluid mixing.) In this work we assume the *droplet-based* model, where the fluids are manipulated in discrete quantities called *droplets*. For convenience, we will identify droplets by their reactant concentrations, which are numbers in the interval $[0, 1]$ with finite binary precision. In particular, a droplet of reactant is denoted by 1 and a droplet of buffer by 0. We focus on the mixing technology that utilizes modules called *1–1 micro-mixers*. A micro-mixer has two inlets and two outlets. It receives two droplets of fluid, one in each inlet, mixes these droplets perfectly, and produces two droplets of the mixed fluid, one on each outlet. (Thus, if the inlet droplets have reactant concentrations a and b, then the two outlet droplets each will have concentration $\frac{1}{2}(a+b)$.) Input droplets are injected into the chip via droplet dispensers and output droplets are collected in droplet collectors. All these components are connected via micro-channels that transport droplets, forming naturally an acyclic graph that we call a *mixing graph*, whose source nodes are fluid dispensers, internal nodes (of in-degree and out-degree 2) are micro-mixers, and sink nodes are droplet collectors. Graph G_1 in Fig. 1 shows an example.

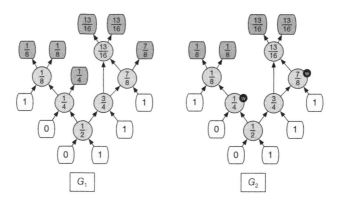

Fig. 1. Mixing graph G_1 produces droplet set $\left\{ \frac{1}{8}, \frac{1}{8}, \frac{1}{4}, \frac{13}{16}, \frac{13}{16}, \frac{7}{8} \right\}$ from input set $I = \{0, 0, 1, 1, 1, 1\}$. Numbers on the internal nodes represent droplet concentrations produced by the corresponding micro-mixers. Mixing graph G_2 produces droplets $\left\{ \frac{1}{8}, \frac{1}{8}, \frac{13}{16}, \frac{13}{16} \right\}$. Small black circles labeled "w" on micro-mixers represent droplets of waste.

Given some target set of droplets with specified reactant concentrations, the objective is to design a mixing graph that produces these droplets from pure reactant and buffer droplets, while optimizing some objective function. Some target sets can be produced only if we allow the mixing graph to also produce some superfluous amount of fluid that we refer to as *waste*; see graph G_2 in Fig. 1. One natural objective function is to minimize the number of waste droplets (or equivalently, the total number of input droplets). As reactant is typically more expensive than buffer, one other common objective is to minimize the reactant usage. Yet another possibility is to minimize the number of micro-mixers or the depth of the mixing graph. There is growing literature on developing techniques and algorithms for designing such mixing graphs that attempt to optimize some of the above criteria.

State-of-the-Art. Most earlier papers on this topic studied designing mixing graphs for single-droplet targets. This line of research was pioneered by Thies *et al.* [11], who proposed an algorithm called `Min-Mix` that constructs a mixing graph for a single-droplet target with the minimum number of mixing operations. Roy *et al.* [10] developed an algorithm called `DMRW` designed to minimize waste. Huang *et al.* [7] considered minimizing reactant usage, and proposed an algorithm called `REMIA`. Another algorithm called `GORMA`, for minimizing reactant usage and based on a branch-and-bound technique, was developed by Chiang *et al.* [3].

The algorithms listed above are heuristics, with no formal performance guarantees. An interesting attempt to develop an algorithm that minimizes waste, for target sets with multiple droplets, was reported by Dinh *et al.* [4]. Their algorithm, that we refer to as `ILP`, is based on a reduction to integer linear programming and, since their integer program could be exponential in the precision d of the target set (and thus also in terms of the input size), its worst-case running time is doubly exponential. Further, as this algorithm only considers mixing graphs of depth at most d, it does not always finds an optimal solution (see [5]). In spite of these deficiencies, for very small values of d it is still likely to produce good mixing graphs.

Additional work regarding the design of mixing graphs for multiple droplets includes Huang *et al.*'s algorithm called `WARA`, which is an extension of Algorithm `REMIA`, that focuses on reactant minimization; see [8]. Mitra *et al.* [9] also proposed an algorithm for multiple droplet concentrations by modeling the problem as an instance of the Asymmetric TSP on a de Bruijn graph.

As discussed in [5], the computational complexity of computing mixing graphs with minimum waste is still open, even in the case of single-droplet targets. In fact, it is not even known whether the minimum-waste function is computable at all, or whether it is decidable to determine if a given target set can be produced without *any* waste. To our knowledge, the only known result that addresses theoretical aspects of designing mixing graphs is a polynomial-time algorithm in [5] that determines whether a given collection of droplets with specified concentrations can be mixed perfectly with a mixing graph.

Our Results. Continuing the line of work in [3,7,10,11], we develop a new efficient algorithm `RPRIS` (for *Recursive Precision Reduction with Initial Shift*) for designing mixing graphs for single-droplet targets, with the objective to minimize waste. Our algorithm was designed to provide improved worst-case waste estimate; specifically to cut it by half for most concentrations. Its main idea is quite natural: recursively, at each step it reduces the precision of the target droplet by 2, while only adding one waste droplet when adjusting the mixing graph during backtracking.

While designed with worst-case performance in mind, RPRIS significantly out-performs algorithms Min-Mix, DMRW and GORMA in our experimental study (see Sect. 6), producing on average about 50% less waste than Min-Mix, between 21 and 25% less waste than DMRW (with the percentage increasing with the precision d of the target droplet), and about 17% less waste than GORMA. (It also produces about 40% less waste than REMIA.) Further, when compared to ILP, RPRIS produces on average only about 7% more waste.

Unlike earlier work in this area, that was strictly experimental, we introduce a performance measure for waste minimization algorithms and show that RPRIS has better worst-case performance than Min-Mix and DMRW. This measure is based on two attributes d and γ of the target concentration t. As defined earlier, d is the precision of t, and γ is defined as the number of equal leading bits in t's binary representation, not including the least-significant bit 1. For example, if $t = .00001011$ then $\gamma = 4$, and if $t = .1111$ then $\gamma = 3$. (Both d and γ are functions of t, but we skip the argument t, as it is always understood from context.) In the discussion below we provide more intuition and motivations for using these parameters.

Algorithm RPRIS produces at most $\frac{1}{2}(d + \gamma) + 2$ droplets of waste (see Theorem 1 in Sect. 5). In comparison, Algorithm Min-Mix from [11] produces exactly d droplets of waste, independently of the value of t. This means that the waste of RPRIS is about half that of Min-Mix for almost all concentrations t. (More formally, for a uniformly chosen random t with precision d the probability that the waste is larger than $(\frac{1}{2} - \epsilon)d$ vanishes when d grows, for any $\epsilon > 0$.) As for Algorithm DMRW, its average waste is better than that of Min-Mix, but its worst-case bound is still $d - O(1)$ even for small values of γ (say, when $t \in [\frac{1}{4}, \frac{3}{4}]$), while Algorithm RPRIS' waste is at most $d/2 + O(1)$ in this range.

Regarding time performance, for the problem of computing mixing graphs it would be reasonable to express the time complexity of an algorithm as a function of its output, which is the size of the produced graph. This is because the output size is at least as large as the input size, which is equal to d – the number of bits of t. Algorithm RPRIS runs in time that is linear in the size of the computed graph, and the graphs computed by Algorithm RPRIS have size $O(d^2)$.

Discussion. To understand better our performance measure for waste, observe that the optimum waste is never smaller than $\gamma + 1$. This is because if the binary representation of t starts with γ 0's then any mixing graph has to use $\gamma + 1$ input droplets 0 and at least one droplet 1. (The case when the leading bits of t are 1's is symmetric.) For this reasons, a natural approach is to express the waste in the form $\gamma + f(d - \gamma)$, for some function $f()$. In Algorithm RPRIS we have $f(x) \approx \frac{1}{2}x$. It is not known whether smaller functions $f()$ can be achieved.

Ideally, one would like to develop efficient "approximation" algorithms for waste minimization, that measure waste performance in terms of the additive or multiplicative approximation error, with respect to the optimum value. This is not realistic, however, given the current state of knowledge, since currently no close and computable bounds for the optimum waste are known.

Note: Due to lack of space, some details of the algorithm and some proofs are omitted here. They can be found in the extended version of this paper [6].

2   Preliminaries

We use notation $\mathsf{prec}(c)$ for the precision of concentration c, that is the number of fractional bits in the binary representation of c. In other words, $\mathsf{prec}(c) = d \in \mathbb{Z}_{\geq 0}$ such that $c = a/2^d$ for an odd $a \in \mathbb{Z}$.

We will deal with sets of droplets, some possibly with equal concentrations. We define a *configuration* as a multiset of droplet concentrations. Let A be an arbitrary configuration. By $|A| = n$ we denote the number of droplets in A. We will often write a configuration as $A = \{f_1 : a_1, f_2 : a_2, ..., f_m : a_m\}$, where each a_i represents a different concentration and f_i denotes the multiplicity of a_i in A. (If $f_i = 1$, then, we will write "a_i" instead of "$f_i : a_i$".) Naturally, $\sum_{i=1}^{m} f_i = n$.

We defined mixing graphs in the introduction. A mixing graph can be thought of as a linear mapping from the source values (usually 0's and 1's) to the sink values. Yet in the paper, for convenience, we will assume that the source concentration vector is part of a mixing graph's specification, and that all sources, micro-mixers, and sinks are labeled by their associated concentration values.

We now define an operation of coupling of mixing graphs G_1 and G_2. Let T_1 be the output configuration (the concentration labels of the sink nodes) of G_1 and I_2 be the input configuration (the concentration labels of the source nodes) for G_2. To construct the *coupling* of G_1 and G_2, denoted $G_2 \bullet G_1$, we identify inlet edges of the sinks of G_1 with labels from $T_1 \cap I_2$ with outlet edges of the corresponding sources in G_2. More precisely, repeat the following steps as long as $T_1 \cap I_2 \neq \emptyset$: (1) choose any $a \in T_1 \cap I_2$, (2) choose any sink node t_1 of G_1 labeled a, and let (u_1, t_1) be its inlet edge, (3) choose any source node s_2 of G_2 labeled a, and let (s_2, v_2) be its outlet edge, (4) remove t_1 and s_2 and their incident

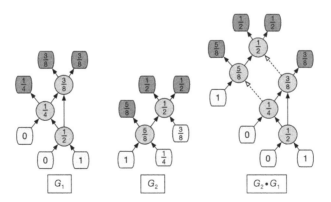

Fig. 2. Coupling of two mixing graphs G_1 and G_2. $G_2 \bullet G_1$ is obtained by identifying inlet edges of two sinks of G_1, one labelled $\frac{1}{4}$ and one $\frac{3}{8}$, with the outlet edges of the corresponding sources of G_2. These new edges are shown as dotted arrows.

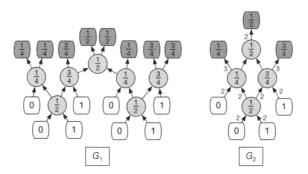

Fig. 3. Graph G_2 is a compact representation of G_1. All nodes in G_2 (except the last mixer node with label $\frac{1}{2}$) represent an aggregation of at least two nodes from G_1.

edges, and finally, (5) add edge (u_1, v_2). The remaining sources of G_1 and G_2 become sources of $G_2 \bullet G_1$, and the remaining sinks of G_1 and G_2 become sinks of $G_2 \bullet G_1$. See Fig. 2 for an example.

We define an $(i : \alpha, j : \beta)$-*converter* as a mixing graph that produces a configuration of the form $T = \{i : \alpha, j : \beta\} \cup W$, where W denotes a set of waste droplets, and whose input droplets have concentration labels either 0 or 1. For example, graph G_2 in Fig. 1 can be interpreted as a $(2 : \frac{1}{8}, 2 : \frac{13}{16})$-converter that produces two waste droplets of concentrations $\frac{1}{4}$ and $\frac{7}{8}$.

If needed, to avoid clutter, sometimes we will use a more compact graphical representation of mixing graphs by aggregating (not necessarily all) nodes with the same concentration labels into a single node, and with edges labeled by the number of droplets that flow through them. (We will never aggregate two micro-mixer nodes if they both produce a droplet of waste.) If the label of an edge is 1, then we will simply omit the label. See Fig. 3 for an example of such a compact representation.

3   Algorithm Description

In this section, we describe our algorithm RPRIS for producing a single-droplet target of concentration t with precision $d = \mathsf{prec}(t)$. We first give the overall strategy and then we gradually explain its implementation. The core idea behind RPRIS is a recursive procedure that we refer to as *Recursive Precision Reduction*, that we outline first. In this procedure, t_s denotes the concentration computed at the sth recursive step with $d_s = \mathsf{prec}(t_s)$; initially, $t_0 = t$. Also, by $\mathcal{B}$ we denote the set of base concentration values with small precision for which we give explicit mixing graphs later in this section.

Procedure RPR(t_s)
   If $t_s \in \mathcal{B}$, let G_s be the base mixing graph (defined later) for t_s, **else:**
   (rpr1) Replace t_s by another concentration value t_{s+1} with $d_{s+1} = d_s - 2$.
   (rpr2) Recursively construct a mixing graph G_{s+1} for t_{s+1}.

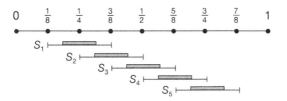

Fig. 4. Graphical representation of intervals $S_1, S_2, \ldots, S_5$. The thick shaded part of each interval $S_k = [l, r]$ marks its "middle section" $[l + \frac{1}{16}, r - \frac{1}{16}]$. Each concentration within interval $[\frac{1}{4}, \frac{3}{4}]$ belongs to the middle section of some S_k.

> (rpr3) Convert G_{s+1} into a mixing graph G_s for t_s, increasing waste by one droplet.
>
> **Return** G_s.

The mixing graph produced by this process is G_0.

When we convert G_{s+1} into G_s in part (rpr3), the precision of the target increases by 2, but the waste only increases by 1, which gives us a rough bound of $d/2$ on the overall waste. However, as it turns out, the above process only works when $t_0 \in [\frac{1}{4}, \frac{3}{4}]$. To deal with values outside this interval, we map t into t_0 so that $t_0 \in [\frac{1}{4}, \frac{3}{4}]$, next we apply Recursive Precision Reduction to t_0, and then we appropriately modify the computed mixing graph. This process is called *Initial Shift*.

We next describe these two processes in more detail, starting with Recursive Precision Reduction, followed by Initial Shift.

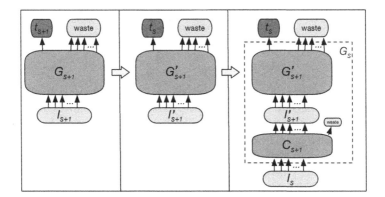

Fig. 5. Conversion from G_{s+1} to G_s. The left image illustrates the computed mixing graph G_{s+1} with input labels I_{s+1} (consisting of only 0's and 1's) that produces t_{s+1} along with some waste. The middle figure illustrates G'_{s+1}, which is obtained from G_{s+1} by changing concentration labels. The last figure illustrates the complete mixing graph $G_s = G'_{s+1} \bullet C_{s+1}$ for t_s, shown within a dotted rectangle.

Recursive Precision Reduction (RPR). We start with concentration t_0 that, by applying Initial Shift (described next), we can assume to be in $[\frac{1}{4}, \frac{3}{4}]$.

Step (rpr1): Computing t_{s+1}. We convert t_s into a carefully chosen concentration t_{s+1} for which $d_{s+1} = d_s - 2$. One key idea is to maintain an invariant so that at each recursive step, this new concentration value t_{s+1} satisfies $t_{s+1} \in [\frac{1}{4}, \frac{3}{4}]$. To accomplish this, we consider five intervals $S_1 = [\frac{1}{8}, \frac{3}{8}]$, $S_2 = [\frac{1}{4}, \frac{1}{2}]$, $S_3 = [\frac{3}{8}, \frac{5}{8}]$, $S_4 = [\frac{1}{2}, \frac{3}{4}]$, and $S_5 = [\frac{5}{8}, \frac{7}{8}]$. We choose an interval S_k that contains t_s "in the middle", that is $S_k = [l, r]$ for k such that $t_s \in [l + \frac{1}{16}, r - \frac{1}{16}]$. (See Fig. 4.) We then compute $t_{s+1} = 4(t_s - l)$. Note that t_{s+1} satisfies both $t_{s+1} \in [\frac{1}{4}, \frac{3}{4}]$ (that is, our invariant) and $d_{s+1} = d_s - 2$.

Step (rpr3): Converting G_{s+1} into G_s. Let G_{s+1} be the mixing graph obtained for t_{s+1} in step (rpr2). We first modify G_{s+1} to obtain a graph G'_{s+1}, which is then coupled with an appropriate converter C_{s+1} to obtain mixing graph $G_s = G'_{s+1} \bullet C_{s+1}$. Figure 5 illustrates this process.

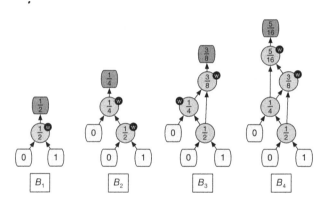

Fig. 6. Mixing graphs B_1, B_2, B_3 and B_4 for concentrations $\frac{1}{2}, \frac{1}{4}, \frac{3}{8}$ and $\frac{5}{16}$, respectively.

    Graph G'_{s+1} consists of the same nodes and edges as G_{s+1}, only the concentration labels are changed. Specifically, every concentration label c from G_{s+1} is changed to $l + c/4$ in G'_{s+1}. Note that this is simply the inverse of the linear function that maps t_s to t_{s+1}. In particular, this will map the 0- and 1-labels of the source nodes in G_{s+1} to the endpoints l and r of the corresponding interval S_k.

    The converter C_{s+1} used in G_s needs to have sink nodes with labels equal to the source nodes for G'_{s+1}. That is, if the labeling of the source nodes of G'_{s+1} is $I'_{s+1} = \{i : l, j : r\}$, then C_{s+1} will be an $(i : l, j : r)$-converter. As a general rule, C_{s+1} should produce at most one waste droplet, but there will be some exceptional cases where it produces two. (Nonetheless, we will show that at most one such "bad" converter is used during the RPR process.) The construction of these converters is somewhat intricate, and is deferred to the next section.

The Base Case. We now specify the set of base concentration values and their mixing graphs. Let $\mathcal{B} = \{\frac{1}{2}, \frac{1}{4}, \frac{3}{4}, \frac{3}{8}, \frac{5}{8}, \frac{5}{16}, \frac{11}{16}\}$. Figure 6 gives the graphs for concentrations $\frac{1}{2}, \frac{1}{4}, \frac{3}{8}$, and $\frac{5}{16}$; the graphs for the remaining concentrations are symmetric.

Initial Shift (IS). We now describe the IS procedure. At the fundamental level, the idea is similar to a single step of RPR. We can assume that $t < \frac{1}{4}$ (because for $t > \frac{3}{4}$ the process is symmetric). Thus the binary representation of t starts with $\gamma \geq 2$ fractional 0's. Since $2^{\gamma-1}t \in [\frac{1}{4}, \frac{1}{2})$, we could use this value as the result of the initial shift, but to improve the waste bound we refine this choice as follows: If $2^{\gamma-1}t \in (\frac{3}{8}, \frac{1}{2})$ then let $t_0 = 2^{\gamma-1}t$ and $\sigma = 1$. Otherwise, we have $2^{\gamma-1}t \in [\frac{1}{4}, \frac{3}{8}]$, in which case we let $t_0 = 2^{\gamma}t$ and $\sigma = 0$. In either case, $t_0 = 2^{\gamma-\sigma}t \in [\frac{1}{4}, \frac{3}{4}]$ and $d_0 = d - \gamma + \sigma$.

Let G_0 be the mixing graph obtained by applying RPR to t_0. It remains to show how to modify G_0 to obtain the mixing graph G for t. This is analogous to the process shown in Fig. 5. We first construct a mixing graph G_0' that consists of the same nodes and edges as G_0, only each concentration label c is replaced by $c/2^{\gamma-\sigma}$. In particular, the label set of the source nodes in G_0' will have the form $I_0' = \{i : 0, j : 1/2^{\gamma-\sigma}\}$. We then construct a $(i : 0, j : 1/2^{\gamma-\sigma})$-converter C_0 and couple it with G_0' to obtain G; that is, $G = G_0' \bullet C_0$. This C_0 is easy to construct: The 0's don't require any mixing, and to produce the j droplets $1/2^{\gamma-\sigma}$ we start with one droplet 1 and repeatedly mix it with 0's, making sure to generate at most one waste droplet at each step. Specifically, after z steps we will have j_z droplets with concentration $1/2^z$, where $j_z = \lceil j/2^{\gamma-\sigma-z} \rceil$. In step z, mix these j_z droplets with j_z 0's, producing $2j_z$ droplets with concentration $1/2^{z+1}$. We then either have $j_{z+1} = 2j_z$, in which case there is no waste, or $j_{z+1} = 2j_z - 1$, in which case one waste droplet $1/2^{z+1}$ is produced. Overall, C_0 produces at most $\gamma - \sigma$ waste droplets.

4   Construction of Converters

We now detail the construction of our converters. We can assume that $t_s \in [\frac{1}{4}, \frac{1}{2}]$, because the case $t_s \in (\frac{1}{2}, \frac{3}{4}]$ is symmetric. Recall that for a t_s in this range, in Step (rpr1) we will chose an appropriate interval S_k, for some $k \in \{1, 2, 3\}$. Let $S_k = [l, r]$ (that is, $l = k \cdot \frac{1}{8}$ and $r = l + \frac{1}{4}$). For each such k and all $i, j \geq 1$ we give a construction of an $(i : l, j : r)$-converter that we will denote $C_{i,j}^k$.

4.1 $(i : \frac{1}{4}, J : \frac{1}{2})$-Converters $C_{i,j}^2$

Here we only show how to construct, for all $i, j \geq 1$, our $(i : \frac{1}{4}, j : \frac{1}{2})$-converter $C_{i,j}^2$. (The construction of converters $C_{i,j}^1$ and $C_{i,j}^3$ is given in the full version of this paper [6].) These converters are constructed via an iterative process. We first give initial converters $C_{i,j}^2$, for some small values of i and j, by providing specific graphs. Other converters are obtained from these initial converters by repeatedly coupling them with other mixing graphs that we refer to as *extenders*.

Let $J^2_{\mathrm{init}} = \{(i,j)\}_{i,j \in \{1,2\}}$. The initial converters $C^2_{i,j}$ are defined for the four index pairs $(i,j) \in J^2_{\mathrm{init}}$. Figure 7 illustrates the initial converters $C^2_{2,1}, C^2_{1,2}$ and two extenders X^2_1, X^2_2. Converter $C^2_{1,2}$ produces one waste droplet and converter $C^2_{2,1}$ does not produce any waste. Converter $C^2_{1,1}$ can be obtained from $C^2_{2,1}$ by designating one of the $\frac{1}{4}$ outputs as waste. Converter $C^2_{2,2}$ is defined as $C^2_{2,2} = X^2_1 \bullet C^2_{2,1}$, and produces one waste droplet of $\frac{1}{2}$. (Thus $C^2_{2,2}$ is simply a disjoint union of $C^2_{2,1}$ and X^2_1 with one output $\frac{1}{2}$ designated as waste.)

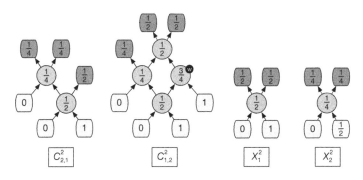

Fig. 7. Initial converters and extenders for the case $I = \{i : \frac{1}{4}, j : \frac{1}{2}\}$.

The construction of other converters $C^2_{i,j}$ is based on the following observation: Suppose that we already have constructed some $C^2_{i,j}$. Then (i) $X^2_1 \bullet C^2_{i,j}$ is a $C^2_{i,j+2}$ converter that produces the same waste as $C^2_{i,j}$, and (ii) provided that $j \geq 2$, $X^2_2 \bullet C^2_{i,j}$ is a $C^2_{i+2,j-1}$ converter that produces the same waste as $C^2_{i,j}$.

Let now $i, j \geq 1$ with $(i,j) \notin J^2_{\mathrm{init}}$ be arbitrary. To construct $C^2_{i,j}$, using and the above observation, express the integer vector (i,j) as $(i,j) = (i',j') + \phi(0,2) + \psi(2,-1)$, for some $i', j' \in J^2_{\mathrm{init}}$ and integers $\psi = \lceil \frac{i}{2} \rceil - 1$ and $\phi = \lceil \frac{i+\psi}{2} \rceil - 1$. Then $C^2_{i,j}$ is constructed by starting with $C^2_{i',j'}$ and coupling it ϕ times with X^2_1 and then ψ times with X^2_2. (This order of coupling is not unique but is also not arbitrary, because each extender X^2_2 requires a droplet of concentration $\frac{1}{2}$ as input.) Since X^2_1 and X^2_2 do not produce waste, $C^2_{i,j}$ will produce at most one waste droplet.

5   Performance Bounds

We now give the analysis of Algorithm RPRIS, including the bounds on produced waste, the size of computed mixing graphs, and the running time.

Waste Bound. We first estimate the number of waste droplets. Let G be the mixing graph constructed by RPRIS for a target concentration t with its corresponding values $d = \mathsf{prec}(t)$ and γ (see Sect. 1). We prove the following theorem.

Theorem 1. *The number of waste droplets in G is at most $\frac{1}{2}(d + \gamma) + 2$.*

To prove Theorem 1, we show that the total number of sink nodes in G is at most $\frac{1}{2}(d + \gamma - \sigma) + 3$. (This is sufficient, as one sink node is used to produce t). Following the algorithm description in Sect. 3, let $G = G'_0 \bullet C_0$. From our construction of C_0 (at the end of Sect. 3), we get that C_0 contributes at most $\gamma - \sigma$ sink nodes to G. (Each waste droplet produced by C_0 represents a sink node in G.) Therefore, to prove Theorem 1 it remains to show that G'_0 contains at most $\frac{1}{2}(d - \gamma + \sigma) + 3$ sink nodes. This is equivalent to showing that G_0 contains at most $\frac{1}{2}d_0 + 3$ sink nodes, where $d_0 = \mathsf{prec}(t_0) = d - \gamma + \sigma$. This fact is established in [6].

Size of Mixing Graphs and Running Time. Let $G = G'_0 \bullet C_0$ be the mixing graph computed by Algorithm RPRIS for t; C_0 is constructed by process IS while G'_0 is obtained from G_0 (constructed by process RPR) by changing concentration labels appropriately. We claim that the running time of Algorithm RPRIS is $O(|G|)$, and that the size of G is $O(d^2)$, for $d = \mathsf{prec}(t)$. We give bounds for G_0 and C_0 individually, then we combine them to obtain the claimed bounds.

First, following the description of process RPR in Sect. 3, suppose that at recursive step s, G_{s+1}, G'_{s+1} and converter $C_{s+1} = C^k_{i,j}$ are computed. The size of $C^k_{i,j}$ is $O(i + j)$ and it takes time $O(i + j)$ to assemble it (as the number of required extenders is $O(i+j)$). Coupling C_{s+1} with G'_{s+1} also takes time $O(i+j)$, since I'_{s+1} (the input for G'_{s+1}) has cardinality $O(i + j)$ as well. In other words, the running time of each recursive RPR step is proportional to the number of added nodes. Thus the overall running time to construct G_0 is $O(|G_0|)$.

Now, let t_0 be the target concentration for the RPR process, with $d_0 = \mathsf{prec}(t_0)$. Then, the size of G_0 is $O(d_0^2)$. This is because the depth of recursion in the RPR process is $O(d_0)$, and each converter used in this process has size $O(d_0)$ as well. The reason for this bound on the converter size is that, from a level of recursion to the next, the number of source nodes increases by at most one (with an exception of at most one step, as explained earlier in this section), and the size of $C^k_{i,j}$ used at this level is asymptotically the same as the number of source nodes at this level.

Regarding the bounds for C_0, we note that the running time to construct C_0 is $O(|C_0|)$, because in step s there are $2j_s$ droplets being mixed, which requires j_s nodes; thus the entire step takes time $O(j_s)$. We next show that the size of C_0 is $O(d^2)$. Let I_0 be the input configuration for G_0. From the analysis for G_0, we get that $|I_0| = O(d_0)$, so the last step in C_0 contains $O(d_0)$ nodes. Therefore, as the depth of C_0 is $\gamma - \sigma$, the size of C_0 is $O(\gamma d_0) = O(d_0^2)$.

Combining the bounds from G_0 and C_0, we get that the running time of Algorithm RPRIS is $O(|G|)$ and the size of G is $O(d^2)$.

6   Experimental Study

In this section, we compare the performance of our algorithm with algorithms Min-Mix, REMIA, DMRW, GURMA and ILP. (A brief description for each of these algorithms is given in [6].) We carried out four experiments. Each experiment consisted on generating all concentration values with precision d, for $d \in \{7, 8, 15, 20\}$, and comparing the outputs of each of the algorithms. The results for GORMA and ILP are shown only for $d \in \{7, 8\}$, since for $d \in \{15, 20\}$ the running time of both GORMA and ILP is prohibitive.

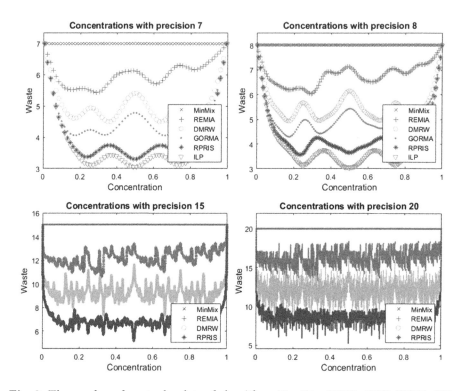

Fig. 8. The number of waste droplets of algorithms Min-Mix, REMIA, DMRW, GORMA, ILP, and our algorithm RPRIS, for all concentrations with precision 7 (top-left), 8 (top-right), 15 (bottom-left) and 20 (bottom-right). All graphs are smoothed using MATLAB's *smooth* function.

Figure 8 illustrates the experiments for concentrations of precision 7, 8, 15 and 20. In all figures, the data was smoothed using MATLAB's *smooth* function to reduce clutter and to bring out the differences in performance between different algorithms. As can be seen from these graphs, RPRIS significantly outperforms algorithm Min-Mix, REMIA, DMRW and GORMA: It produces on average about 50% less waste than Min-Mix (consistently with our bound of $\frac{1}{2}(d + \gamma) + 4$ on waste produced by RPRIS), and 40% less waste than REMIA. It also produces on average between 21 and 25% less waste than DMRW, with this percentage increasing with d. Additionally, for $d = 7, 8$, RPRIS produces on average about 17% less waste than GORMA and only about 7% additional waste than ILP.

Among all of the target concentration values used in our experiments, there is not a single case where RPRIS is worse than either Min-Mix or REMIA. When compared to DMRW, RPRIS never produces more waste for precision 7 and 8. For precision 15, the percentage of concentrations where RPRIS produces more waste than DMRW is below 2%, and for precision 20 it is below 3.5%. Finally, when compared to GORMA, the percentage of concentrations where RPRIS produces more waste is below 4%.

7   Final Comments

Many questions about mixing graphs remain open. We suspect that our bound on waste can be significantly improved. It is not clear whether waste linear in d is needed for concentrations not too close to 0 or 1, say in $[\frac{1}{4}, \frac{3}{4}]$. In fact, we do not know even a *super-constant* lower bound on waste for concentrations in this range.

For single-droplet targets it is not known whether minimum-waste mixing graphs can be effectively computed. The most fascinating open question, in our view, is whether it is decidable to determine if a given multiple-droplet target set can be produced without any waste.

Another interesting problem is about designing mixing graphs for producing multiple droplets of the same concentration. Using perfect-mixing graphs from [5], it can be shown that if the number of droplets exceeds a certain threshold then such target sets can be produced with at most one waste droplet. However, this threshold value is very large and the resulting algorithm very complicated.

It would also be interesting to extend our proposed worst-case performance measure to reactant minimization. It is quite possible that our general approach of recursive precision reduction could be adapted to this problem.

References

1. Bhattacharjee, S., Bhattacharya, B.B., Chakrabarty, K.: Algorithms for Sample Preparation with Microfluidic Lab-on-Chip. River Publishers, Delft (2019)
2. Bhattacharya, B.B., Roy, S., Bhattacharjee, S.: Algorithmic challenges in digital microfluidic biochips: protocols, design, and test. In: Gupta, P., Zaroliagis, C. (eds.) ICAA 2014. LNCS, vol. 8321, pp. 1–16. Springer, Cham (2014). https://doi.org/10.1007/978-3-319-04126-1_1
3. Chiang, T.W., Liu, C.H., Huang, J.D.: Graph-based optimal reactant minimization for sample preparation on digital microfluidic biochips. In: 2013 International Symposium on VLSI Design, Automation and Test (VLSI-DAT), pp. 1–4. IEEE (2013)
4. Dinh, T.A., Yamashita, S., Ho, T.Y.: A network-flow-based optimal sample preparation algorithm for digital microfluidic biochips. In: 19th Asia and South Pacific Design Automation Conference (ASP-DAC), pp. 225–230. IEEE (2014)

5. Coviello Gonzalez, M., Chrobak, M.: Towards a theory of mixing graphs: a characterization of perfect mixability (extended abstract). In: Heggernes, P. (ed.) CIAC 2019. LNCS, vol. 11485, pp. 187–198. Springer, Cham (2019). https://doi.org/10.1007/978-3-030-17402-6_16

6. Gonzalez, M.C., Chrobak, M.: A waste-efficient algorithm for single-droplet sample preparation on microfluidic chips. CoRR abs/1908.09618 (2019)

7. Huang, J.D., Liu, C.H., Chiang, T.W.: Reactant minimization during sample preparation on digital microfluidic biochips using skewed mixing trees. In: Proceedings of the International Conference on Computer-Aided Design, pp. 377–383. ACM (2012)

8. Huang, J.D., Liu, C.H., Lin, H.S.: Reactant and waste minimization in multitarget sample preparation on digital microfluidic biochips. IEEE Trans. Comput. Aided Des. Integr. Circuits Syst. **32**(10), 1484–1494 (2013)

9. Mitra, D., Roy, S., Chakrabarty, K., Bhattacharya, B.B.: On-chip sample preparation with multiple dilutions using digital microfluidics. In: IEEE Computer Society Annual Symposium on VLSI (ISVLSI), pp. 314–319. IEEE (2012)

10. Roy, S., Bhattacharya, B.B., Chakrabarty, K.: Optimization of dilution and mixing of biochemical samples using digital microfluidic biochips. IEEE Trans. Comput. Aided Des. Integr. Circuits Syst. **29**(11), 1696–1708 (2010)

11. Thies, W., Urbanski, J.P., Thorsen, T., Amarasinghe, S.: Abstraction layers for scalable microfluidic biocomputing. Nat. Comput. **7**(2), 255–275 (2008)

Shortest Covers of All Cyclic Shifts
of a String

Maxime Crochemore[1] , Costas S. Iliopoulos[1] , Jakub Radoszewski[2] ,
Wojciech Rytter[2] , Juliusz Straszyński[2] , Tomasz Waleń[2] ,
and Wiktor Zuba[2(✉)]

[1] Department of Informatics, King's College London, London, UK
{maxime.crochemore,c.iliopoulos}@kcl.ac.uk
[2] Institute of Informatics, University of Warsaw, Warsaw, Poland
{jrad,rytter,jks,walen,w.zuba}@mimuw.edu.pl

Abstract. A factor W of a string X is called a cover of X, if X can
be constructed by concatenations and superpositions of W. Breslauer
(IPL, 1992) proposed a well-known $\mathcal{O}(n)$-time algorithm that computes
the shortest cover of every prefix of a string of length n. We show an
$\mathcal{O}(n \log n)$-time algorithm that computes the shortest cover of every
cyclic shift of a string and an $\mathcal{O}(n)$-time algorithm that computes the
shortest among these covers. A related problem is the number of different
lengths of shortest covers of cyclic shifts of the same string of length n.
We show that this number is $\Omega(\log n)$.

1 Introduction

We consider strings as finite sequences of letters from an integer alphabet Σ. The
notion of periodicity in strings and its many variants have been well-studied in
many fields like combinatorics on words, pattern matching, data compression,
automata theory, formal language theory, and molecular biology. A typical regu-
larity, the *period* U of a given string X, grasps the repetitiveness of X since X is
a prefix of a string constructed by concatenations of U. If $X = AWB$, for some,
possibly empty, strings A, W, B, then W is called a *factor* of X and, respectively,
X is a superstring of W. A factor W of X is called a *cover* of X, if X can be constructed by concatenations and superpositions of W. A factor W of X is called
a *seed* of X, if there exists a superstring of X which is constructed by concatenations and superpositions of W. For example, abc is a period of $abcabcabca, abca$
is a cover of $abcabcaabca$, and $abca$ is a seed of $bcabcaabc$. The notions "cover"
and "seed" are generalizations of periods in the sense that superpositions as well
as concatenations are considered to define them, whereas only concatenations
are considered for periods.

In computation of covers, two problems have been considered in the literature.
The shortest-cover problem (also known as the superprimitivity test) is that of

J. Radoszewski, J. Straszyński, T. Waleń and W. Zuba—Supported by the Polish
National Science Center, grant no. 2018/31/D/ST6/03991.

M. S. Rahman et al. (Eds.): WALCOM 2020, LNCS 12049, pp. 69–80, 2020.
https://doi.org/10.1007/978-3-030-39881-1_7

computing the shortest cover of a given string of length n, and the all-covers problem is that of computing all the covers of a given string. Apostolico et al. [1] introduced the notion of covers and gave a linear-time algorithm for the shortest-cover problem. Breslauer [4] proposed an on-line algorithm for computing the shortest cover that works in linear time. In particular, his algorithm computes the shortest cover of every prefix of a string. The other direction was taken by Moore and Smyth [20, 21] and by Li and Smyth [19] who computed all the covers of a string and a representation of all the covers of all prefixes of a string, respectively.

Covers of circular strings were also considered. It is implicit in [13] that covers of a circular string S are exactly seeds of S^2 (see also [15]).

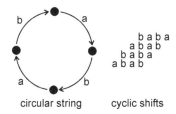

circular string   cyclic shifts

Fig. 1. The string aba is a cover of the string $S = abab$ treated as a single circular string, but is not a cover of any of cyclic shifts of S.

All the seeds of a string of length n can be represented in $\mathcal{O}(n)$ space as a collection of a linear number of disjoint paths in the suffix trees of the string and of its reversal. This representation can be computed in $\mathcal{O}(n \log n)$ time [13] and even in $\mathcal{O}(n)$-time [16]. Recently it was also shown in [17] that all the seeds can also be represented as a linear number of disjoint paths in just the suffix tree of the string. This implies the following fact:

Lemma 1. *The problem of computing the shortest cover of a circular string can be solved in linear time.*

We say that a string Y is a *cyclic shift* of a string X if $X = AB$ and $Y = BA$ for some strings A and B; in this case we also write $Y = rot_{|A|}(X)$. It seems that the problem of computing shortest covers of all cyclic shifts of a string is harder than that of computing the shortest cover of a circular string. A straightforward application of any of the aforementioned algorithms for computing covers of a string yields an $\mathcal{O}(n^2)$-time solution to the problem. One should note that covers of circular strings are a different notion than that of covers of cyclic shifts of a string; see Fig. 1.

The shortest covers of cyclic shifts of a string can behave rather irregularly. For example, the length of the shortest cover of $S = abaababababababababa$ equals 3, whereas the shortest cover of $rot_1(S)$ has length 18.

We consider the following problem.

SHORTEST COVERS OF ALL CYCLIC SHIFTS OF A STRING

Input: A string S of length n.

Output: The lengths of the shortest covers of all cyclic shifts of S.

Let S be a string of length n and $ShCov(S)$ denote the shortest cover of S. We introduce an array $CyCo_S$ of length n such that $CyCo_S[i] = |ShCov(rot_i(S))|$. Our main result is computing this array. We also denote

$$CyCoSet(S) = \{CyCo_S[i] : i = 0, \ldots, n - 1\}.$$

Example 2. For the Fibonacci strings $S_1 = abaab, S_2 = abaababaabaab$ we have:

$$CyCo_{S_1} = [5, 5, 5, 3, 5], \quad CyCo_{S_2} = [5, 5, 13, 3, \ldots]$$
$$CyCoSet(S_1) = \{3, 5\}, \quad CyCoSet(S_2) = \{3, 5, 8, 13\}.$$

Our Results. We show that the whole array $CyCo_S$ and $\min_i CyCo_S[i]$ for a string S of length n can be computed in $\mathcal{O}(n \log n)$ time and $\mathcal{O}(n)$ time, respectively. For this we use a characterization of covers of cyclic shifts of a string by seeds and squares, i.e., strings of the form W^2, and the suffix tree data structure. We also show that there exists a (known) infinite family of strings for which $|CyCoSet(S)| = \Theta(\log |S|)$.

Structure of the Paper. In Sect. 2 we recall the definition and basic properties of a suffix tree of a string. Then in Sect. 3 we present characterizations of shortest covers of cyclic shifts of a string, which lead us to the main algorithmic results in Sect. 4. The lower bound on the size of the $CyCoSet$ set is shown using Fibonacci strings in Sect. 5. We conclude and mention some open problems in Sect. 6.

2   Applications of the Suffix Tree

Recall that a *suffix tree* of a string S is a compact trie of all the suffixes of $S\#$, where $\#$ is a special end marker. The root, branching nodes, and leaves of the tree are *explicit*. All the remaining nodes are *implicit* in the tree. Each leaf is labeled with the starting position of the corresponding suffix. Every factor of S is represented as an explicit or implicit node of the tree. A suffix tree of a string of length n over an integer alphabet can be constructed in $\mathcal{O}(n)$ time [10].

Observation 1. *Let S be a string of length n. After $\mathcal{O}(n)$-time preprocessing, all the occurrences of a factor of S, represented as a node in the suffix tree of S, can be reported in linear time w.r.t. the number of these occurrences.*

Proof. It suffices to store a list L of leaves of the suffix tree in a left-to-right order. Then for every explicit node v of the tree, we precompute the endpoints of the sublist of L that corresponds to the occurrences of the string v. This precomputation is done bottom-up in $\mathcal{O}(n)$ time.   □

We also use the following lemma.

Lemma 3 ([17]). *Given a collection of factors $U_1, \ldots, U_k$ of a string S of length n, each represented by an occurrence in S, in $\mathcal{O}(n + k)$ time we can compute the implicit or explicit node in the suffix tree of S that corresponds to each factor U_i. Moreover, all these nodes can be made explicit in $\mathcal{O}(n + k)$ time.*

The set (possibly of a quadratic size) of all seeds of a string can be represented as a collection of linearly many disjoint paths in the suffix tree [17]. It can be assumed that each path belongs to a single edge of the suffix tree. The endpoints of the paths can be implicit nodes. For an example, see Fig. 2.

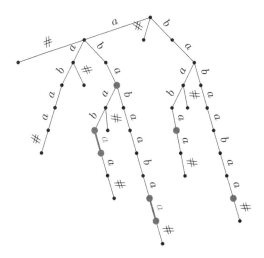

Fig. 2. The string $S = ababaabaa$ has the following seeds: *aba, abaab, baaba, abaaba, ababaaba, babaabaa, ababaabaa*. They can be represented on the (uncompressed) suffix tree of S as shown in the figure. Each seed is a path from root to marked node. In some cases, e.g. *abaab, abaaba*, multiple seeds are represented on a single path.

3   Covers of Cyclic Shifts

A string X is called *primitive* if $X = Y^k$ for positive integer k implies that $k = 1$. A string of the form Z^2 is called a *square*; it is called *primitively rooted* if Z is primitive. We denote by *Squares*(S) the set of factors Z of S such that the square Z^2 is also a factor of S and by *PSquares*(S) the subset of *Squares*(S) that consists only of primitive strings. We further denote by *Seeds*(S) the set of factors which are seeds of S. We use these sets of S^3 in order to characterize covers of all cyclic shifts of S.

Lemma 4. *Let S be a string of length n and C be a string of length up to n. Then C is a cover of $rot_i(S)$ if and only if $C \in Seeds(S^3) \cap Squares(S^3)$ and C^2 occurs with its center at position $j \equiv i \pmod{n}$ in S^3.*

Moreover, if C is the shortest cover of $rot_i(S)$, then $C \in Seeds(S^3) \cap PSquares(S^3)$.

Proof.
($\Rightarrow$) String C is a cover of $(rot_i(S))^4$, and thus a seed of its factor S^3. Moreover, $S^3[j - |C|, j + |C| - 1]$, that is, the factor of S^3 of length $2|C|$ with center at position j, is equal to C^2 for $j = i + n$.
($\Leftarrow$) The square C^2 occurs in S^3 with its center at position $j \equiv i \pmod{n}$. Thus C is a prefix and a suffix of $rot_j(S) = rot_i(S)$ as $|C| < n$. C is also a seed of $rot_i(S)$ which is a factor of S^3, hence it is a cover of $rot_i(S)$.

As for the "moreover" part, it suffices to note that the shortest cover of a string is obviously primitive. $\square$

Example 5. In the above lemma, one could not take S^2 instead of S^3. Indeed, for $S = abaaaaba$ we have that $rot_4(S) = aabaabaa$ has the shortest cover $aabaa$, but $S^2 = abaaaabaabaaaaba$ does not contain the square $(aabaa)^2$ (Fig. 3).

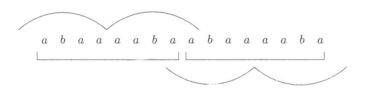

Fig. 3. Illustration of Example 5.

Let $T(S^3)$ be the suffix tree of S^3 in which we distinguish the nodes v corresponding to strings Z^2 for $Z \in Seeds(S^3) \cap PSquares(S^3)$. These nodes are called *candidate nodes*. Some of these nodes could be implicit nodes in the suffix tree. Then they are made explicit. Denote by $CandAnc(v)$ the set of ancestor nodes of v in $T(S^3)$ which are candidate nodes. Let $|v|$ be the length of the string corresponding to the node v.

We can reformulate Lemma 4 as follows (see Fig. 4):

Lemma 6. $CyCo_S[i]$, *i.e., the length of the shortest cover of $rot_i(S)$, equals*

$$\min_{j,v} \{ k \; : \; k = |v|/2, \; i = (j + k) \bmod n, \; j \in Leaves(T(S^3)), \; v \in CandAnc(j) \}.$$

Clearly if C is a cover of $rot_i(S)$, then C is a cover of S treated as a circular string. As we have already noted in Fig. 1, the converse is not necessarily true. However, we show that every shortest cover of the circular string S is a cover of the corresponding cyclic shift of S.

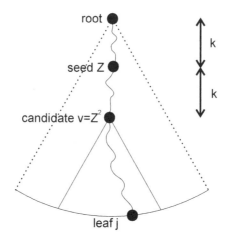

Fig. 4. Illustration of Lemma 6. The situation when $CyCo_S[i] = k$. We have that $i = (j + k) \bmod n$ and Z^2 is a primitively rooted square of length $2k$; it corresponds to the node v which is possibly inside an edge of the suffix tree.

Lemma 7. *A shortest cover of a circular string is always a (shortest) cover of some cyclic shift.*

Proof. We need the following claim.

Claim (See [13]). String C is a cover of S considered as a circular string iff it is a seed of S^2, hence also iff it is a seed of S^3.

Consider a cover C of a circular string S, such that C^2 does not occur in it. Consider the last position covered by any occurrence of C in the string.

The position must be also covered by another occurrence of C (the next position must be covered and cannot be the first position of some C). Thus by erasing the last position of C we obtain a shorter cover. Hence if C is a shortest cover then C^2 must appear in the circular string S.

By Lemma 4 it is a cover of some cyclic shift of the string.   □

By computing the shortest cover of the circular string S using Lemma 1 we obtain the following preliminary result.

Corollary 8. *For a string S of length n, $\min CyCo_S$ can be computed in $\mathcal{O}(n)$ time.*

4   Main Algorithm

First we have to show how to compute efficiently the tree $T(S^3)$. We denote by $OccPSquares(S)$ the set of all occurrences of primitively rooted squares in S. Each occurrence is represented in $\mathcal{O}(1)$ space as a factor of S. A direct consequence of the Three-square-prefix Lemma, see [8], is that a string of length n has no more than $\log n$ prefixes that are primitively rooted squares.

Lemma 9 ([8]). *For a string S of length n, $|OccPSquares(S)| = \mathcal{O}(n \log n)$.*

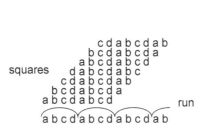

squares

run

Fig. 5. Primitive squares can be derived from runs (maximal repetitions), knowing the shortest periods of runs.

Lemma 10. *For a string S of length n, $|PSquares(S)| = \mathcal{O}(n)$ and this set can be computed in $\mathcal{O}(n)$ time.*

Proof. Let us start with efficient computation of squares.

Claim ([7,9,11,12]). For a string S of length n, $|Squares(S)| = \mathcal{O}(n)$ and this set can be computed in $\mathcal{O}(n)$ time.

By the claim, $|PSquares(S)| = \mathcal{O}(n)$ since $PSquares(S) \subseteq Squares(S)$.

The set $PSquares(S)$ can be computed by filtering out the factors from $Squares(S)$ that are not primitive. This can be done in $\mathcal{O}(1)$ time per factor after $\mathcal{O}(n)$-space and time preprocessing using so-called Two-Period queries [3,18]. A more direct approach would be to (effortlessly) adapt the algorithm for computing different square factors from [7] using relations between primitive squares and *runs* (maximal repetitions); see Fig. 5. □

Lemma 11. *The tree $T(S^3)$ can be computed in $\mathcal{O}(n)$ time.*

Proof. We use a version of a minimal augmented suffix tree (MAST, in short), a data structure that was initially introduced in [2].

Let us recall that Lemma 3 can be used to augment the suffix tree with nodes that correspond to a set of factors of S^3. We first apply the lemma to the collection of factors $PSquares(S^3)$, which can be efficiently computed due to the previous lemma.

Then we compute a representation of all the seeds of S^3 in the suffix tree using the algorithm from [17]. The representation consists of a collection of disjoint paths, each located on a single edge in the suffix tree; see Fig. 2. The endpoints of the paths that are implicit nodes can also be made explicit using the lemma.

For every node v corresponding to an element in $PSquares(S^3)$, we check if it is located on some path that belongs to the representation of $Seeds(S^3)$. Finally, we again use Lemma 3 for the original suffix tree of S^3 and set of factors Z^2 that correspond to all such elements $Z \in PSquares(S^3) \cap Seeds(S^3)$ to obtain the set of candidate nodes. This completes the proof. □

The number of integers $(j + k)$ equal *modulo* n is constant, hence we can forget about computing modulo n and for each $0 \le i < 3n$ we are to compute:

$$\min_{j,v} \{ k \; : \; k = |v|/2, \; i = j + k, \; j \in Leaves(T(S^3)), \; v \in CandAnc(j)\}. \quad (1)$$

Algorithm ComputeCyCo

1. Initialize each entry of $CyCo$ to $+\infty$
2. Compute $T(S^3)$
3. **for each** candidate node v in $T(S^3)$ **do**
    for each occurrence $S^3[j, j + |v| - 1]$ of v in S^3 **do**
    $i := (j + |v|) \bmod n$
    $CyCo[i] := \min(CyCo[i], |v|)$
4. **return** $CyCo$

Theorem 12. *The algorithm* ComputeCyCo *computes the lengths of shortest covers for all cyclic shifts of a string in* $\mathcal{O}(n \log n)$ *time.*

Proof. In the third paragraph we simply implement (1). This proves correctness. The occurrences of a node are computed using Observation 1. The required complexity follows from Lemmas 9 and 11. □

5   Strings with Arbitrarily Large Size of $CyCoSet(S)$

We show that the size of $CyCoSet(S)$, for a binary alphabet, is not bounded by a constant. It grows at least logarithmically.

Recall that the Fibonacci strings are defined as $Fib_0 = b$, $Fib_1 = a$, $Fib_k = Fib_{k-1} Fib_{k-2}$ for $k \ge 2$. In other words, $Fib_k = \phi^k(Fib_0)$, where ϕ is a morphism

$$\phi(a) = ab, \quad \phi(b) = a.$$

Hence

$$Fib_2 = ab, \; Fib_3 = aba, \; Fib_4 = abaab, \; Fib_5 = abaababa, \ldots$$

We denote $F_k = |Fib_k|$, the k-th Fibonacci number.

We use the following well known properties of Fibonacci strings.

Observation 2. *Fib_k does not contain the factors aaa and bb.*

Fact 1 (see [14,22]). *For every non-empty factor U^2 of Fib_k, U is a cyclic shift of Fib_m for some m.*

An example for the theorem below can be found in Fig. 6.

Theorem 13. *For $k \ge 3$, $CyCoSet(Fib_k) = \{F_3, \ldots, F_k\}$.*

Fib$_2$

cyclic shift	shortest cover	length
ab	ab	2
ba	ba	2

Fib$_3$

aba	aba	3
baa	baa	3
aab	aab	3

Fib$_4$

abaab	abaab	5
baaba	baaba	5
aabab	aabab	5
ababa	aba	3
babaa	babaa	5

Fib$_5$

abaababa	aba	3
baababaa	baababaa	8
aababaab	aababaab	8
ababaaba	aba	3
babaabaa	babaabaa	8
abaabaab	abaab	5
baabaaba	baaba	5
aabaabab	aabaabab	8

Fib$_6$

cyclic shift	shortest cover	length
abaababaabaab	abaab	5
baababaabaaba	baaba	5
aababaabaabab	aababaabaabab	13
ababaabaababa	aba	3
babaabaababaa	babaabaababaa	13
abaabaababaab	abaab	5
baabaababaaba	baaba	5
aabaababaabab	aabaababaabab	13
abaababaababa	aba	3
baababaababaa	baababaa	8
aababaababaab	aababaab	8
ababaababaaba	aba	3
babaababaabaa	babaababaabaa	13

Fig. 6. Shortest covers of cyclic shifts of Fibonacci strings.

Proof. We show two inclusions.

Proof of the inclusion $\{F_3, \ldots, F_k\} \supseteq CyCoSet(Fib_k)$.

By Lemma 4, every element of the set $CyCoSet(Fib_k)$ is a square half of length at most F_k in Fib_k^3. By Fact 1, the lengths of square halves in a Fibonacci string are Fibonacci numbers. It suffices to note that, for $k \geq 3$, Fib_k^3 is a factor of Fib_{k+5} since

$$Fib_8 = abaababaabaababaab\underline{abaabaababa}abaab$$

contains a cube Fib_3^3 (underlined). It can be readily verified that no string of length $F_1 = 1$ and $F_2 = 2$ covers Fib_k for $k \geq 3$.

Proof of the inclusion $\{F_3, \ldots, F_k\} \subseteq CyCoSet(Fib_k)$.

We will prove that for every $i = 3, \ldots, k$, there exists a cyclic shift S of Fib_k such that $|ShCov(S)| = F_i$ and $S \neq aaXbab$ for a binary string X. The proof goes by induction over k. For $k = 3$ the conclusion is straightforward. Assume now that the conclusion holds for $k - 1$.

Let us consider $i \in \{3, \ldots, k-1\}$ and let S be a cyclic shift of Fib_{k-1} such that $|ShCov(S)| = F_i$ and S is not of the form $aaXbab$. We use the following observation.

Observation 3. Let $S_{1,1} = abXb$, $S_{1,2} = bXba$, $S_{2,1} = baaXaa$, and $S_{2,2} = aaXaab$ be cyclic shifts of a Fibonacci string, where X is a binary string. For every $i = 1, 2$ and string C, C is a cover of $S_{i,1}$ if and only if $rot_1(C)$ is a cover of $S_{i,2}$.

Proof. Any cover C of $S_{1,1}$ starts with the letter a and ends with the letter b. By Observation 2, each of its occurrences except for the occurrence as a suffix is followed by the letter a. Hence, $rot_1(C)$ is a cover of $S_{1,2}$. Similarly, a cover C' of $S_{1,2}$ starts with the letter b and ends with the letter a, so each of its occurrences except for the occurrence as a prefix is preceded by the letter a. Hence, $rot_{|C'|-1}(C')$ is a cover of $S_{1,1}$.

The proof for $S_{2,1}$ and $S_{2,2}$ is analogous.   □

If S ends with the letter b, we either move its first letter to its end to obtain a string that matches $S_{1,2}$ or the last letter to the start to obtain a string that matches $S_{2,1}$. We use the additional property that S is not of the form $aaXbab$, in which case none of the shifts would be possible. After the potential transformation, $|ShCov(S)|$ did not change and S starts with the letter b.

Now $S' = \phi(S)$ is a cyclic shift of Fib_k that ends with $\phi(a) = ab$.

Observation 4. Assume that $S' = \phi(S)$ and S' ends with the letter b. Let C' be a cover of S'. Then there exists a unique cover C of S such that $\phi(C) = C'$.

By the observation, if $C = ShCov(S)$, then $C' = \phi(C)$ is the shortest cover of S'. By Lemma 4, C'^2 occurs in Fib_{k-1}^3. Hence, Fact 1 implies that C is a cyclic shift of Fib_i, so C' is a cyclic shift of Fib_{i+1} and $|C'| = F_{i+1}$.

Thus we have obtained that $\{F_4, \ldots, F_k\} \subseteq CyCoSet(Fib_k)$. To conclude, we notice that $rot_3(Fib_k)$ starts and ends with $Fib_2 = aba$, so aba is its cover by Observation 2 (and the shortest cover by the first inclusion). Thus $F_3 \in CyCoSet(Fib_k)$.

Consequently, $\{F_3, F_4, \ldots, F_k\} \subseteq CyCoSet(Fib_k)$.   □

6   Conclusions and Open Problems

Breslauer [4] proposed a linear-time algorithm for computing the shortest cover of every prefix of a string. We have proposed an $\mathcal{O}(n \log n)$-time algorithm for computing the shortest cover of every cyclic shift of a string. It remains an open problem if these values can be computed in $\mathcal{O}(n)$ time.

$\mathcal{O}(n)$, $\mathcal{O}(n \log n)$ and $\mathcal{O}(n^2)$-time algorithms for computing the shortest left seed, right seed, and seed, respectively, of all the prefixes of a string are known; see [5,6]. Here left and right seed are notions that are intermediate between cover and seed. It remains an open problem if the shortest left seed, right seed, and seed can be computed efficiently for all the cyclic shifts of a string.

Based on computer experiments we make the following conjecture.

Conjecture 1. For a string S of length n, $|CyCoSet(S)| = \mathcal{O}(\log n)$.

References

1. Apostolico, A., Farach, M., Iliopoulos, C.S.: Optimal superprimitivity testing for strings. Inf. Process. Lett. **39**(1), 17–20 (1991). https://doi.org/10.1016/0020-0190(91)90056-N
2. Apostolico, A., Preparata, F.P.: Data structures and algorithms for the string statistics problem. Algorithmica **15**(5), 481–494 (1996). https://doi.org/10.1007/BF01955046
3. Bannai, H., Tomohiro, I., Inenaga, S., Nakashima, Y., Takeda, M., Tsuruta, K.: The "runs" theorem. SIAM J. Comput. **46**(5), 1501–1514 (2017). https://doi.org/10.1137/15M1011032
4. Breslauer, D.: An on-line string superprimitivity test. Inf. Process. Lett. **44**(6), 345–347 (1992). https://doi.org/10.1016/0020-0190(92)90111-8
5. Christou, M., Crochemore, M., Guth, O., Iliopoulos, C.S., Pissis, S.P.: On left and right seeds of a string. J. Discrete Algorithms **17**, 31–44 (2012). https://doi.org/10.1016/j.jda.2012.10.004
6. Christou, M., et al.: Efficient seed computation revisited. Theor. Comput. Sci. **483**, 171–181 (2013). https://doi.org/10.1016/j.tcs.2011.12.078
7. Crochemore, M., Iliopoulos, C.S., Kubica, M., Radoszewski, J., Rytter, W., Waleń, T.: Extracting powers and periods in a word from its runs structure. Theor. Comput. Sci. **521**, 29–41 (2014). https://doi.org/10.1016/j.tcs.2013.11.018
8. Crochemore, M., Rytter, W.: Squares, cubes, and time-space efficient string searching. Algorithmica **13**(5), 405–425 (1995). https://doi.org/10.1007/BF01190846
9. Deza, A., Franek, F., Thierry, A.: How many double squares can a string contain? Discrete Appl. Math. **180**, 52–69 (2015). https://doi.org/10.1016/j.dam.2014.08.016
10. Farach, M.: Optimal suffix tree construction with large alphabets. In: 38th Annual Symposium on Foundations of Computer Science, FOCS 1997, Miami Beach, Florida, USA, 19–22 October 1997, pp. 137–143. IEEE Computer Society (1997). https://doi.org/10.1109/SFCS.1997.646102
11. Fraenkel, A.S., Simpson, J.: How many squares can a string contain? J. Comb. Theory Ser. A **82**(1), 112–120 (1998). https://doi.org/10.1006/jcta.1997.2843
12. Gusfield, D., Stoye, J.: Linear time algorithms for finding and representing all the tandem repeats in a string. J. Comput. Syst. Sci. **69**(4), 525–546 (2004). https://doi.org/10.1016/j.jcss.2004.03.004
13. Iliopoulos, C.S., Moore, D.W.G., Park, K.: Covering a string. Algorithmica **16**(3), 288–297 (1996). https://doi.org/10.1007/BF01955677
14. Iliopoulos, C.S., Moore, D.W.G., Smyth, W.F.: A characterization of the squares in a Fibonacci string. Theor. Comput. Sci. **172**(1–2), 281–291 (1997). https://doi.org/10.1016/S0304-3975(96)00141-7
15. Iliopoulos, C.S., Moore, D.W.G., Smyth, W.F.: The covers of a circular Fibonacci string. J. Comb. Math. Comb. Comput. **26**, 227–236 (1998)
16. Kociumaka, T., Kubica, M., Radoszewski, J., Rytter, W., Waleń, T.: A linear time algorithm for seeds computation. In: Rabani, Y. (ed.) Proceedings of the Twenty-Third Annual ACM-SIAM Symposium on Discrete Algorithms, SODA 2012, Kyoto, Japan, 17–19 January 2012, pp. 1095–1112. SIAM (2012). https://doi.org/10.1137/1.9781611973099.86
17. Kociumaka, T., Kubica, M., Radoszewski, J., Rytter, W., Waleń, T.: A linear time algorithm for seeds computation. CoRR abs/1107.2422v2 (2019). http://arxiv.org/abs/1107.2422v2

18. Kociumaka, T., Radoszewski, J., Rytter, W., Waleń, T.: Internal pattern matching queries in a text and applications. In: Indyk, P. (ed.) Proceedings of the Twenty-Sixth Annual ACM-SIAM Symposium on Discrete Algorithms, SODA 2015, San Diego, CA, USA, 4–6 January 2015, pp. 532–551. SIAM (2015). https://doi.org/10.1137/1.9781611973730.36

19. Li, Y., Smyth, W.F.: Computing the cover array in linear time. Algorithmica **32**(1), 95–106 (2002). https://doi.org/10.1007/s00453-001-0062-2

20. Moore, D.W.G., Smyth, W.F.: An optimal algorithm to compute all the covers of a string. Inf. Process. Lett. **50**(5), 239–246 (1994). https://doi.org/10.1016/0020-0190(94)00045-X

21. Moore, D.W.G., Smyth, W.F.: A correction to "an optimal algorithm to compute all the covers of a string". Inf. Process. Lett. **54**(2), 101–103 (1995). https://doi.org/10.1016/0020-0190(94)00235-Q

22. Séébold, P.: Propriétés combinatoires des mots infinis engendrés par certains morphismes. Report no. 85-16, LITP, Paris (1985)

Packing Trees into 1-Planar Graphs

Felice De Luca[1][ID], Emilio Di Giacomo[2][ID], Seok-Hee Hong[3][ID],
Stephen Kobourov[1][ID], William Lenhart[4][ID], Giuseppe Liotta[2][ID], Henk Meijer[5],
Alessandra Tappini[2(✉)][ID], and Stephen Wismath[6][ID]

[1] Department of Computer Science, University of Arizona, Tucson, USA
[2] Dipartimento di Ingegneria, Università degli Studi di Perugia, Perugia, Italy
alessandra.tappini@studenti.unipg.it
[3] School of Computer Science, University of Sydney, Sydney, Australia
[4] Department of Computer Science, Williams College, Williamstown, USA
[5] Department of Computer Science, University College Roosevelt,
Middelburg, The Netherlands
[6] Department of Computer Science, University of Lethbridge, Lethbridge, Canada

Abstract. We introduce and study the *1-planar packing problem*: Given
k graphs with n vertices $G_1, \ldots, G_k$, find a 1-planar graph that contains
the given graphs as edge-disjoint spanning subgraphs. We mainly focus
on the case when each G_i is a tree and $k = 3$. We prove that a triple
consisting of three caterpillars or of two caterpillars and a path may not
admit a 1-planar packing, while two paths and a special type of caterpillar always have one. We then study 1-planar packings with few crossings
and prove that three paths (resp. cycles) admit a 1-planar packing with
at most seven (resp. fourteen) crossings. We finally show that a quadruple consisting of three paths and a perfect matching with $n \geq 12$ vertices
admits a 1-planar packing, while such a packing does not exist if $n \leq 10$.

1 Introduction

In the *graph packing problem* we are given a collection of n-vertex graphs
$G_1, \ldots, G_k$ and we are requested to find a graph G that contains the given
graphs as edge-disjoint spanning subgraphs. Various settings of the problem can
be defined depending on the type of graphs that have to be packed and on the
restrictions put on the packing graph G. The most general case is when G is the
complete graph on n vertices and there is no restriction on the input graphs.
Sauer and Spencer [17] prove that any two graphs with at most $n - 2$ edges
can be packed into K_n; Woźniak and Wojda [19] give sufficient conditions for
the existence of a packing of three graphs. The setting when G is K_n and each
G_i is a tree ($i = 1, 2, \ldots, k$) has been intensively studied. Hedetniemi et al. [10]
show that two non-star trees can always be packed into K_n. Notice that, the

This work started at the Bertinoro Workshop on Graph Drawing 2019 and it is par-
tially supported by: (*i*) MIUR grant 20174LF3T8, (*ii*) Dipartimento di Ingegneria -
Università degli Studi di Perugia grants RICBASE2017WD and RICBA18WD, (*iii*)
NFS grants CCF-1740858, CCF-1712119, DMS-1839274, DMS-1839307.

hypothesis that the trees are not stars is necessary for the existence of the packing because each vertex must have degree at least one in each tree, which is not possible if a vertex is adjacent to every other vertex as it is the case for a star. Wang and Sauer [18] give sufficient conditions for the existence of a packing of three trees into K_n, while Mahéo et al. [13] characterize the triples of trees that admit such a packing.

García et al. [7] consider the *planar packing problem*, that is the case when the graph G is required to be planar. They conjecture that the result of Hedetniemi et al. extends to this setting, i.e., that every pair of non-star trees can be packed into a planar graph. Notice that, when G is required to be planar, two is the maximum number of trees that can be packed (because three trees have more than $3n - 6$ edges). García et al. prove their conjecture for some restricted cases, namely when one of the trees is a path and when the two trees are isomorphic. In a series of subsequent papers the conjecture has been proved true for other pairs of trees. Oda and Ota [14] prove it when one tree is a caterpillar or it is a spider of diameter four. Frati et al. [6] extend the last result to any spider, while Frati [5] considers the case when both trees have diameter four. Geyer et al. show that a planar packing always exists for a pair of binary trees [8] and for a pair of non-star trees [9], thus finally settling the conjecture.

In the present paper we initiate the study of the *1-planar packing problem*, i.e., the problem of packing a set of graphs into a 1-planar graph. A 1-planar graph is a graph that can be drawn so that each edge has at most one crossing. 1-planar graphs have been introduced by Ringel [16] and have received increasing attention in the last years in the research area called *beyond planarity* (see, e.g., [4,11]). Since any two non-star trees admit a planar packing, a natural question is whether we can pack more than two trees into a 1-planar graph. On the other hand, since each 1-planar graph has at most $4n - 8$ edges edges [15], it is not possible to pack more than three trees into a 1-planar graph. Thus, our main question is whether any three trees with maximum vertex degree $n - 3$ admit a 1-planar packing. The restriction to trees of degree at most $n - 3$ is necessary because a vertex of degree larger than $n - 3$ in one tree cannot have degree at least one in the other two trees. Our results can be listed as follows.

- We show that there exist triples of structurally simple trees that do not admit a 1-planar packing (Sect. 3). These triples consist of three caterpillars with at least 10 vertices and of two caterpillars and a path with 7 vertices.
- Motivated by the above results, we study triples consisting of two paths and a caterpillar (Sect. 4). We characterize the triples consisting of two paths and a 5-legged caterpillar (a caterpillar where each vertex of the spine has no leaves attached or it has at least five) that admit such a packing. We also characterize the triples that admit a 1-planar packing and that consist of two paths and a caterpillar whose spine has exactly two vertices.
- The packing technique of the results above is constructive and it gives rise to 1-plane graphs (i.e., 1-planar embedded graphs) with a linear number of crossings. This naturally raises the question about the number of edge crossings required by a 1-planar packing. We show that any three paths with at

least six vertices can be packed into a 1-plane graph with seven edge cross-ings in total (Sect. 5). We also extend this technique to three cycles obtaining 1-plane graphs with fourteen crossings in total.

- We finally consider the 1-planar packing problem for quadruples of acyclic graphs (Sect. 6). Since, as already observed, four paths cannot be packed into a 1-planar graph, we consider three paths and a perfect matching. We show that when $n \geq 12$ such a quadruple admits a 1-planar packing and that when $n \leq 10$ a 1-planar packing does not exist.

Preliminary definitions are given in Sect. 2 and open problems are listed in Sect. 7. Some proofs are sketched or removed and can be found in [3].

2 Preliminaries

Given a graph G and a vertex v of G, $\deg_G(v)$ denotes the vertex degree of v in G. Let $G_1, \ldots, G_k$ be k graphs with n vertices; a *packing* of $G_1, \ldots, G_k$ is an n-vertex graph G that has $G_1, \ldots, G_k$ as edge-disjoint spanning subgraphs. We consider the case when G is a 1-planar graph; in this case we say that G is a *1-planar packing* of $G_1, \ldots, G_k$. If $G_1, \ldots, G_k$ admit a (1-planar) packing G, we also say that $G_1, \ldots, G_k$ *can be packed into G*. We mainly concentrate on the case when each G_i is a tree $(1 \leq i \leq k)$. In this case (and generally when each G_i is connected), we have restrictions on the values of k and n for which a packing exists.

Property 1. A 1-planar packing of k connected n-vertex graphs $G_1, \ldots, G_k$ exists only if $k \leq 3$ and $n \geq 2k$. Moreover, $\deg_{G_i}(v) \leq n - k$ for each vertex v.

A *caterpillar* T is a tree such that removing all the leaves results in a path called the *spine*. A *backbone* of T is a path $v_0, v_1, v_2, \ldots, v_k, v_{k+1}$ of T where $v_1, v_2, \ldots, v_k$ is the spine of T and v_0 and v_{k+1} are two leaves adjacent in T to v_1 and v_k, respectively. T is *h-legged* if every vertex of its spine has degree either 2 or at least $h + 2$ in T.

3 Trees that Do Not Admit 1-Planar Packings

In this section we describe triples of trees that do not admit a 1-planar packing.

Theorem 1. *For every $n \geq 10$, there exists a triple of caterpillars that does not admit a 1-planar packing.*

Proof. The triple consists of three isomorphic caterpillars T_1, T_2, T_3 with $n \geq 10$ vertices. Each T_i has a backbone of length 5 and $n - 5$ leaves all adjacent to the middle vertex of the spine, which we call the *center* of T_i. First, notice that each T_i satisfies Property 1, i.e., $\deg_{T_i}(v) \leq n - 3$. Namely, the vertex with largest degree in T_i is its center, which has degree $n - 3$. Let G be any packing of T_1, T_2, and T_3 and let v_1, v_2, and v_3 be the three vertices of G where the three centers of T_1, T_2, T_3, respectively, are mapped. The three vertices v_1, v_2, and v_3 must be distinct because otherwise they would have degree larger than $n - 1$ in G, which is impossible. For each v_i we have $\deg_{T_i}(v_i) = n - 3$ and $\deg_{T_j}(v_i) \geq 1$, for $j \neq i$. This implies that $\deg_G(v_i) = n - 1$ for each v_i. In other words, each v_i is adjacent to all the other vertices of G. Thus, G contains $K_{3,n-3}$ as a subgraph. Since $n \geq 10$ and $K_{3,7}$ is not 1-planar [2], G is not 1-planar. □

Motivated by Theorem 1, we consider triples where one of the caterpillars is a path. Also in this case there exist triples that do not have a 1-planar packing.

Theorem 2. *There exists a triple consisting of a path and two caterpillars with $n = 7$ vertices that does not admit a 1-planar packing.*

Proof. Let T_i $(i = 1, 2)$ be a caterpillar with a backbone of length four such that one of the two internal vertices has degree three and the other one has degree four. Let G be a packing of T_1, T_2 and a path P of 7 vertices. Let v_1, v_2, v_3, and v_4 be the four vertices of G where the internal vertices of the backbones of T_1 and T_2 are mapped to. We first observe that v_1, v_2, v_3, and v_4 must be distinct. Suppose, as a contradiction, that two of them coincide, say v_1 and v_2; then $\deg_{T_1}(v_1) + \deg_{T_2}(v_1) \geq 6$. On the other hand $\deg_P(v_1) \geq 1$, and therefore $\deg_G(v_1) \geq 7$, which is impossible (since G has only 7 vertices). Denote by $G_{1,2}$ the subgraph of G containing only the edges of T_1 and T_2. Two vertices among v_1, v_2, v_3, and v_4, say v_1 and v_2, have degree 5 in $G_{1,2}$, while the other two have degree 4 in $G_{1,2}$. Consider now the edges of P. Since the maximum vertex degree in a graph of seven vertices is six, v_1 and v_2 must be the end-vertices of P, while v_3 and v_4 are internal vertices. This means that they all have degree 6 in G. The vertices distinct from v_1, v_2, v_3, and v_4 have degree 2 in $G_{1,2}$ and degree 4 in G. Thus in G there are four vertices of degree 6 and three vertices of degree 4. The only graph of seven vertices with this degree distribution is the graph obtained from K_7 by deleting all the edges of a 3-cycle, which is known to be non-1-planar [12]. □

4   1-Planar Packings of Two Paths and a Caterpillar

In this section we prove that a triple consisting of two paths P_1 and P_2 and a 5-legged caterpillar T with at least six vertices admits a 1-planar packing. Let P be the backbone of T and let P_1' and P_2' be two paths with the same length as P. We first show how to construct a 1-planar packing of P, P_1' and P_2'. We then modify the computed packing to include the leaves of the caterpillar; this requires transforming some edges of P_1' and P_2' to sub-paths that pass through the added leaves. The resulting packing is a 1-planar packing of P_1, P_2 and T.

Let Γ be a 1-planar drawing, possibly with parallel edges, and let e be an edge of Γ. If e has one crossing c, then each of the two parts in which e is divided by c are called *sub-edges* of e; if e has no crossing, e itself is called a *sub-edge* of e. Let v be a vertex of Γ; a *cutting curve* of v is a Jordan arc γ such that: (i) γ has v as an end-point; (ii) γ intersects two edges $e_1 = (u_1, v_1)$ and $e_2 = (u_2, v_2)$ (possibly $u_1 = u_2$ and/or $v_1 = v_2$); (iii) γ does not intersect any other edge of

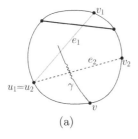 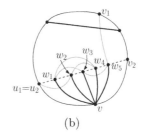

(a)   (b)

Fig. 1. A 5-leaf addition operation. The cutting curve is shown with a zig-zag pattern on it.

Γ; (iv) e_1 and e_2 do not cross each other; (v) if e_1 and e_2 are parallel edges (i.e., $u_1 = u_2$ and $v_1 = v_2$), they have no crossings. The *stub* of e_i with respect to γ is the sub-edge of e_i intersected by γ ($i = 1, 2$). Given a cutting curve γ of a vertex v, and an integer $k \geq 5$, a *k-leaf addition* operation adds k vertices $w_1, w_2, \ldots, w_k$ and the edges $(v, w_1), (v, w_2), \ldots, (v, w_k)$ to Γ in such a way that: (i) the added vertices subdivide the stubs of both e_1 and e_2 with respect to γ; (ii) the subgraph induced by $u_1, u_2, v_1, v_2, w_1, w_2, \ldots, w_k$ has no multiple edges (see Fig. 1 for an example). In other words, a leaf addition adds a set of vertices adjacent to v and replaces the stubs of e_1 and e_2 with two edge-disjoint paths. This operation will be used to modify the 1-planar packing of P, P_1' and P_2' to include the leaves of the caterpillar. When the value of k is not relevant, a k-leaf addition will be simply called a *leaf addition*.

Lemma 1. *Let Γ be a 1-planar drawing possibly with parallel edges, let v be a vertex of Γ and let γ be a cutting curve of v. It is possible to execute a k-leaf addition for every $k \geq 5$ in such a way that the resulting drawing is still 1-planar.*

Proof. Denote by e_1 and e_2 the two edges crossed by γ. If one of them or both are crossed in Γ replace their crossing points with dummy vertices. Let e_i' be the stub of e_i with respect to γ (if e_i is not crossed in Γ, e_i' coincides with e_i). After the replacement of the crossings with the dummy vertices the two stubs e_1' and e_2' have no crossing. Since γ does not cross any edge distinct from e_1 and e_2, the drawing Γ' obtained by removing e_1' and e_2' has a face f whose boundary contains the vertex v and all the end-vertices of e_1' and of e_2' (there are at least two and at most four such vertices). The idea now is to insert into the face f, without creating any crossing, a gadget that realizes the k-leaf addition for the desired value of $k \geq 5$. A gadget has k vertices that will be added to Γ, a vertex that will be identified with v, and four vertices a, b, c, and d that will be identified with the end-vertices of e_1' and e_2'. The four vertices a, b, c, and d will be called *attaching vertices* and the edges incident to them will be called *attaching edges*. In order to guarantee that the leaf addition is valid and that the drawing Γ'' obtained by the insertion of the gadget inside f is 1-planar, we have to pay attention to two aspects: (i) if an attaching edge is crossed in the gadget,

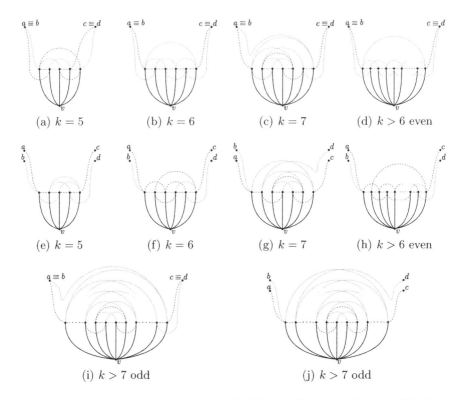

Fig. 2. Gadgets for the proof of Lemma 1. (a)–(d) and (i) are used for parallel edges; (e)–(h) and (j) are used for non-parallel edges.

then its attaching vertex cannot be identified with a dummy vertex (otherwise when we remove the dummy vertex we obtain an edge that is crossed twice); (ii) if two attaching vertices of the gadget are coincident (because two end-vertices of e_1' and e_2' coincide), then the corresponding attaching edges must not have the second end-vertex in common in the gadget (otherwise the leaf addition is not valid because it creates multiple edges). We use different gadgets depending on whether e_1 and e_2 are parallel edges or not. If they are parallel edges, we use the gadgets of Figs. 2(a)–(d) and (i). Notice that in this case, e_1 and e_2 are not crossed by definition of cutting curve. It follows that f has no dummy vertex and (i) is guaranteed. On the other hand, both end-vertices of e_1 and e_2 coincide and therefore the end-vertices of the attaching edges that are not attaching vertices must be distinct. This is true for the gadgets used in this case. If e_1 and e_2 are non-parallel, we use the gadgets of Figs. 2(e)–(h) and (j). All these gadgets have only one attaching edge that is crossed (labeled d in the figure); also, vertex d can be identified with vertex c without creating multiple edges. If e_1 and e_2 are non-parallel, at most two end-vertices of e_1' and e_2' are dummy; they cannot belong to the same stub, and they cannot coincide (because e_1 and e_2 do not

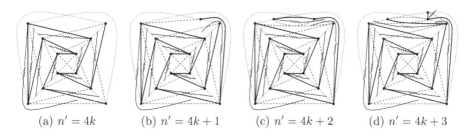

(a) $n' = 4k$ (b) $n' = 4k + 1$ (c) $n' = 4k + 2$ (d) $n' = 4k + 3$

Fig. 3. 1-planar packings of three paths with $n' \geq 8$ vertices (case $k = 3$); A cutting curve is shown (zig-zag pattern) for each internal vertex of the black path.

cross each other). Thus we can identify d with a non-dummy vertex and we can identify c and d if needed.   □

We are ready to describe our construction of a 1-planar packing of P_1, P_2, and T. We use different techniques for different lengths of the backbone of T.

Lemma 2. *Two paths and a 5-legged caterpillar whose backbone contains $n' \geq 6$ vertices admit a 1-planar packing.*

Proof. We start with the construction of a 1-planar packing of the three paths P_1', P_2' and P. Let n' be the number of vertices of P_1', P_2' and P, assume first that $n' \geq 8$ and $n' \equiv 0 \pmod 4$. A 1-planar packing of P_1', P_2' and P for this case is shown in Fig. 3(a) for $n' = 16$ and it is easy to see that it can be extended to any n' multiple of 4. Assume that the backbone P of T is the path shown in black in Fig. 3(a). To add the leaves of T to the construction we define a cutting curve for each vertex that has some leaves attached; we then execute a leaf addition operation for each such vertex. By Lemma 1, it is possible to execute each leaf addition so to guarantee the 1-planarity of the resulting drawing. The cutting curve for each internal vertex of P is shown in Fig. 3(a) with a zig-zag pattern. Note that, regardless of the order in which the leaf additions are executed, the cutting curves remain valid.

Suppose that $n' \geq 8$ and $n' \not\equiv 0 \pmod 4$. We first construct a 1-planar packing of three paths with $n'' = 4k$ vertices (with $k = \lfloor \frac{n'}{4} \rfloor$) using the same construction as in the previous case and then we add one, two or three vertices as shown in Figs. 3(b)–(d), which also show the cutting curves for each internal vertex of P. If n' is 6 or 7, we use the same approach; the difference is in the construction of the 1-planar packing of P_1', P_2' and P. The construction for such a packing and the cutting curves for the internal vertices of P are in Figs. 4(a)–(b).   □

Lemma 3. *Two paths and a 5-legged caterpillar T whose backbone contains $n' = 5$ vertices admit a 1-planar packing, unless T is a path.*

Proof. If T is a path, then P_1, P_2 and T are all paths of length five, and by Property 1, a 1-planar packing of P_1, P_2 and T does not exist. Suppose therefore that at least one internal vertex of the backbone P of T has some leaves attached.

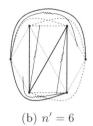
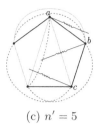

| (a) $n' = 7$ | (b) $n' = 6$ | (c) $n' = 5$ |

Fig. 4. 1-planar packings of three paths with $n' \in \{5, 6, 7\}$ vertices, with a cutting curve (zig-zag pattern) for each internal vertex of the black path.

We use an approach similar to the one of Lemma 2. However, as just explained, a 1-planar packing of P'_1, P'_2 and P does not exist in this case. We start with a 1-planar packing with two pairs of parallel edges. For each pair, one edge belongs to P'_1 and the other one to P'_2. We will remove the parallel edges by performing the leaf addition operations. To this aim we must guarantee that there is a cutting curve for each pair of parallel edges. The 1-planar packing P'_1, P'_2 and P and the cutting curves for the internal vertices of P are shown in Fig. 4(c), for the case when at least two vertices have leaves attached. Indeed, if only two vertices have leaves attached, they are either consecutive along the backbone or not. In the first case, these two vertices are mapped to the vertices labeled a and b in Fig. 4(c) and the depicted cutting curves will remove the parallel edges; in the second case, the two vertices are mapped to the vertices labeled a and c and also in this case the depicted cutting curves will remove the parallel edges.

If only one vertex of P has leaves attached, we have only one cutting curve and thus it is not possible to intersect both pairs of parallel edges. To handle this case we use an ad-hoc technique which can be found in [3].   □

The next theorem gives a complete characterization for the case in which the backbone of T has length four.

Theorem 3. *Two paths and a caterpillar T whose backbone contains $n' = 4$ vertices admit a 1-planar packing if and only if $n \geq 6$ and $\deg_T(v) \leq n - 3$ for every vertex v.*

Lemmas 2 and 3, together with Theorem 3 imply the next theorem.

Theorem 4. *Two paths and a 5-legged caterpillar T with n vertices admit a 1-planar packing if and only if $n \geq 6$ and $\deg_T(v) \leq n - 3$ for every vertex v.*

5   1-Planar Packings with Constant Edge Crossings

The technique described in the previous section constructs 1-planar drawings that have a linear number of crossings. A natural question is whether it is possible to compute a 1-planar packing with a constant number of crossings. In this section we prove that seven (resp. fourteen) crossings suffice for packing three

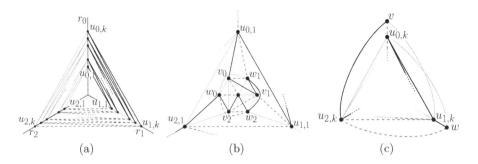

Fig. 5. Illustration for the proof of Theorem 5.

paths (resp. cycles). It is worth remarking that a 1-planar packing of three paths has at least three crossings (because it has $3n-3$ edges), while a 1-planar packing of three cycles has at least six crossings (because it has $3n$ edges).

Theorem 5. *Three paths with $n \geq 6$ vertices can be packed into a 1-plane graph with at most 7 edge crossings.*

Proof. We prove the statement by showing how to construct a 1-planar drawing with at most 7 crossings of a graph that is the union of three paths. Suppose first that $n = 7 + 3k$ for $k \in \mathbb{N}$. If $k = 0$, we draw the union of the three paths with 7 vertices as shown in Fig. 4(a). The drawing is 1-planar and has three crossings in total. Suppose now that $k > 0$. We consider three rays r_0, r_1, r_2 with a common origin pairwise forming a 120° angle and we place k vertices on each line. We denote by $u_{i,1}, u_{i,2}, \ldots, u_{i,k}$ the vertices of line r_i ($i = 0, 1, 2$) in the order they appear along r_i starting from the origin (see Fig. 5(a)). In the following, indices will be taken modulo 3 when working with the indices of the rays r_i. To draw path P_i ($i = 0, 1, 2$) we draw the edges $(u_{i,1}, u_{i+1,1})$, $(u_{i,j}, u_{i+1,j-1})$, and $(u_{i,j}, u_{i+1,j})$ (for $j = 2, \ldots, k$) as straight-line segments. Notice that, these edges form a zig-zagging path between the vertices of rays r_i and r_{i+1}, so P_i passes through all vertices of r_i and r_{i+1} but not through the vertices of r_{i+2}. To include these missing vertices in P_i, we add to P_i edges $(u_{i+2,j}, u_{i+2,j+1})$ (for $j = 1, 2, \ldots, k - 1$). In this way we draw two disjoint sub-paths for each path P_i, namely a zig-zagging path between r_i and r_{i+1} and a straight-line path along r_{i+2}. Moreover, we only draw $3k$ edges and therefore there are still 7 missing vertices (and 8 missing edges) in each path. To add the missing vertices and edges and to connect the two sub-paths of each path, we construct a drawing Γ_0 of three paths P'_0, P'_1, P'_2 with seven vertices as in the case when $k = 0$. Denote with v_i and w_i the end-vertices of P'_i in Γ_0. We place Γ_0 inside the triangle $u_{0,1}, u_{1,1}, u_{2,1}$ and add the edges $(v_i, u_{i,1})$ and $(w_i, u_{i+2,1})$. It is easy to see (see also Fig. 5(b)) that these six edges can be added so that the drawing is still 1-planar and so that the total number of crossings is 6. This concludes the proof for $n = 7 + 3k$. If $n = 7 + 3k + 1$ we start with the same construction as in the previous case and then add an extra vertex v outside the triangle $u_{1,k}, u_{2,k}, u_{3,k}$. Notice that each of these three vertices is the end-vertex of two of the three paths with $7 + 3k$ vertices. Thus we can extend each path to include v by connecting it to each of the three vertices $u_{1,k}, u_{2,k}, u_{3,k}$ in a planar way (see Fig. 5(c) ignoring vertex w). If $n = 7 + 3k + 2$, then we add two extra vertices outside the triangle $u_{0,k}, u_{1,k}, u_{2,k}$ and connect both of them to the three vertices $u_{0,k}, u_{1,k}, u_{2,k}$ (recall that each of these three vertices is the end-vertex of two distinct paths with $7 + 3k$ vertices). In this case however the addition of the two extra vertices causes the creation of one crossing. Thus the final drawing is 1-planar and the total number of crossings is at most 7 (see Fig. 5(c)). This concludes the proof for $n \geq 7$. If $n = 6$ we construct a 1-planar packing of three paths with three crossings in total as shown in Fig. 4(b).   □

The construction of Theorem 5 can be extended to three cycles.

Theorem 6. *Three cycles with $n \geq 20$ vertices can be packed into a 1-plane graph with at most 14 edge crossings.*

6   From Triples to Quadruples

In this section we extend the study of 1-planar packings from triples of graphs to quadruples of graphs. By Property 1, a 1-planar packing of four graphs does not exist if all graphs are connected, because the number of edges of the four graphs is higher than the number of edges allowed in a 1-planar graph. We consider therefore a quadruple consisting of three paths and a perfect matching. Notice that, in this case the number of vertices n has to be even.

Theorem 7. *Three paths and a perfect matching with $n \geq 12$ vertices admit a 1-planar packing. If $n \leq 10$, the quadruple does not admit a 1-planar packing.*

Proof. Three paths and a perfect matching have a total of $3(n-1) + \frac{n}{2} = \frac{7n}{2} - 3$ edges. Since a 1-planar graph has at most $4n - 8$ edges, a 1-planar packing of three paths and a perfect matching exists only if $\frac{7n}{2} - 3 \leq 4n - 8$, i.e., if $n \geq 10$. If $n = 10$, we have $\frac{7n}{2} - 3 = 32$ and $4n - 8 = 32$, which means that any 1-planar packing of three paths and a perfect matching with $n = 10$ vertices is an optimal 1-planar graph. It is known that every optimal 1-planar graph has at least eight vertices of degree exactly six [1]. On the other hand, in any 1-planar packing of three paths and a perfect matching all vertices, except the at most six end-vertices of the three paths, have degree seven, which implies that a 1-planar packing of three paths and a perfect matching does not exist.

We now prove that a 1-planar packing exists if $n \geq 12$. We only discuss here the case when $n \geq 24$; the cases in which $12 \leq n \leq 22$ are described in [3]. Based on the fact that in any 1-planar packing of three paths and a perfect matching at least $n - 6$ vertices have degree seven, we construct the desired 1-planar packing starting from a 1-planar graph G such that at least $n - 6$ vertices have degree at least seven; we then partition the edges of G into five sets; three of these sets form a spanning path each, the fourth one forms a perfect matching, and the fifth one

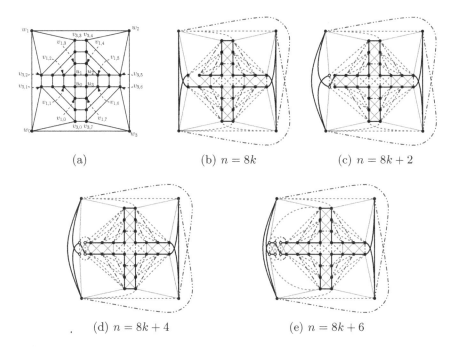

Fig. 6. (a) Graph G' used in the proof of Theorem 7 ($n = 8k, k = 3$). (b)–(e) 1-planar packings of three paths and a perfect matching obtained starting from G'.

contains edges that will not be part of the 1-planar packing. For every $n = 8k$ and $k \geq 3$ it is possible to construct a 1-planar graph with n vertices each having degree at least seven as follows. We start with $k - 1$ cycles $C_1, C_2, \ldots, C_{k-1}$. Each cycle C_i ($1 \leq i \leq k - 1$) has eight vertices $v_{i,j}$ with $0 \leq j \leq 7$. Cycle C_i, for $1 \leq i \leq k - 2$, is embedded inside C_{i+1} and is connected to it with edges $(v_{i,j}, v_{i+1,j})$ for each $0 \leq j \leq 7$. We have a cycle with four vertices u_0, u_1, u_2, u_3 embedded inside C_1 and connected to it with edges $(u_j, v_{1,2j})$ and $(u_j, v_{1,2j+1})$. Finally, we have a cycle with four vertices w_0, w_1, w_2, w_3 embedded outside C_{k-1} and connected to it with edges $(w_j, v_{k-1,2j})$ and $(w_j, v_{k-1,2j+1})$. The graph G' described so far has n vertices, is planar, all its vertices have degree four, and each vertex is incident to at most one face of size three (see Fig. 6(a)). By adding two crossing edges inside each face of size four, we obtain a 1-planar graph G with n vertices having degree at least seven. The graph G and the partition of the edges of G in five sets defining three paths and a matching is shown in Fig. 6(b). If n is not a multiple of 8, then it will be $n = 8k + r$, with $0 < r < 8$ and r even (because n is even). In this case we construct G' as explained above and then we extend the paths $u_0, v_{1,1}, \ldots, v_{k-1,1}$ and $u_1, v_{1,2}, \ldots, v_{k-1,2}$ to the left with 1, 2 or 3 vertices each; we then suitably rearrange the edges of G'. The graph G is then obtained, as in the previous case, by adding a pair of crossing edges inside each face of size four. The resulting graph G and a partition of its edges in five sets defining three paths and a matching is shown in Figs. 6(c), (d), and (e), for the cases when $r = 2$, $r = 4$, and $r = 6$, respectively.   □

7   Open Problems

We find that the 1-planar packing problem is a fertile and still largely unexplored research subject. We conclude the paper with a list of open problems.(i) Theorem 2 holds only for $n = 7$. Do two caterpillars (or more general trees) and a path admit a 1-planar packing if they have more than 7 vertices? (ii) Can Theorem 4 be extended to general caterpillars? What about two paths and a tree more complex than a caterpillar, for example a binary tree? (iii) Is it possible to compute a 1-planar packing of three paths or cycles with the minimum number of crossings (three and six, respectively)? Can we compute 1-planar packings with few crossings for triples of other types of trees?

References

1. Brandenburg, F.J.: Recognizing optimal 1-planar graphs in linear time. Algorithmica **80**(1), 1–28 (2018)
2. Czap, J., Hudák, D.: 1-planarity of complete multipartite graphs. Discrete Appl. Math. **160**(4), 505–512 (2012)
3. De Luca, F., et al.: Packing trees into 1-planar graphs. CoRR abs/1911.01761 (2019)
4. Didimo, W., Liotta, G., Montecchiani, F.: A survey on graph drawing beyond planarity. ACM Comput. Surv. **52**(1), 4:1–4:37 (2019)
5. Frati, F.: Planar packing of diameter-four trees. In: Proceedings of the 21st Annual Canadian Conference on Computational Geometry, pp. 95–98 (2009)
6. Frati, F., Geyer, M., Kaufmann, M.: Planar packing of trees and spider trees. Inf. Process. Lett. **109**(6), 301–307 (2009)
7. García Olaverri, A., Hernando, M.C., Hurtado, F., Noy, M., Tejel, J.: Packing trees into planar graphs. J. Graph Theory **40**(3), 172–181 (2002)
8. Geyer, M., Hoffmann, M., Kaufmann, M., Kusters, V., Tóth, C.D.: Planar packing of binary trees. In: Dehne, F., Solis-Oba, R., Sack, J.R. (eds.) WADS 2013. LNCS, vol. 8037, pp. 353–364. Springer, Heidelberg (2013). https://doi.org/10.1007/978-3-642-40104-6_31
9. Geyer, M., Hoffmann, M., Kaufmann, M., Kusters, V., Tóth, C.D.: The planar tree packing theorem. JoCG **8**(2), 109–177 (2017)
10. Hedetniemi, S., Hedetniemi, S., Slater, P.: A note on packing two trees into K_n. Ars Combin. **11**, 149–153 (1981)
11. Kobourov, S.G., Liotta, G., Montecchiani, F.: An annotated bibliography on 1-planarity. Comput. Sci. Rev. **25**, 49–67 (2017)
12. Korzhik, V.P.: Minimal non-1-planar graphs. Discrete Math. **308**(7), 1319–1327 (2008)
13. Mahéo, M., Saclé, J., Wozniak, M.: Edge-disjoint placement of three trees. Eur. J. Comb. **17**(6), 543–563 (1996)
14. Oda, Y., Ota, K.: Tight planar packings of two trees. In: 22nd European Workshop on Computational Geometry (2006)

15. Pach, J., Tóth, G.: Graphs drawn with few crossings per edge. Combinatorica **17**(3), 427–439 (1997)
16. Ringel, G.: Ein sechsfarbenproblem auf der kugel. Abh. Math. Semin. Univ. Hamburg **29**(1–2), 107–117 (1965)
17. Sauer, N., Spencer, J.: Edge disjoint placement of graphs. J. Comb. Theory Ser. B **25**(3), 295–302 (1978)
18. Wang, H., Sauer, N.: Packing three copies of a tree into a complete graph. Eur. J. Comb. **14**(2), 137–142 (1993)
19. Wozniak, M., Wojda, A.P.: Triple placement of graphs. Graphs Comb. **9**(1), 85–91 (1993)

Angle Covers:
Algorithms and Complexity

William Evans[1] , Ellen Gethner[2], Jack Spalding-Jamieson[1(✉)],
and Alexander Wolff[3]

[1] Department of Computer Science, University of British Columbia,
Vancouver, B.C., Canada
will@cs.ubc.ca,jacketsj@alumni.ubc.ca
[2] Department of Computer Science and Engineering, University of Colorado, Denver,
CO, U.S.A.
ellen.gethner@ucdenver.edu
[3] Universität Würzburg, Würzburg, Germany

Abstract. Consider a graph with a rotation system, namely, for every
vertex, a circular ordering of the incident edges. Given such a graph,
an *angle cover* maps every vertex to a pair of consecutive edges in the
ordering – an *angle* – such that each edge participates in at least one
such pair. We show that any graph of maximum degree 4 admits an angle
cover, give a poly-time algorithm for deciding if a graph with no degree-3
vertices has an angle-cover, and prove that, given a graph of maximum
degree 5, it is NP-hard to decide whether it admits an angle cover. We
also consider extensions of the angle cover problem where every vertex
selects a fixed number $a > 1$ of angles or where an angle consists of more
than two consecutive edges. We show an application of angle covers to
the problem of deciding if the 2-blowup of a planar graph has isomorphic
thickness 2.

1 Introduction

A well-known problem in combinatorial optimization is *vertex cover*: given an
undirected graph, select a subset of the vertices such that every edge is incident to
at least one of the selected vertices. The aim is to select as few vertices as possible.
The problem is one of Karp's 21 NP-complete problems [7] and remains NP-hard
even for graphs of maximum degree 3 [1]. Moreover, vertex cover is APX-hard [2]
and while it is straightforward to compute a 2-approximation (take all endpoints
of a maximal matching), the existence of a $(2 - \varepsilon)$-approximation for any $\varepsilon > 0$
would contradict the so-called *Unique Games Conjecture* [8]. Vertex cover is the
"book example" of a fixed-parameter tractable problem.

Note that in vertex cover, a vertex covers *all* its incident edges. In this paper
we alter the problem by restricting the covering abilities of the vertices. We

The full version of this article is available at ArXiv [3]. The authors acknowledge
support by NSERC Discovery Grant (W.E.), Simons Foundation Collaboration Grant
for Mathematicians #311772 (E.G.), and DFG grant WO 758/10-1 (A.W.).

© Springer Nature Switzerland AG 2020
M. S. Rahman et al. (Eds.): WALCOM 2020, LNCS 12049, pp. 94–106, 2020.
https://doi.org/10.1007/978-3-030-39881-1_9

assume that the input graph has a given *rotation system*, that is, for every vertex, a circular ordering of its incident edges. In the basic version of our problem, the *angle cover problem*, at each vertex we can cover one pair of its incident edges that are consecutive in the ordering (i.e., form an *angle* at the vertex), and every edge must be covered. An example of a planar graph with a vertex cover and an angle cover is shown in Fig. 1.

In this paper, we mainly treat the decision version of angle cover, but various optimization versions are interesting as well; see Sect. 5. Clearly, a graph that admits an angle cover cannot have too many edges, both locally and globally and we formalize this notion as follows: we say that a graph G has *low edge density* if, for all k, every k-vertex subgraph has at most $2k$ edges. Observe that only graphs of low edge density can have an angle cover. Whether G has low

Fig. 1. A graph with a minimum vertex cover (black vertices) and an angle cover (gray arcs)

edge density can be easily checked by testing whether the following bipartite auxiliary graph G_{mat} has a matching of size at least $|E|$. The graph G_{mat} has one vertex for each edge of G, two vertices for each vertex of G, and an edge for every pair (v, e) where v is a vertex of G and e is an edge of G incident to v.

Example classes of graphs with low edge density are outerplanar graphs and maximum-degree-4 graphs (both of which always admit angle covers; see below), Laman graphs (graphs such that for all k, every k-vertex subgraph has at most $2k - 3$ edges), and pointed pseudo-triangulations. Given a set P of points in the plane, a *pseudo-triangulation* is a plane graph with vertex set P and straight-line edges that partitions the convex hull of P into pseudo-triangles, that is, simple polygons with exactly three convex angles. A pseudo-triangulation is *pointed* if the edges incident to a vertex span an angle less than π. Thus every vertex has one large angle (greater than π) but these angles do not necessarily form an angle cover. It is known that every pointed pseudo-triangulation is a planar Laman graph [10] and that every planar Laman graph can be realized as a pointed pseudo-triangulation [5]. In the full version [3], we show that not all Laman graphs have an angle cover.

Our interest in angle covers arose from the study of graphs that have isomorphic thickness 2. The *thickness*, $\theta(G)$, of a graph G is the minimum number of planar graphs whose union is G. By allowing each edge to be a polygonal line with bends, we can draw G on $\theta(G)$ parallel planes where each vertex appears in the same position on each plane and within each plane the edges do not cross [6]. The *isomorphic thickness*, $\iota(G)$, of a graph G is the minimum number of *isomorphic* planar graphs whose union is G. The k-*blowup* of a graph $G = (V, E)$ is the graph $B^k(G)$ with $k|V|$ vertices $\bigcup_{a=1}^{k} V_a$ and edges $\bigcup_{1 \le a, b \le k} E_{ab}$, where $V_a = \{v_a \colon v \in V\}$ and $E_{ab} = \{(u_a, v_b) \colon (u, v) \in E\}$. As we will show, it is NP-hard to determine if a graph has isomorphic thickness 2, but all graphs that are the 2-blowup of a plane graph with an angle cover have isomorphic thickness 2.

As a warm-up, we observe that every outerplane graph has an angle cover. (Recall that an outerplane graph is an outerplanar graph with a given embedding and, hence, fixed rotation system.) The statement can be seen as follows. Any n-vertex outerplane graph has at most $2n - 4$ edges, and outerplanarity is hereditary, so outerplane graphs have low edge density. Additionally, every such graph has an *ear decomposition*, that is, an ordering $(v_1, v_2, \ldots, v_n)$ of the vertex set $V = \{v_1, v_2, \ldots, v_n\}$ such that, for $i = n, n - 1, \ldots, 2$, vertex v_i is incident to at most two edges in $G[v_1, \ldots, v_i]$. Due to outerplanarity, the ear decomposition can be chosen such that for each vertex the at most two "ear edges" are consecutive in the ordering around the vertex. This shows the existence of an angle cover for any outerplane graph.

Our Contribution. We first consider a few concrete examples that show that not every planar graph of low edge density admits an angle cover. We then show that any graph of maximum degree 4 does admit an angle cover and give a polytime algorithm to decide if a graph with no degree 3 vertex admits an angle cover (Sect. 3). Then we prove that, given a graph of maximum degree 5, it is NP-hard to decide whether it admits an angle cover (Sect. 4). We also consider two extensions of the angle cover problem where (i) every vertex is associated with a fixed number $a > 1$ of angles or (ii) an angle consists of more than two consecutive edges (Sect. 5). Finally, we show that the 2-blowup of any plane graph with an angle cover has isomorphic thickness 2 (Sect. 6).

2   Preliminaries and Examples

In this section we show that even graphs with several seemingly nice properties do not always admit an angle cover. All our examples are *plane* graphs, that is, they are planar and their rotation system corresponds to a planar drawing.

Observation 1. *There is a plane graph (Fig. 2a) of maximum degree 5 and with low edge density that does not admit an angle cover.*

Proof. Consider the graph in Fig. 2a. We have oriented its edges so that each vertex has outdegree 2. Hence, the graph has low edge density. The graph has $n = 21$ vertices and $2n$ edges. Due to the way the two degree-2 vertices (filled black) are arranged around the central vertex (square) of degree 4, one of the horizontal edges incident to the central vertex is covered twice. This implies that the n vertices cover at most $2n - 1$ edges. Thus, there is no angle cover.   □

Note that the counterexample critically exploits the use of degree-2 vertices. Next we show that there are also counterexamples without such vertices.

Observation 2. *There is a plane graph (Fig. 2b) of low edge density with vertex degrees in $\{3, 4, 5\}$ that does not admit an angle cover.*

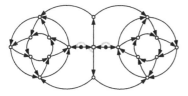
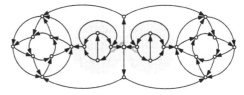

(a) example with vertex degrees 2–5   (b) example with vertex degrees 3–5

Fig. 2. Two plane graphs that do not admit angle covers. Edges are oriented such that each vertex has outdegree 2 (hence both graphs have low edge density) (Color figure online).

Proof. Consider the graph in Fig. 2b. Again, we oriented the edges such that every vertex has outdegree 2, which shows that the graph has low edge density. The graph is the same as the one in Fig. 2a except we replaced the two degree-2 vertices by copies of K_4 (light blue). Since the number of edges is $2n$ and K_4 has four vertices and six edges, the two edges that connect each copy of K_4 to the rest of the graph, must necessarily be directed away from K_4. As a result, the two copies of K_4 behave like the degree-2 vertices in Fig. 2a: they cover one of the horizontal edges incident to the central vertex twice. Thus, there is no angle cover.   □

Note that the counterexamples so far were not strongly connected (w.r.t. the chosen edge orientation). But strongly connected counterexamples exist, too.

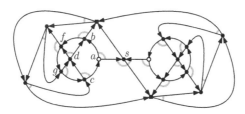

Fig. 3. A plane graph of low edge density that has a strongly connected orientation of its edges but does not admit an angle cover.

Observation 3. *There is a plane graph (Fig. 3) of low edge density with vertex degrees in $\{3, 4, 5\}$ and a strongly connected edge orientation that does not admit an angle cover.*

Proof. Consider the graph depicted in Fig. 3. We assume that it has an angle cover and show that this yields a contradiction. Clearly, one of the horizontal edges incident to s must be covered by s. Due to symmetry, we can assume that it is the edge to vertex a on the left. Since a has degree 3, it must cover its other two edges, to vertices b up and to c down. Let d be the vertex adjacent to both b and c and let b, f, g, c be the neighbors of d in counterclockwise order. We have the following four cases, each of which leads to a contradiction.

1. d covers db and dc: Then b covers bf, f covers fg (since f must cover fd), and g covers gc (since g must cover gd). But then c has only one edge to cover.
2. d covers db and df: Again, b covers bf and (since only two edges remain uncovered at f) f covers fg and (as before) g covers gc. But then c has two non-consecutive edges to cover.
3. d covers df and dg: Then b must cover bd and bf, f must cover fg, g must cover gc. But then c has two non-consecutive edges to cover.
4. d covers dg and dc: Then c must cover cg, g must cover gf, f must cover fb. But then b has two non-consecutive edges to cover.

□

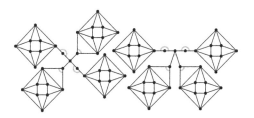

Fig. 4. A graph with two embeddings; one without and one with an angle cover.

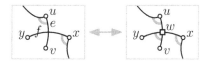

Fig. 5. Any topological graph (left) admits an angle cover if and only if its planarization (right) admits an angle cover.

Observation 4. *There is a planar maximum-degree-5 graph (Fig. 4) with two embeddings such that one admits an angle cover, but the other does not.*

Thus, when determining whether a graph (of maximum degree greater than 4) has an angle cover, we must consider a particular embedding, which determines a rotation system. This applies to non-planar graphs as well. However, if we have a topological embedding of a non-planar graph, we can decide whether it has an angle cover by considering its planarization. By a *topological graph* we mean a graph together with a drawing of that graph where any pair of edges (including their endpoints) has at most one (crossing not touching) point in common and any point of the plane is contained in at most two edges. By the *planarization* of a topological graph we mean the plane graph that we get if we replace, one by one, in arbitrary order, each crossing by a new vertex that is incident exactly to the four pieces of the two edges that defined the crossing. We define the order of the four new edges around the new vertex to be the same as the order of the four endpoints of the old edges around the crossing.

Proposition 1. *Any topological graph admits an angle cover if and only if its planarization admits an angle cover.*

Proof. We show the equivalence for the first step of the planarization procedure defined above. Then, induction proves our claim.

Let G be a topological graph, and let G' be the graph that we obtain from G by replacing an arbitrary crossing of two edges $e = uv$ and $f = xy$ by a new vertex w that is incident to u, v, x, and y; see Fig. 5. (Let the order of the endpoints around the crossing in G and around the new vertex in G be $\langle u, x, v, y \rangle$.)

Suppose that G has an angle cover α. Edges e and f, say, must be covered by angles incident to vertices u and x. Then it is simple to extend α to G' by mapping w to the angle $\{wv, wy\}$ incident to w.

Now suppose that G' has an angle cover α' with, say, $\alpha'(w) = \{vw, yw\}$. Since w does not cover uw and xw, u must cover uw and x must cover xw. Now we restrict α' to G: we replace uw by uv and xw by xy. Hence, both uv and xy are covered. Finally, we remove w (with vw and yw). Clearly, the resulting map is an angle cover for G.   □

3   Algorithms for Graphs with Restricted Degrees

Theorem 1. *Any maximum-degree-4 n-vertex graph with any rotation system admits an angle cover, and such a cover can be found in $O(n)$ time.*

Proof. We can assume that the given graph is connected since we can treat each connected component independently. If the given graph is not 4-regular, we arbitrarily add dummy edges between vertices of degree less than 4 until the resulting (multi)graph is 4-regular or there is a single vertex, say v, of degree less than 4. If one vertex remains with degree less than 4, we add self-loops to that vertex until it has degree 4. This is always possible since all other vertices have degree 4, which is even, hence the last vertex must also have even degree. An angle cover in the new graph implies an angle cover in the original, where the assigned angle at a vertex in the original graph is the angle that contains the assigned angle at the same vertex in the new graph.

We find a collection of directed cycles in the now 4-regular graph, similarly to the algorithm for finding an Eulerian cycle. We follow the rule to exit a degree-4 vertex always on the opposite edge from where we enter it. Whenever we close a cycle and there are still edges that we have not traversed yet, we start a new cycle from one of these edges. In this way we never visit an edge of the input graph twice, which establishes the linear running time.

The algorithm yields a partition of the edge set into (directed) cycles with the additional property that pairs of cycles may cross each other (or themselves), but they never touch without crossing. Hence, in every vertex the two outgoing edges are always consecutive in the circular ordering around the vertex. We assign to each vertex the angle formed by this pair of edges.   □

Theorem 2. *The angle cover problem for n-vertex graphs with no vertices of degree 3 and any rotation system can be solved in $O(n^2)$ time.*

Proof. Given a graph $G = (V, E)$ with no vertices of degree 3, at most $2n$ edges, and a corresponding rotation system, we create a 2SAT instance in conjunctive normal form. For each vertex v and edge e adjacent to v, create a variable x_{ve}. In our potential assignments, the variable x_{ve} is true if and only if in a corresponding angle cover, the edge e is covered by the angle at the vertex v. For each edge $e = (u, v)$, add a clause $(x_{ve} \lor x_{ue})$, where the clause is true if and only if the edge is covered. For each vertex v with incident edges $e_1, e_2, \ldots, e_d$ $(d > 3)$, create clauses $(\neg x_{ve_i} \lor \neg x_{ve_j})$ for any pair e_i, e_j that are not adjacent in the circular ordering around v. This guarantees that there can be at most 2 true variables for v, and that they must be adjacent in the circular ordering. Furthermore, given any satisfying assignment, we can force v to have exactly 2 such variables, which then specify an angle for v in an angle cover. A vertex v with degree ≤ 2 can always cover all of its edges. Thus, G with its rotation system has an angle cover if and only if the constructed 2SAT instance is satisfiable. The number of clauses is bounded above by $|E| + |E|^2$ where $|E| \in O(n)$. Since 2SAT can be solved in linear time [4], the algorithm takes $O(n^2)$ time. $\qquad\square$

4 NP-Hardness for Graphs of Maximum Degree 5

Theorem 3. *The angle cover problem is NP-hard even for graphs of maximum degree 5.*

Proof. We reduce from 3-colouring. Given a graph $G = (V, E)$, we construct a graph $H = (U, F)$ with a rotation system $f \colon U \to F^*$ such that (H, f) admits an angle cover if and only if G has a 3-colouring. Note that f is a function that, given a vertex u, will provide a circular order of the edges around u.

For each vertex $v \in V$, let $E_1(v), \ldots, E_{\deg(v)}(v)$ be its adjacent edges in some arbitrary order. We create a graph, called a *gadget*, for vertex v that contains $1 + 9 \deg(v)$ vertices. The centre of the gadget is a vertex $c(v)$ that is adjacent to the first vertex in three paths, $e_1^k(v), e_2^k(v), \ldots, e_{\deg(v)}^k(v)$, one for each of the three colours $k \in \{0, 1, 2\}$. Each vertex $e_j^k(v)$, for $j = 1, \ldots, \deg(v)$, is adjacent to two degree-1 vertices $a_j^k(v)$ and $b_j^k(v)$ (as well as its neighbours in the path) that are part of the gadget (see Fig. 6). In addition, if $E_i(u) = E_j(v)$, that is, (u, v) is an edge in G and is the ith edge adjacent to u and the jth edge adjacent to v, then the vertex $e_j^k(v)$ is adjacent to $e_i^k(u)$ (see Fig. 7). The circular order of edges around $e_j^k(v)$ is $[e_{j-1}^k(v), a_j^k(v), e_i^k(u), e_{j+1}^k(v), b_j^k(v)]$, where $e_{j-1}^k(v)$ is $c(v)$ if $j = 1$ and $e_{j+1}^k(v)$ does not exist if $j = \deg(v)$. The *separator* edges $(e_j^k(v), a_j^k(v))$ and $(e_j^k(v), b_j^k(v))$ prevent an angle cover at $e_j^k(v)$ from (i) covering both $(e_j^k(v), e_{j-1}^k(v))$ and $(e_j^k(v), e_{j+1}^k(v))$ or (ii) covering both $(e_j^k(v), e_{j-1}^k(v))$ and $(e_j^k(v), e_i^k(u))$. The graph containing all of the gadgets and the edges between them along with the specified rotation system is then $(H = (U, F), f)$. Observe that the maximum degree of H is 5 and that the construction takes polynomial time.

It remains to show that G is 3-colourable if and only if $(H = H(G), f)$ has an angle cover. We start with the "only if" direction.

"$\Rightarrow$": *G is 3-colourable implies that (H, f) has an angle cover:*
Let $t \colon V \to \{0, 1, 2\}$ be a 3-colouring of G. We construct an angle cover $\alpha \colon U \to F \times F$. For each vertex $v \in V$, if $t(v) = k$, then set $\alpha(c(v)) = ((c(v), e_1^{k+1}(v)),$

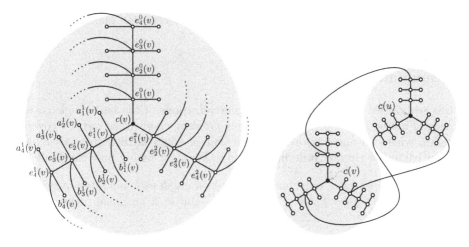

Fig. 6. Gadget for a degree-4 vertex v; the edge incident to $c(v)$ that is not covered by the angle cover corresponds to the colour of v.

Fig. 7. The edge (u, v) of G is represented by the three curved edges in H. Here, (u, v) is the third edge of u and the second edge of v; $\deg(u) = 3$ and $\deg(v) = 4$.

$(c(v), e_1^{k+2}(v)))$, where all superscripts are taken modulo 3. Also, for $j = 1, \ldots,$ $\deg(v)$, set $\alpha(e_j^k(v)) = \left((e_j^k(v), e_{j-1}^k(v)), (e_j^k(v), a_j^k(v))\right)$, where $e_{j-1}^k(v)$ is $c(v)$ for $j = 1$. Furthermore, for $\ell \neq k$ (i.e., $\ell \in \{k + 1, k + 2\}$), and $(u, v) \in E$, set $\alpha(e_j^\ell(v)) = \left((e_j^\ell(v), e_{j+1}^\ell(v)), (e_j^\ell(v), e_i^\ell(u))\right)$, where $E_i(u) = E_j(v)$. Since any vertex of degree at most 2 covers all its adjacent edges, all edges in the construction, except for possibly $(e_j^k(v), e_i^k(u))$ where $E_i(u) = E_j(v)$, are covered. Since t is a 3-colouring, we know that $t(u) \neq t(v) = k$. Therefore, the edge $(e_j^k(v), e_i^k(u))$ is covered by $e_i^k(u)$ by the construction above, so all edges are covered by the constructed angle cover.

"$\Leftarrow$": (H, f) *has an angle cover implies that G is 3-colourable:*
For every vertex $v \in V$, if $c(v)$ covers edges $(c(v), e_1^{k+1}(v))$ and $(c(v), e_1^{k-1}(v))$ in the angle cover then set $t(v) = k$ (i.e., the colour given by the edge not covered by $c(v)$). Suppose for the sake of contradiction that an edge (u, v) in G is not properly coloured and $t(u) = t(v) = k$. The edge $(c(u), e_1^k(u))$ is not covered by $c(u)$ and the edge $(c(v), e_1^k(v))$ is not covered by $c(v)$. Then, those edges must be covered by $e_1^k(u)$ and $e_1^k(v)$, respectively, so $e_1^k(u)$ and $e_1^k(v)$ cannot cover the edges $(e_1^k(u), e_2^k(u))$ and $(e_1^k(v), e_2^k(v))$, respectively, nor the edges $(e_1^k(u), c_1^k(w))$ where $E_1(u) = E_i(w) = (u, w)$ and $(e_1^k(v), e_j^k(x))$ where $E_1(v) = E_j(x) = (v, x)$, respectively. Let $E_{i^*}(u) = E_{j^*}(v) = (u, v)$. Repeating this argument, we see that the edge $(e_{i^*}^k(u), e_{j^*}^k(v))$ is neither covered by $e_{i^*}^k(u)$ nor by $e_{j^*}^k(v)$. This is a contradiction since we assumed that (H, f) has an angle cover. □

Now we apply Proposition 1 to a drawing of the graph in the above reduction.

Corollary 1. *The angle cover problem is NP-hard even for* planar *graphs of maximum degree 5.*

5   Generalizations

In this section we consider two natural generalizations of the basic angle cover problem. First we consider the *a-angle cover problem* where every vertex v covers a angles, where an angle is (as before) a pair of edges incident to v that are consecutive in the circular ordering around v. We start with a positive result.

Theorem 4. *For even d, any maximum-degree-d graph with any rotation system admits an a-angle cover for $a \geq d/2 - \lfloor d/6 \rfloor$, and such an angle cover can be found in linear time.*

Proof. The proof is similar to that of Theorem 1. As in that proof, we can assume that the given graph is connected and d-regular.

We again direct the edges of the graph to form a directed Eulerian cycle. To this end, we number the edges incident to each vertex from 0 to $d - 1$ in circular order. For $i = 0, \ldots, \lfloor d/6 \rfloor - 1$, we call the group of edges $6i, \ldots, 6i + 5$ a *sextet*. In creating the directed cycle, when entering a vertex v, our goal is to exit (directing an outgoing edge) in such a way that we obtain two consecutive outgoing edges in each sextet at v. This proves the theorem since every vertex v is then able to cover all of its $d/2$ outgoing edges (and thus all edges in the graph are covered) using at most $d/2 - \lfloor d/6 \rfloor$ angles.

The rule we follow to ensure that every sextet contains two consecutive outgoing edges is when an incoming edge to v enters a sextet for the first time, we exit v on an edge e in the same sextet that is consecutive with two undirected edges (edges that are not part of the cycle yet). Since the sextet contains six edges, such an edge exists. When the cycle next enters v on an edge in this sextet, we exit on one of the remaining undirected edges consecutive to e.   □

Theorem 4 above yields, for $d = 4, 6, 8$, a-angle covers with $a = 2, 2, 3$, respectively. For $d = 4$, Theorem 1 shows that $a = 1$ suffices. For $d = 6$, $a = 1$ certainly does not suffice, so $a = 2$ is optimal. For $d = 8$, Theorem 6 shows that we cannot decide efficiently (unless $P = NP$) whether a 2-angle cover exists, so $a = 3$ is optimal.

Theorem 5. *For any $a \geq 1$ the a-angle cover problem is NP-hard even for graphs of maximum degree $4a + 1$.*   □

Proof. If every vertex can select $a > 1$ angles, we can use the same NP-hardness reduction described in the proof of Theorem 3 but attach $2(a - 1)$ adjacent edges, each connected to its own copy of K_{4a+1}, to every vertex $e_j^k(v)$ to force the vertex to "waste" $a - 1$ of its angles on these edges. These edges precede the edge $(e_j^k(v), a_j^k(v))$ in the circular order at vertex $e_j^k(v)$. Now, $d(e_j^k(v)) = 2a + 3 < 4a + 1$. For the centre vertex $c(v)$, we similarly attach $2(a - 1)$ edges, each connected to its own copy of K_{4a+1}, to $c(v)$ so that these edges lie between $(c(v), e_1^0(v))$ and $(c(v), e_1^1(v))$. Now, $d(c(v)) = 2a + 1 < 4a + 1$.

Finally, since the maximum degree in each of these attached K_{4a+1} copies is $4a + 1$, the maximum degree of the graph is $4a + 1$. □

Theorem 6. *The 2-angle cover problem is NP-hard even for graphs of maximum degree* 8.

Proof. The maximum degree given by the construction described in the proof of Theorem 5 is as a result of the use of K_9. To decrease the maximum degree to 8, instead of attaching two copies of K_9 to a vertex, we attach one copy of the following graph, T, using two edges . The graph T contains a copy of K_7 and two vertices b_1 and b_2 that are connected to all seven vertices of the copy of K_7. The graph T is attached to an external vertex v (not in T) by edges (b_1, v) and (b_2, v). Thus every vertex in T has degree 8. Including these edges, T contains 37 edges and nine vertices. Since each vertex can cover at most four edges, at least one edge must be covered by v in any valid angle cover. Furthermore, all edges except for one of the external outgoing edges can be covered; see Fig. 8.

For each vertex $e_j^k(v)$ in the original NP-hardness reduction, we now instead attach three new edges, using a copy of T, and another isolated vertex x. Similarly to the high-degree construction, these edges, in the order $(e_j^k(v), b_1)$, $(e_j^k(v), b_2)$, $(e_j^k(v), x)$, directly precede the edge $(e_j^k(v), a_j^k(v))$ in the circular order at vertex $e_j^k(v)$. As a result, $e_j^k(v)$ must "waste" one of its angles on either $(e_j^k(v), b_1)$ or $(e_j^k(v), b_2)$, and hence, it cannot use this angle on any other edge connected to a non-isolated vertex. Now $d(e_j^k(v)) = 8$. For the centre vertex $c(v)$, we similarly attach four edges using two copies of T, to $c(v)$ so that their edges lie between $(c(v), e_1^0(v))$ and $(c(v), e_1^1(v))$. Now, $d(c(v)) = 7$. If $c(v)$ covers

Fig. 9. The three cases of $c(v)$ with the new added edges to copies of T.

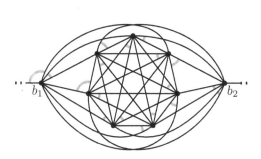

Fig. 8. The graph T with a 2-angle cover that covers all edges except the outgoing edge from b_2. A symmetric cover leaves only the outgoing edge from b_1 uncovered.

Fig. 10. An example of how the existence of an angle cover on a graph G implies that the 2-blowup of G has isomorphic thickness 2.

two of these new edges, then $c(v)$ cannot cover all three of its original edges, since it can only cover a total of four. Furthermore, assuming that the edges to each copy of T are consecutive in the edge-ordering of $c(v)$, we can cover any two of the original edges alongside one edge from each copy of T. All three cases are depicted in Fig. 9.   □

Another obvious generalization of angle covers is to consider "wider" angles. In the *m-wide angle cover problem* every vertex v can cover m consecutive edges in their circular order around v. In the full version [3], we show the following.

Theorem 7. *For $m \geq 3$, the m-wide angle cover problem is NP-hard even for graphs of maximum degree $3m - 3$.*

We now turn to a relaxation of angle cover where each vertex can select a different number of (2-wide) angles, and still, all edges must be covered. We call this an *angle allocation* and it is *optimal* if it uses the minimum number of angles among all allocations.

Theorem 8. *Given a graph $G = (V, E)$ with a rotation system, an optimal angle allocation can be computed in $O(|E|^{3/2})$ time.*

Proof. Consider the *medial graph* $G_{med} = (E, A)$ associated with the given graph $G = (V, E)$ and its rotation system. The vertices of G_{med} are the edges of G, and two vertices of G_{med} are adjacent if the corresponding edges of G are incident to the same vertex of G and consecutive in the circular ordering around that vertex. The medial graph is always 4-regular. If G has no degree-1 vertices, G_{med} has no loops. If G has minimum degree 3, G_{med} is simple.

Find a maximum matching M in G_{med} using $O(\sqrt{|E|}|A|)$ time [9]. The edges in M correspond to *independent* angles that collectively single-cover $2|M|$ edges of G (i.e., these edges are covered by only one of their adjacent vertices). Add additional angles, one for every uncovered edge of G, to obtain an allocation $\alpha \colon V \to 2^{E \times E}$ of total size $|M| + (|E| - 2|M|) = |E| - |M|$. We claim that α minimizes the number of angles. Indeed, suppose that there were an optimal angle allocation with fewer angles that covered all edges of G, and then, since no angle will double-cover two edges, there would be a larger set of independent angles (a set of angles that do not double-cover any edge), contradicting the maximality of M.   □

6   Isomorphic Thickness

Our motivation for considering angle covers was the observation that an angle cover of a plane graph can be used to place the duplicate vertices in its 2-blowup to show that the original graph is the union of two isomorphic planar graphs.

Theorem 9. *If a plane graph G has an angle cover then the 2-blowup of G has isomorphic thickness at most 2.*

Proof. Let V and E be the vertices and edges of G. Let $V_a = \{v_a \colon v \in V\}$ for $a = 1, 2$. Let $E_{ab} = \{(u_a, v_b) \colon (u, v) \in E\}$ for all $a, b \in \{1, 2\}$. Let λ be an angle cover for G. Let H be the graph with vertices $V_1 \cup V_2$ and edges $E_{11} \cup \bigcup_{v \in V}\{(v_2, x_1), (v_2, y_1) \colon \lambda(v) = \{(v, x), (v, y)\}\}$, however, if for some edge (u, v) in E, both $\lambda(u)$ and $\lambda(v)$ contain (u, v) then add only (u_2, v_1) or (u_1, v_2) (not both) to H. Note that since λ is an angle cover for G, for every edge $(u, v) \in E$, either (u_2, v_1) or (u_1, v_2) is an edge in H. Let $\widetilde{H}$ be the graph isomorphic to H where vertex v_a maps to v_b with $b = 3 - a$. We claim that H and $\widetilde{H}$ are planar graphs whose union is the 2-blowup of G.

To see that H (and hence $\widetilde{H}$) is planar, fix a planar straight-line drawing of G realizing the embedding for which λ is the angle cover for G. Place v_2 close enough to v in the angle formed by the edges $\lambda(v) = \{(v, x), (v, y)\}$, so that the segments (v_2, x) and (v_2, y) do not cross any edge segments of G. Such a placement exists since the straight lines (v, x) and (v, y) do not cross any edge segments of G. Relabel each vertex v in G as v_1 and add the edges (u_2, v_1) in H to the drawing. This creates a planar drawing of H (see Fig. 10).

The graph H contains all edges (u_1, v_1) for $(u, v) \in E$. It also contains the edge (u_1, v_2) or (u_2, v_1) for every $(u, v) \in E$ since λ is an angle cover. Hence, $\widetilde{H}$ contains, for every $(u, v) \in E$, the edge (u_2, v_2) and the edge (u_1, v_2) or (u_2, v_1) that is not in H. Thus the union of H and $\widetilde{H}$ equals the 2-blowup of G. □

As one would expect, not every planar graph whose 2-blowup has isomorphic thickness 2 has a plane embedding that admits an angle cover. In the full version [3], we provide a small example.

7   Conclusion and Open Problems

For even d, we have shown that every maximum-degree-d graph with a rotation system admits an a-angle cover for $a \approx \lceil d/3 \rceil$; see Theorem 4. This is optimal for $d = 6$ and $d = 8$ (see Theorem 6). For $d = 4$, however, we need only *one* angle per vertex, so we pose the following questions: For even d, does every graph of maximum degree d with a rotation system admit an a-angle cover for $a \approx \lceil \delta d \rceil$, where $\delta < 1/3$? What about graphs of maximum degree d if d is odd?

References

1. Alimonti, P., Kann, V.: Some APX-completeness results for cubic graphs. Theor. Comput. Sci. **237**(1), 123–134 (2000). https://doi.org/10.1016/S0304-3975(98)00158-3
2. Dinur, I., Safra, S.: On the hardness of approximating vertex cover. Ann. Math. **162**(1), 439–485 (2005). https://doi.org/10.4007/annals.2005.162.439
3. Evans, W., Gethner, E., Spalding-Jamieson, J., Wolff, A.: Angle covers: algorithms and complexity. Arxiv (2019). http://arxiv.org/abs/1911.02040
4. Even, S., Itai, A., Shamir, A.: On the complexity of timetable and multicommodity flow problems. SIAM J. Comput. **5**(4), 691–703 (1976). https://doi.org/10.1137/0205048

5. Haas, R., et al.: Planar minimally rigid graphs and pseudo-triangulations. Comput. Geom. **31**(1), 31–61 (2005)
6. Kainen, P.C.: Thickness and coarseness of graphs. Abh. Math. Semin. Univ. Hambg. **39**(1), 88–95 (1973). https://doi.org/10.1007/BF02992822
7. Karp, R.M.: Reducibility among combinatorial problems. In: Miller, R., Thatcher, J., Bohlinger, J. (eds.) Complexity of Computer Computations, IBM Research Symposia, pp. 85–103. Springer, Boston (1972). https://doi.org/10.1007/978-1-4684-2001-2
8. Khot, S., Regev, O.: Vertex cover might be hard to approximate to within $2 - \epsilon$. J. Comput. Syst. Sci. **74**(3), 335–349 (2008). https://doi.org/10.1016/j.jcss.2007.06.019
9. Micali, S., Vazirani, V.V.: An $O(\sqrt{|V|}|E|)$ algorithm for finding maximum matching in general graphs. In: Proceedings 21st Annual Symposium on Foundations of Computer Science (FOCS), pp. 17–27. IEEE (1980). https://doi.org/10.1109/SFCS.1980.12
10. Streinu, I.: A combinatorial approach to planar non-colliding robot arm motion planning. In: Proceedings 41st Annual Symposium on Foundations of Computer Science (FOCS), pp. 443–453. IEEE (2000). https://doi.org/10.1109/SFCS.2000.892132

Fast Multiple Pattern Cartesian Tree Matching

Geonmo Gu[1], Siwoo Song[1], Simone Faro[2], Thierry Lecroq[3],
and Kunsoo Park[1(✉)]

[1] Seoul National University, Seoul, Korea
{gmgu,swsong,kpark}@theory.snu.ac.kr
[2] University of Catania, Catania, Italy
faro@dmi.unict.it
[3] Normandie University, Rouen, France
thierry.lecroq@univ-rouen.fr

Abstract. Cartesian tree matching is the problem of finding all substrings in a given text which have the same Cartesian trees as that of a given pattern. In this paper, we deal with Cartesian tree matching for the case of multiple patterns. We present two fingerprinting methods, i.e., the parent-distance encoding and the binary encoding. By combining an efficient fingerprinting method and a conventional multiple string matching algorithm, we can efficiently solve multiple pattern Cartesian tree matching. We propose three practical algorithms for multiple pattern Cartesian tree matching based on the Wu-Manber algorithm, the Rabin-Karp algorithm, and the Alpha Skip Search algorithm, respectively. In the experiments we compare our solutions against the previous algorithm [18]. Our solutions run faster than the previous algorithm as the pattern lengths increase. Especially, our algorithm based on Wu-Manber runs up to 33 times faster.

Keywords: Multiple pattern Cartesian tree matching ·
Parent-distance encoding · Binary encoding · Fingerprinting methods

1 Introduction

Cartesian tree matching is the problem of finding all substrings in a given text which have the same Cartesian trees as that of a given pattern. For instance, given text $T = (6, 1, 5, 3, 6, 5, 7, 4, 2, 3, 1)$ and pattern $P = (1, 4, 3, 4, 1)$ in Fig. 1a, P has the same Cartesian tree as the substring $(3, 6, 5, 7, 4)$ of T. Among many generalized matchings, Cartesian tree matching is analogous to order-preserving

A full version of this paper is available at https://arxiv.org/abs/1911.01644. Gu, Song, and Park were supported by Collaborative Genome Program for Fostering New Post-Genome industry through the National Research Foundation of Korea (NRF) funded by the Ministry of Science ICT and Future Planning (No. NRF-2014M3C9A3063541).

M. S. Rahman et al. (Eds.): WALCOM 2020, LNCS 12049, pp. 107–119, 2020.
https://doi.org/10.1007/978-3-030-39881-1_10

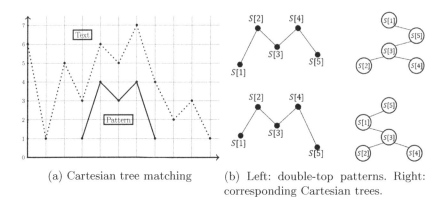

(a) Cartesian tree matching   (b) Left: double-top patterns. Right: corresponding Cartesian trees.

Fig. 1. Cartesian tree matching: multiple Cartesian trees are required for the double-top pattern.

matching [5,9,13,15] in the sense that they deal with relative order between numbers. Accordingly, both of them can be applied to time series data such as stock price analysis, but Cartesian tree matching can be sometimes more appropriate than order-preserving matching in finding patterns [18].

In this paper, we deal with Cartesian tree matching for the case of multiple patterns. Although finding multiple different patterns is interesting by itself, multiple pattern Cartesian tree matching can be applied in finding one meaningful pattern when the meaningful pattern is represented by multiple Cartesian trees: Suppose we are looking for the double-top pattern [17]. Two Cartesian trees in Fig. 1b are required to identify the pattern, where the relative order between $S[1]$ and $S[5]$ causes the difference. In general, the more complex the pattern is, the more Cartesian trees having the same lengths are required. (e.g., the head-and-shoulder pattern [17] requires four Cartesian trees.)

Recently, Park et al. [18] introduced (single pattern) Cartesian tree matching, multiple pattern Cartesian tree matching, and Cartesian tree indexing with their respective algorithms. They proposed the parent-distance representation that has a one-to-one mapping with Cartesian trees, and gave linear-time solutions for the problems, utilizing the representation and existing string algorithms, i.e., KMP algorithm, Aho-Corasick algorithm, and suffix tree construction algorithm. Song et al. [19] proposed new representations about Cartesian trees, and proposed practically fast algorithms for Cartesian tree matching based on the framework of filtering and verification.

Extensive works have been done to develop algorithms for multiple pattern matching, which is one of the fundamental problems in computer science [11,16,20]. Aho and Corasick [1] presented a linear-time algorithm based on an automaton. Commentz-Walter [6] presented an algorithm that combines the Aho-Corasick algorithm and the Boyer-Moore technique [3]. Crochemore et al. [8] proposed an algorithm that combines the Aho-Corasick automaton and a Directed Acyclic Word Graph, which runs linear in the worst case and runs in

$O((n/m)\log m)$ time in the average case, where m is the length of the shortest pattern. Rabin and Karp [12] proposed an algorithm that runs linear on average and $O(nM)$ in the worst case, where M is the sum of lengths of all patterns. Charras et al. [4] proposed an algorithm called Alpha Skip Search, which can efficiently handle both single pattern and multiple patterns. Wu and Manber [22] presented an algorithm that uses an extension of the Boyer-Moore-Horspool technique.

In this paper we present practically fast algorithms for multiple pattern Cartesian tree matching. We present three algorithms based on Wu-Manber, Rabin-Karp, and Alpha Skip Search. All of them use the filtering and verification approach, where filtering relies on efficient fingerprinting methods of a string. Two fingerprinting methods are presented, i.e., the parent-distance encoding and the binary encoding. By combining an efficient fingerprinting method and a conventional multiple string matching algorithm, we can efficiently solve multiple pattern Cartesian tree matching. In the experiments we compare our solutions against the previous algorithm [18] which is based on the Aho-Corasick algorithm. Our solutions run faster than the previous algorithm. Especially, our algorithm based on Wu-Manber runs up to 33 times faster.

2   Problem Definition

2.1   Notation

A *string* is a sequence of characters drawn from an alphabet Σ, which is a set of integers. We assume that a comparison between any two characters can be done in constant time. For a string S, $S[i]$ represents the i-th character of S, and $S[i..j]$ represents the substring of S starting from i and ending at j.

A *Cartesian tree* [21] is a binary tree derived from a string. Specifically, the Cartesian tree $CT(S)$ for a string S can be uniquely defined as follows:

- If S is an empty string, $CT(S)$ is an empty tree.
- If S is not empty and $S[i]$ is the minimum value in $S[1..n]$, $CT(S)$ is the tree with $S[i]$ as the root, $CT(S[1..i-1])$ as the left subtree, and $CT(S[i+1..n])$ as the right subtree. If there is more than one minimum value, we choose the leftmost one as the root.

Given two strings $T[1..n]$ and $P[1..m]$, where $m \leq n$, we say that P *matches* T at position i if $CT(T[i-m+1..i]) = CT(P[1..m])$. For example, given $T = (6, 1, 5, 3, 6, 5, 7, 4, 2, 3, 1)$ and $P = (1, 4, 3, 4, 1)$ in Fig. 1a, P matches T at position 8. We also say that $T[4..8]$ is *a match* of P in T.

Cartesian tree matching is the problem of finding all the matches in the text which have the same Cartesian trees as a given pattern.

Definition 1. *(Cartesian tree matching [18]) Given two strings text $T[1..n]$ and pattern $P[1..m]$, find every $m \leq i \leq n$ such that $CT(T[i-m+1..i]) = CT(P[1..m])$.*

2.2   Multiple Pattern Cartesian Tree Matching

Cartesian tree matching can be extended to the case of multiple patterns. *Multiple pattern Cartesian tree matching* is the problem of finding all the matches in the text which have the same Cartesian trees as at least one of the given patterns.

Definition 2. *(Multiple pattern Cartesian tree matching [18]) Given a text $T[1..n]$ and patterns $P_1[1..m_1], P_2[1..m_2], ..., P_k[1..m_k]$, find every position in the text which matches at least one pattern, i.e., it has the same Cartesian tree as that of at least one pattern.*

3   Fingerprinting Methods

Fingerprinting is a technique that maps a string to a much shorter form of data, such as a bit string or an integer. In Cartesian tree matching, we can use fingerprints to filter out unpromising matching positions with low computational cost.

In this section we introduce two fingerprinting methods, i.e., the parent-distance encoding and the binary encoding, for the purpose of representing information about Cartesian tree as an integer. The two encodings make use of the parent-distance representation and the binary representation, respectively, both of which are strings that represent Cartesian trees.

3.1   Parent-Distance Encoding

In order to represent Cartesian trees efficiently, Park et al. proposed the *parent-distance representation* [18], which is another form of the all nearest smaller values [2].

Definition 3. *(Parent-distance representation) Given a string $S[1..n]$, the parent-distance representation of S is an integer string $PD(S)[1..n]$, which is defined as follows:*

$$PD(S)[i] = \begin{cases} i - \max_{1 \leq j < i}\{j : S[j] \leq S[i]\} & \text{if such } j \text{ exists} \\ 0 & \text{otherwise} \end{cases} \quad (1)$$

Intuitively, $PD(S)[i]$ stores the distance between $S[i]$ and the parent of $S[i]$ in $CT(S[1..i])$. For example, the parent-distance representation of string $S = (11, 14, 13, 15, 12)$ is $PD(S) = (0, 1, 2, 1, 4)$, where $PD(S)[3] = 3 - 1 = 2$ stores the distance between $S[3]$ and $S[1]$ ($S[1]$ is the parent of $S[3]$ in $CT(S[1..3])$). The parent-distance representation has a one-to-one mapping to the Cartesian tree [18], and so if two strings have the same parent-distance representations, the two strings also have the same Cartesian trees. The parent-distance representation of a string can be computed in linear time [18]. Note that $PD(S)[i]$ holds a value between 0 to $i - 1$ by definition, and $PD(S)[1] = 0$ at all times.

With the parent-distance representation, we can define a fingerprint encoding function that maps a string to an integer, using the factorial number system [14].

Definition 4. *(Parent-distance Encoding) Given a string $S[1..n]$, the encoding function $f(S)$, which maps S into an integer within the range $[0..n! - 1]$, is defined as follows:*

$$f(S) = \sum_{i=2}^{n}(PD(S)[i]) \cdot (i - 1)!. \tag{2}$$

The parent-distance encoding maps a string into a unique integer according to its parent-distance representation. That is, given two strings S_1 and S_2, $CT(S_1) = CT(S_2)$ if and only if $f(S_1) = f(S_2)$. This is because if $PD(S_1) \neq PD(S_2)$ then $f(S_1) \neq f(S_2)$ due to the fact that $PD(S)[i] < i$. The encoding function $f(S[1..n])$ can be computed in $O(n)$ time, since $PD(S)$ can be computed in linear time. For a long string, the fingerprint may not fit in a word size, so we select a prime number by which we divide the fingerprint, and use the residue instead of the actual fingerprint. A similar encoding function was used to solve the multiple pattern order-preserving matching problem [10].

3.2 Binary Encoding

For order-preserving matching, the representation of a string as a binary string is first presented by Chhabra and Tarhio [5]. Recently, Song et al. make use of the *binary representation* for Cartesian tree matching as follows [19].

Definition 5. *(Binary representation) Given an n-length string S, binary representation $\beta(S)$ of length $n - 1$ is defined as follows: for $1 \leq i \leq n - 1$,*

$$\beta(S)[i] = \begin{cases} 1 & \text{if } S[i] \leq S[i + 1] \\ 0 & \text{otherwise.} \end{cases} \tag{3}$$

Given two strings $S_1[1..n]$ and $S_2[1..n]$, the binary representations $\beta(S_1)$ and $\beta(S_2)$ are the same if the Cartesian trees $CT(S_1)$ and $CT(S_2)$ are the same [19]. Obviously, the Cartesian tree has a many-to-one mapping to the binary representation. Thus, two strings whose binary representations are the same may not have the same Cartesian trees, but two strings whose Cartesian trees are the same have the same binary representations.

A fingerprint encoding function $f(S)$ can be defined using the binary representation.

Definition 6. *(Binary Encoding) Given a string $S[1..n]$, encoding function $f(S)$, which maps S into an integer within the range $[0..2^{n-1} - 1]$, is defined as follows:*

$$f(S) = \sum_{i=1}^{n-1}(\beta(S)[i] \cdot 2^{n-1-i}). \tag{4}$$

Since $f(S)$ is a polynomial, it can be efficiently computed in linear time using Horner's rule [7]. Moreover, a fingerprint computed by the binary encoding can be reused when two strings overlap, which is discussed in the full version of this paper. Like the parent-distance encoding, in case the fingerprint does not fit in a word size, we select a prime number by which we divide the fingerprint, and use the residue instead of the actual fingerprint.

4 Fast Multiple Pattern Cartesian Tree Matching Algorithms

In this section we introduce three algorithms for multiple pattern Cartesian tree matching. Each of them consists of preprocessing and search. In the preprocessing step, hash tables are built using fingerprints of patterns. In the search step, the filtering and verification approach is adopted. To filter out unpromising matching positions, a fingerprinting method is applied to either length-m substrings of the text, where m is the length of the shortest pattern, or much shorter length-b substrings of the text (we will discuss how to set b in Sect. 4.4). Then each candidate pattern is verified by an efficient comparison method (which is described in the full version of this paper).

Algorithm 1. Algorithm based on Wu-Manber

1: **input:** text $T[1..n]$ and patterns $P_1[1..m_1], P_2[1..m_2], ..., P_k[1..m_k]$
2: **output:** every position in T that matches at least one of the patterns
3: **procedure** PREPROCESSING
4:     $m \leftarrow \min(m_1, m_2, ..., m_k)$
5:     $b \leftarrow \log_2(km)$
6:     Initialize each entry of SHIFT to $m - b + 1$
7:     **for** $i \leftarrow 1$ to k **do**
8:         **for** $j \leftarrow b$ to $m - 1$ **do**
9:             $fp \leftarrow f(P_i[j - b + 1..j])$
10:             **if** SHIFT$[fp] > m - j$ **then**
11:                 SHIFT$[fp] \leftarrow m - j$
12:         $fp \leftarrow f(P_i[m - b + 1..m])$
13:         HASH$[fp].add(i)$
14: **procedure** SEARCH
15:     $index \leftarrow m$
16:     **while** $index \leq n$ **do**
17:         $fp \leftarrow f(T[index - b + 1..index])$
18:         **for** $i \in$ HASH$[fp]$ **do**
19:             **if** P_i matches $T[index - m + 1..index - m + m_i]$ **then**
20:                 output $index - m + m_i$
21:         $index \leftarrow index +$ SHIFT$[fp]$

4.1 Algorithm Based on Wu-Manber

Algorithm 1 shows the pseudo-code of an algorithm for multiple pattern Cartesian tree matching based on the Wu-Manber algorithm [22]. The algorithm uses
two hash tables, HASH and SHIFT. Both tables use a fingerprint of length-b
string, called a *block*. Either the parent-distance encoding or the binary encoding is used to compute the fingerprint. Given patterns $P_1, P_2, ..., P_k$, let m be
the length of the shortest pattern. HASH maps a fingerprint fp of a block to the
list of patterns P_i such that the fingerprint of the last block in P_i's length-m
prefix is the same as fp. For a block $B[1..b]$ and a fingerprint encoding function
f, HASH is defined as follows:

$$\text{HASH}[f(B)] = \{i : f(P_i[m - b + 1..m]) = f(B), 1 \leq i \leq k\} \tag{5}$$

SHIFT maps a fingerprint fp of a block to the amount of a valid shift when
the block appears in the text. The shift value is determined by the rightmost
occurrence of a block in terms of the fingerprint among length-$(m - 1)$ prefixes
of the patterns. For a block $B[1..b]$ and a fingerprint encoding function f, we
define the rightmost occurrence r_B as follows:

$$r_B = \begin{cases} \max_{b \leq j \leq m-1}\{j : f(P_i[j - b + 1..j]) = f(B), 1 \leq i \leq k\} & \text{if such } j \text{ exists} \\ 0 & \text{otherwise} \end{cases} \tag{6}$$

Then SHIFT is defined as follows:

$$\text{SHIFT}[f(B)] = m - r_B \tag{7}$$

In the preprocessing step, we build HASH and SHIFT (as described in
Algorithm 1). In the search step, we scan the text from left to right, computing the fingerprint of a length-b substring of the text to get a list of patterns
from HASH. Let *index* be the current scanning position of the text. We compute
fingerprint fp of $T[index - b + 1..index]$, and get a list of patterns in the entry
HASH[fp]. If the list is not empty, each pattern is verified by an efficient comparison method (see the full version). Consider $P_i[1..m_i]$ in the list. The comparison
method verifies whether P_i matches $T[index - m + 1..index - m + m_i]$. After
verifying all patterns in the list, the text is shifted by SHIFT[fp].

The worst case time complexity of Algorithm 1 is $O((M + b)n)$, where M
is the total pattern length, b is the block size, and n is the length of the text
(consider $T = 1^n$ and the patterns of which prefixes are 1^m). On the other hand,
the best case time complexity of Algorithm 1 is $O(\frac{bn}{m-b})$.

4.2 Algorithm Based on Rabin-Karp

The second algorithm for multiple pattern Cartesian tree matching is based
on the Rabin-Karp algorithm [12]. The algorithm uses one hash table, namely
HASH. HASH is similarly defined as in Algorithm 1 except that we consider

length-m prefixes instead of blocks and we use only binary encoding for fingerprinting. For a string $S[1..m]$ and the binary encoding function f, HASH is defined as follows:

$$\text{HASH}[f(S)] = \{i : f(P_i[1..m]) = f(S), 1 \leq i \leq k\} \tag{8}$$

In the preprocessing step, we build HASH. In the search step, we shift one by one, and compute the fingerprint of a length-m substring of the text to get candidate patterns by using HASH. Again, each candidate pattern is verified by an efficient comparison method.

Given a fingerprint at position i of the text, the next fingerprint at position $i + 1$ can be computed in constant time if we use the binary encoding as a fingerprinting method. Let the former fingerprint be $fp_i = f(T[i - m + 1..i])$ and the latter one be $fp_{i+1} = f(T[i - m + 2..i + 1])$. Then,

$$fp_{i+1} = 2(fp_i - 2^{m-2}\beta(T)[i - m + 1]) + \beta(T)[i] \tag{9}$$

Subtracting $2^{m-2}\beta(T)[i - m + 1]$ removes the leftmost bit from fp_i, multiplying the result by 2 shifts the number to the left by one position, and adding $\beta(T)[i]$ brings in the appropriate rightmost bit.

The worst case time complexity of the algorithm is $O(Mn)$ (consider $T = 1^n$ and patterns of which prefixes are 1^m). The best case time complexity is $O(n)$ since fingerprint f_i at position i, $m + 1 \leq i \leq n$, can be computed in $O(1)$ time using Eq. (9).

4.3   Algorithm Based on Alpha Skip Search

The third algorithm for multiple pattern Cartesian tree matching is based on Alpha Skip Search [4]. Recall that a length-b string is called a block. The algorithm uses a hash table POS that maps the fingerprint of a block to a list of occurrences in all length-m prefixes of the patterns. Either the parent-distance encoding or the binary encoding is used for fingerprinting. For a block $B[1..b]$ and a fingerprint encoding function f, POS is defined as follows:

$$\text{POS}[f(B)] = \{(i, j) : f(P_i[j - b + 1..j]) = f(B), 1 \leq i \leq k, b \leq j \leq m\} \tag{10}$$

In the preprocessing step, we build POS. In the search step, we scan the text from left to right, computing the fingerprint of a length-b substring of the text to get the list of pairs (i, j), meaning that the fingerprint of $P_i[j - b + 1..j]$ is the same as that of the substring of the text. Verification using an efficient comparison method is performed for each pair in the list. Note that the algorithm always shifts by $m - b + 1$.

The worst case time complexity of the algorithm is $O((M + b)n)$, where M is the total pattern length, b is the block size, and n is the length of the text (consider $T = 1^n$ and patterns of which prefixes are 1^m). On the other hand, the best case time complexity of the algorithm is $O(\frac{bn}{m-b})$ since the algorithm always shifts by $m - b + 1$.

4.4   Selecting the Block Size

The size of the block affects the running time of the presented algorithms based on Wu-Manber and Alpha Skip Search. A longer block size leads to a lower probability of candidate pattern occurrences, so it decreases verification time. On the other hand, a longer block size increases the overhead required for computing fingerprints. Thus, it is important to set a block size appropriate for each algorithm.

In order to set a block size, we first study the matching probability of two strings, in terms of Cartesian trees. Assume that numbers are independent and identically distributed, and there are no identical numbers within any length-n string.

Lemma 1. *Given two strings $S_1[1..n]$ and $S_2[1..n]$, the probability $p(n)$ that S_1 and S_2 have the same Cartesian tree can be defined by the recurrence formula, where $p(0) = 1$ and $p(1) = 1$, as follows:*

$$p(n) = \frac{p(0)p(n-1) + p(1)p(n-2) + \cdots + p(n-1)p(0)}{n^2} \tag{11}$$

We have the following upper bound on the matching probability.

Theorem 1. *Assume that numbers are independent and identically distributed, and there are no identical numbers within any length-n string. Given two strings $S_1[1..n]$ and $S_2[1..n]$, the probability that the two strings match, in terms of Cartesian trees, is at most $\frac{1}{2^{n-1}}$, i.e., $p(n) \leq \frac{1}{2^{n-1}}$.*

We set the block size $b = \log_2(km)$ if $\log_2(km) \leq m$; otherwise we set $b = m$, where k is the number of patterns and m is the length of the shortest pattern, in order to get a low probability of match and a relatively short block size with respect to m. By Theorem 1, if we set $b = \log_2(km)$, $p(b) \leq \frac{2}{km}$.

5   Experiments

We conduct experiments to evaluate the performances of the proposed algorithms against the previous algorithm. We compare algorithms based on Aho-Corasick (AC) [18], Wu-Manber (WM), Rabin-Karp (RM), and Alpha Skip Search (AS). By default, all our algorithms use optimization techniques described in the full version of this paper, except the min-index filtering method which is evaluated in the experiments. Particularly, in order to compare the fingerprinting methods and see the effect of min-index filtering method, we compare variants of our algorithms. The following algorithms are evaluated.

- AC: multiple Cartesian tree matching algorithm based on Aho-Corasick [18].
- WMP: algorithm based on Wu-Manber that uses the parent-distance encoding as a fingerprinting method.
- WMB: algorithm based on Wu-Manber that uses the binary encoding as a fingerprinting method. The algorithm reuses fingerprints when adjacent blocks overlap $b - 1$ characters (i.e., when the text shifts by one position), where b is the block size.

- WMBM: WMB that exploits additional min-index filtering.
- RK: algorithm based on Rabin-Karp that uses the binary encoding as a fingerprinting method.
- ASB: algorithm based on Alpha Skip Search that uses the binary encoding as a fingerprinting method. The algorithm reuses fingerprints when adjacent blocks overlap $b - 1$ characters.

All algorithms are implemented in C++. Experiments are conducted on a machine with Intel Xeon E5-2630 v4 2.20 GHz CPU and 128 GB memory running CentOS Linux.

The total time includes the preprocessing time for building data structures and the search time. To evaluate an algorithm, we run it 100 times and measure the average total time in milliseconds.

We randomly build a text of length 10,000,000 where the alphabet size is 1,000. A pattern is extracted from the text at a random position.

5.1   Evaluation on the Equal Length Patterns

We first conduct experiments with sets of patterns of the same length. Figures 2a, 2c, and 2e show the results, where k is the number of patterns and x-axis represents the length of the patterns, i.e., m. As the length of the patterns increases, WMB, WMBM, and ASB become the fastest algorithms due to a long shift length, low verification time, and light fingerprinting method. WMBM and WMB outperforms AC up to 33 times ($k = 100$ and $m = 256$). ASB outperforms AC up to 28 times ($k = 10$ and $m = 256$). RK outperforms AC up to 3 times ($k = 50, 100$ and $m = 16$). When the length of the patterns is extremely short, however, AC is the clear winner ($m = 4$). In this case, other algorithms naïvely compare the greatest part of patterns for each position of the text. WMP works visibly worse when $m = 8$ due to the extreme situation and overhead of the fingerprinting method. Since short patterns are more likely to have the same Cartesian trees, the proposed algorithms are sometimes faster when $m = 4$ than when $m = 8$ due to the grouping technique described in the full version of this paper. Comparing WMB and WMBM, the min-index filtering method is more effective when there are many short patterns ($k = 100$ and $m = 4, 8$).

5.2   Evaluation on the Different Length Patterns

We compare algorithms with sets of patterns of different lengths. Figures 2b, 2d, and 2f show the results. The length is randomly selected in an interval, i.e., [8, 32], [16, 64], [32, 128], and [64, 256]. After a length is selected, a pattern is extracted from the text at a random position. When there are many short patterns, i.e., $k = 100$ and patterns of length 8–32, AC is the fastest due to the short minimum pattern length.

When the length of the shortest pattern is sufficiently long, however, the proposed algorithms outperform AC. Specifically WMB outperforms AC up to 20 times ($k = 10, 50, 100$ and patterns of length 64–256). ASB outperforms AC

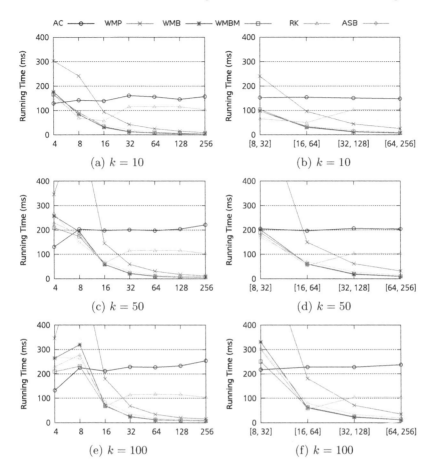

Fig. 2. Evaluation on the length of pattern. Left: patterns of equal length. Right: patterns of different lengths.

up to 14 times ($k = 10$ and patterns of length 64–256). RK outperforms AC up to 4 times ($k = 100$ and patterns of length 16–64).

5.3   Evaluation on the Real Dataset

We conduct experiment on a real dataset, which is a time series of Seoul temperatures. The Seoul temperatures dataset consists of 658,795 integers referring to the hourly temperatures in Seoul (multiplied by ten) in the years 1907–2019 [19]. In general, temperatures rise during the day and fall at night. Therefore, the Seoul temperatures dataset has more matches than random datasets when patterns are extracted from the text. Figure 3 shows the results on the Seoul temperatures dataset with sets of patterns of the same length. As the pattern length grows, the proposed algorithms run much faster than AC. For short patterns ($m = 4, 8$), AC is the fastest algorithm, and AC is up to twice times faster

118   G. Gu et al.

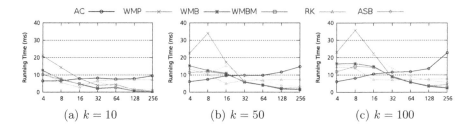

(a) $k = 10$ (b) $k = 50$ (c) $k = 100$

Fig. 3. Evaluation on the Seoul temperatures dataset.

than WMBM ($m = 4$ and $k = 100$) and 1.7 times faster than RK ($m = 8$ and $k = 100$). For moderate-length patterns ($m = 16, 32$), RK is up to 2.8 times faster than AC ($m = 16$ and $k = 10$), and WMB is up to 4 times faster than AC ($m = 32$ and $k = 10$). For relatively long patterns ($m = 64, 128, 256$), all the proposed algorithms outperform AC. Specifically, WMB, WMBM, ASB, and WMP outperform AC up to 28, 26, 11, and 10 times, respectively ($m = 256$ and $k = 10$), and RK outperforms AC up to 2.9 times ($m = 256$ and $k = 100$).

References

1. Aho, A.V., Corasick, M.J.: Efficient string matching: an aid to bibliographic search. Commun. ACM **18**(6), 333–340 (1975)
2. Berkman, O., Schieber, B., Vishkin, U.: Optimal doubly logarithmic parallel algorithms based on finding all nearest smaller values. J. Algorithms **14**(3), 344–370 (1993)
3. Boyer, R.S., Moore, J.S.: A fast string searching algorithm. Commun. ACM **20**(10), 762–772 (1977)
4. Charras, C., Lecroq, T., Pehoushek, J.D.: A very fast string matching algorithm for small alphabets and long patterns. In: Farach-Colton, M. (ed.) CPM 1998. LNCS, vol. 1448, pp. 55–64. Springer, Heidelberg (1998). https://doi.org/10.1007/BFb0030780
5. Chhabra, T., Tarhio, J.: Order-preserving matching with filtration. In: Gudmundsson, J., Katajainen, J. (eds.) SEA 2014. LNCS, vol. 8504, pp. 307–314. Springer, Cham (2014). https://doi.org/10.1007/978-3-319-07959-2_26
6. Commentz-Walter, B.: A string matching algorithm fast on the average. In: Maurer, H.A. (ed.) ICALP 1979. LNCS, vol. 71, pp. 118–132. Springer, Heidelberg (1979). https://doi.org/10.1007/3-540-09510-1_10
7. Cormen, T.H., Leiserson, C.E., Rivest, R.L., Stein, C.: Introduction to algorithms second edition. The Knuth-Morris-Pratt Algorithm (2001)
8. Crochemore, M., Czumaj, A., Gasieniec, L., Lecroq, T., Plandowski, W., Rytter, W.: Fast practical multi-pattern matching. Inf. Process. Lett. **71**(3–4), 107–113 (1999)
9. Ganguly, A., Hon, W.K., Sadakane, K., Shah, R., Thankachan, S.V., Yang, Y.: Space-efficient dictionaries for parameterized and order-preserving pattern matching. In: 27th Annual Symposium on Combinatorial Pattern Matching (CPM), pp. 2:1–2:12. LIPIcs (2016)

10. Han, M., Kang, M., Cho, S., Gu, G., Sim, J.S., Park, K.: Fast multiple order-preserving matching algorithms. In: Lipták, Z., Smyth, W.F. (eds.) IWOCA 2015. LNCS, vol. 9538, pp. 248–259. Springer, Cham (2016). https://doi.org/10.1007/978-3-319-29516-9_21

11. Hua, N., Song, H., Lakshman, T.: Variable-stride multi-pattern matching for scalable deep packet inspection. In: IEEE INFOCOM 2009, pp. 415–423. IEEE (2009)

12. Karp, R.M., Rabin, M.O.: Efficient randomized pattern-matching algorithms. IBM J. Res. Dev. **31**(2), 249–260 (1987)

13. Kim, J., et al.: Order-preserving matching. Theor. Comput. Sci. **525**, 68–79 (2014)

14. Knuth, D.E.: The Art of Computer Programming, volume 2: Seminumerical algorithms. Addison-Wesley Professional, Boston (2014)

15. Kubica, M., Kulczyński, T., Radoszewski, J., Rytter, W., Waleń, T.: A linear time algorithm for consecutive permutation pattern matching. Inf. Process. Let. **113**(12), 430–433 (2013)

16. Liao, H.J., Lin, C.H.R., Lin, Y.C., Tung, K.Y.: Intrusion detection system: a comprehensive review. J. Netw. Comput. Appl. **36**(1), 16–24 (2013)

17. Liu, J.N., Kwong, R.W.: Automatic extraction and identification of chart patterns towards financial forecast. Appl. Soft Comput. **7**(4), 1197–1208 (2007)

18. Park, S., Amir, A., Landau, G.M., Park, K.: Cartesian tree matching and indexing. In: 30th Annual Symposium on Combinatorial Pattern Matching (CPM), pp. 16:1–16:14. LIPIcs (2019)

19. Song, S., Ryu, C., Faro, S., Lecroq, T., Park, K.: Fast cartesian tree matching. In: Brisaboa, N.R., Puglisi, S.J. (eds.) SPIRE 2019. LNCS, vol. 11811, pp. 124–137. Springer, Cham (2019). https://doi.org/10.1007/978-3-030-32686-9_9

20. Song, T., Zhang, W., Wang, D., Xue, Y.: A memory efficient multiple pattern matching architecture for network security. In: IEEE INFOCOM 2008-The 27th Conference on Computer Communications, pp. 166–170. IEEE (2008)

21. Vuillemin, J.: A unifying look at data structures. Commun. ACM **23**(4), 229–239 (1980)

22. Wu, S., Manber, U.: A fast algorithm for multi-pattern searching. Technical report. TR-94-17, Department of Computer Science, University of Arizona (1994)

Generalized Dictionary Matching Under Substring Consistent Equivalence Relations

Diptarama Hendrian[(⊠)]

Graduate School of Information Sciences, Tohoku University, Sendai, Japan
diptarama@tohoku.ac.jp

Abstract. Given a set of patterns called a dictionary and a text, the dictionary matching problem is a task to find all occurrence positions of all patterns in the text. The dictionary matching problem can be solved efficiently by using the Aho-Corasick algorithm. Recently, Matsuoka et al. [TCS, 2016] proposed a generalization of pattern matching problem under substring consistent equivalence relations and presented a generalization of the Knuth-Morris-Pratt algorithm to solve this problem. An equivalence relation ≈ is a substring consistent equivalence relation (SCER) if for two strings X, Y, $X \approx Y$ implies $|X| = |Y|$ and $X[i : j] \approx Y[i : j]$ for all $1 \leq i \leq j \leq |X|$. In this paper, we propose a generalization of the dictionary matching problem and present a generalization of the Aho-Corasick algorithm for the dictionary matching under SCER. We present an algorithm that constructs SCER automata and an algorithm that performs dictionary matching under SCER by using the automata. Moreover, we show the time and space complexity of our algorithms with respect to the size of input strings.

Keywords: Dictionary matching · Aho-Corasick algorithm · Substring consistent equivalence relation

1 Introduction

The pattern matching problem is one of the most fundamental problems in string processing and extensively studied due to its wide range of applications [7,8]. Given a text T of length n and a pattern P of length m, the pattern matching problem is to find all occurrence positions of P in T. A naive approach to solve this problem is by comparing all substrings of T whose length is m to P which takes $O(nm)$ time. One of the algorithms that can solve this problem in linear time and space is the *Knuth-Morris-Pratt (KMP) algorithm* [15]. The KMP algorithm constructs an $O(m)$ space array as a failure function by preprocessing the pattern in $O(m)$ time, then uses the failure function to perform pattern matching in $O(n)$ time.

This research was supported by JSPS KAKENHI Grant Number JP19K20208.

M. S. Rahman et al. (Eds.): WALCOM 2020, LNCS 12049, pp. 120–132, 2020.
https://doi.org/10.1007/978-3-030-39881-1_11

Many variants of pattern matching problems are studied for various applications such as parameterized pattern matching [6] for detecting duplication in source code, order-preserving pattern matching [14,16] for numerical analysis, permuted pattern matching [13] for multi sensor data, and so on [18]. In order to solve these problems efficiently, the KMP algorithm is extended for the above mentioned pattern matching problems [3,9,11,14,16,18].

Table 1. The time complexity of the proposed algorithm on some dictionary matching problems.

	$\xi(n)$	$\phi(\ell)$	π	Preprocessing	Searching
Exact	$O(1)$	$O(1)$	$\|\Sigma\|$	$O(m \log \|\Sigma\|)$	$O(n \log \|\Sigma\| + occ)$
Parameterized	$O(n)$	$O(1)$	$\|\Sigma\|$	$O(m \log \|\Sigma\|)$	$O(n \log \|\Sigma\| + occ)$
Order-preserving	$O(1)$	$O(\log \ell)$	ℓ	$O(m \log \ell)$	$O(n \log \ell + occ)$

Recently, Matsuoka et al. [17] defined a general pattern matching problem under a substring consistent equivalence relation. An equivalence relation $\approx$ for two strings $X \approx Y$ is a *substring consistent equivalence relation (SCER)* [17] if for two strings X, Y, $X \approx Y$ implies $|X| = |Y|$ and $X[i : j] \approx Y[i : j]$ for all $1 \leq i \leq j \leq |X|$. The equivalence relations used in parameterized pattern matching, order-preserving pattern matching, and permuted pattern matching are SCERs. Matsuoka et al. proposed a generalized KMP algorithm that can solve any pattern matching under SCER and showed the time complexity of the algorithm. They also show periodicity properties on strings under SCERs.

The *dictionary matching problem* is a task to find all occurrence positions of multiple patterns in a text. Given a set of patterns called a dictionary $\mathcal{D}$, we can find the occurrence positions of all patterns in a text T by performing pattern matching for each pattern in the dictionary. However, we need to read the text multiple times in this approach. Aho and Corasick [1] proposed an algorithm that can perform dictionary matching in linear time by extending the failure function of the KMP algorithm. The Aho-Corasick (AC) algorithm constructs an automaton (we call this automaton as an *AC-automaton*) from $\mathcal{D}$ and then uses this automaton to find the occurrences of all patterns in the text. The AC-automaton of $\mathcal{D}$ uses $O(m)$ space and can be constructed in $O(m \log |\Sigma|)$ time, where m is the sum of the length of all patterns in $\mathcal{D}$ and $|\Sigma|$ is the alphabet size. By using an AC-automaton, all occurrences of patterns in T can be found only by reading T once, which takes $O(n \log |\Sigma|)$ time. Similarly to the KMP algorithm, the AC algorithm is also extended to perform dictionary matching on some variant of strings [10–14]. In order to perform dictionary matching efficiently, the extended AC algorithms encode the patterns in a dictionary and create an automaton from the encoded patterns instead of the patterns itself.

In this paper, we propose a generalization of the Aho-Corasick algorithm for dictionary matching under SCER. The proposed algorithm encodes the patterns in the dictionary, and then constructs an automaton with a failure function called

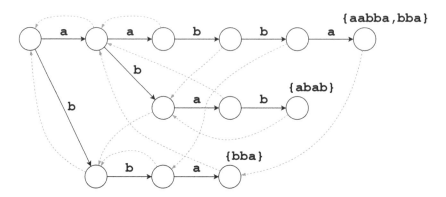

Fig. 1. The AC-automaton of $\mathcal{D} = \{\mathtt{aabba}, \mathtt{abab}, \mathtt{bba}\}$. The solid arrows represent the goto function, the dashed blue arrows represent the failure function, and the sets of strings represent the output function.

a *substring consistent equivalence relation automaton* (*SCER automaton*) from the encoded strings. We present an algorithm to construct SCER automata and show how to perform dictionary matching by using SCER automata. Suppose we can encode a string X in $\xi(|X|)$ time and re-encode $X[i : j][j - i + 1]$ in $\phi(|X[i : j]|)$ time, we show that the size of SCER automaton is $O(m)$ and can be constructed in $O(\xi(m) + m \cdot (\phi(\ell) + \log \pi))$ time, where m is the sum of the length of all patterns in the dictionary, ℓ is the length of the longest patterns in the dictionary, and π is the maximum number of possible outgoing transitions from any state. Moreover, we show that the dictionary matching under SCER can be performed in $O(\xi(n) + n \cdot (\phi(\ell) + \log \pi))$ time by using SCER automata, where n is the length of the text. By using our algorithm, we can perform dictionary matching under any SCER. Table 1 shows the time complexity of our algorithm on some dictionary matching problems.

2   Preliminaries

Let Σ and Π be integer alphabets, and Σ^* (resp. Π^*) be the set of all strings over Σ (resp. Π). The empty string ε is the string of length 0. We assume that the size of any symbol in Σ and Π is constant and a comparison of any two symbols in Σ or Π can be done in constant time in word RAM model. For a string $T \in \Sigma^*$, $|T|$ denotes the length of T. For $1 \leq i \leq j \leq |T|$, $T[i]$ denotes the i-th character of T and $T[i : j]$ denotes the *substring* of T that starts at i and ends at j. Let $T[: j] = T[1 : j]$ denote the *prefix* of T that ends at j and $T[i :] = T[i : |T|]$ denote the *suffix* of T that starts at i. For convenience we define $T[i : j] = \varepsilon$ if $j < i$. Note that ε is a substring, a prefix, and a suffix of any string. For a string T, let $\mathsf{Pref}(T)$ denote the set of all prefixes of P. For two strings X and Y, we denote by XY or $X \cdot Y$ the concatenation of X and Y.

Let $\mathcal{D} = \{P_1, P_2, ..., P_d\}$ be a set of patterns, called a *dictionary*. Let $|\mathcal{D}|$ denote the number of patterns in $\mathcal{D}$ and $\|\mathcal{D}\| = \Sigma_{k=1}^{|\mathcal{D}|} |P_k|$ denote the total length of the patterns in $\mathcal{D}$. For a dictionary $\mathcal{D}$, $\mathsf{Pref}(\mathcal{D}) = \bigcup_{k=1}^{|\mathcal{D}|} \mathsf{Pref}(P_k)$ is the set of all prefixes of the patterns in $\mathcal{D}$.

The Aho-Corasick automaton [1] $\mathsf{AC}(\mathcal{D})$ of $\mathcal{D}$ consists of a set of states and three functions: goto, failure, and output functions. Each state of $\mathsf{AC}(\mathcal{D})$ corresponds to a prefix in $\mathsf{Pref}(\mathcal{D})$. The goto function $\delta_{\mathcal{D}}$ of $\mathsf{AC}(\mathcal{D})$ is defined so that $\delta_{\mathcal{D}}(P_k[: i], P_k[i + 1]) = P_k[: i + 1]$, for any $P_k \in \mathcal{D}$ and $1 \le i < |P_k|$. The failure fuction $\mathsf{fail}_{\mathcal{D}}$ of $\mathsf{AC}(\mathcal{D})$ is defined so that $\mathsf{fail}_{\mathcal{D}}(P_k[: i]) = P_k[j : i]$ where $j = \min\{l \mid l > 1, P_k[l : i] \in \mathsf{Pref}(\mathcal{D})\}$. The output fuction $\mathsf{out}_{\mathcal{D}}$ of $\mathsf{AC}(\mathcal{D})$ is defined as $\mathsf{out}_{\mathcal{D}}(P_k[: j]) = \{P \in \mathcal{D} \mid P = P_k[i : j] \text{ for some } 1 \le i \le j\}$. Figure 1 shows an example of AC-automaton. We will define a generalization of AC-automata later in Sect. 3.

Next, we define the class of equivalence relations that we consider in this paper called *substring consistent equivalence relations*.

Definition 1 (Substring consistent equivalence relation (SCER) [17]). *An equivalence relation $\approx \subseteq \Sigma^* \times \Sigma^*$ is a substring consistent equivalence relation (SCER) if for two string X and Y, $X \approx Y$ implies $|X| = |Y|$ and $X[i : j] \approx Y[i : j]$ for all $1 \le i \le j \le |X|$.*

We say X $\approx$-matches Y iff $X \approx Y$. For instance, the matching relations in parameterized pattern matching [6], order-preserving pattern matching [14,16], and permuted pattern matching [13] are SCERs, while the matching relations in indeterminate string pattern matching [5] and function matching [2] are not.

Matsuoka et al. [17] define occurrences of a pattern in a text under an SCER $\approx$, which is used to define the pattern matching under SCERs as follows.

Definition 2 ($\approx$-occurrence). *For two strings T and P, a position i, $1 \le i \le |T| - |P| + 1$, is an $\approx$-occurrence of P in T iff $P \approx T[i : i + |P| - 1]$.*

By using the above definition we define the *dictionary matching under SCERs*.

Definition 3 (Dictionary matching under SCERs). *Given a dictionary $\mathcal{D} = \{P_1, P_2, \ldots, P_d\}$ and a text T, the dictionary matching with respect to an SCER $\approx$ ($\approx$-dictionary matching) is a task to find all $\approx$-occurrences of P_k in T for all $P_k \in \mathcal{D}$.*

In order to perform some variants of dictionary matching fast, encodings are used on strings. For instance, the prev-encoding is used for parameterized pattern matching [12] and the nearest neighbor encoding is used for order-preserving pattern matching [14]. Following the previous research, we generalize these encodings for SCERs as follows.

Definition 4 ($\approx$-prefix encoding). *Let Σ and Π be alphabets. We say an encoding function $f : \Sigma^* \to \Pi^*$ is a prefix encoding with respect to an SCER $\approx$ ($\approx$-prefix encoding) if (1) for a string X, $|X| = |f(X)|$, (2) $f(X[: i]) = f(X)[: i]$, and (3) for two strings X and Y, $f(X) = f(Y)$ iff $X \approx Y$.*

We can easily confirm that both the prev-encoding [6] and the nearest neighbor encoding [14] are prefix encodings. Amir and Kondratovsky [4] show that there exists a prefix encoding for any SCER. By using a $\approx$-prefix encoding, if $X[: i] \approx Y[: i]$ we can check whether $X[: i + 1] \approx Y[: i + 1]$ just by checking whether $f(X)[i + 1] = f(Y)[i + 1]$. Therefore, $\approx$-dictionary matching can be performed fast by using prefix encoded strings.

For a string P and prefix encoding f, let we denote $f(P)$ by $\langle P \rangle$ for simplicity. For a dictionary $\mathcal{D} = \{P_1, P_2, \ldots, P_d\}$, let $\langle \mathcal{D} \rangle = \{\langle P_1 \rangle, \langle P_2 \rangle, \ldots, \langle P_d \rangle\}$ and $\langle \mathsf{Pref}(\mathcal{D}) \rangle = \mathsf{Pref}(\langle \mathcal{D} \rangle) = \bigcup_{k=1}^{d} \mathsf{Pref}(\langle P_k \rangle)$.

Throughout the paper, let T be a text of length n, $\mathcal{D}$ be a dictionary, $d = |\mathcal{D}|$, $m = \|\mathcal{D}\|$, and $\ell = \max\{|P_k| \mid P_k \in \mathcal{D}\}$. Let Π^* be the co-domain of a $\approx$-prefix encoding. For a string X, suppose that $\langle X \rangle$ can be computed in $\xi(|X|)$ time. Assuming that $\langle X \rangle$ has been computed, suppose we can re-encode $\langle X[i : j] \rangle[j - i + 1]$ in $\phi(|X[i : j]|)$ time.

3   SCER Automata

In this section, we propose automata for the $\approx$-dictionary matching problem called *substring consistent equivalence relation automata (SCERAs)*. First, we describe the definition and properties of the SCERA for a dictionary $\mathcal{D}$, then show the size of the SCERA of $\mathcal{D}$ with respect to the size of $\mathcal{D}$. After that, we propose a $\approx$-dictionary matching algorithm by using SCERAs and show the time complexity of the proposed algorithm. Last, we present an algorithm to construct SCERAs and show its time complexity.

3.1   Definition and Properties

For a dictionary $\mathcal{D} = \{P_1, P_2, \ldots, P_d\}$, the substring consistent equivalence relation automaton $\mathsf{SCERA}(\mathcal{D})$ of $\mathcal{D}$ consists of a set of states and three functions, namely *goto*, *failure*, and *output* functions.

The set of states $\mathcal{S}_\mathcal{D}$ of $\mathsf{SCERA}(\mathcal{D})$ defined as follows.

$$\mathcal{S}_\mathcal{D} = \{S \mid S \in \mathsf{Pref}(\langle \mathcal{D} \rangle)\}$$

Each state of $\mathsf{SCERA}(\mathcal{D})$ corresponds to a prefix of $\langle P_k \rangle$ for some $P_k \in \mathcal{D}$, thus we can identify each state by the corresponded prefix. Since the number of prefixes of P_k is $|P_k| + 1$ and $|P_k| = |\langle P_k \rangle|$, the number of states of $\mathsf{SCERA}(\mathcal{D})$ is as follows.

Lemma 1. *For a dictionary $\mathcal{D}$, the number of states $|\mathcal{S}_\mathcal{D}|$ of $\mathsf{SCERA}(\mathcal{D})$ is $O(m)$.*

Next, we define the functions in $\mathsf{SCERA}(\mathcal{D})$. First, the goto function $\delta_\mathcal{D}$ of $\mathsf{SCERA}(\mathcal{D})$ is defined as follows.

Definition 5 (Goto function). *The goto function* $\delta_{\mathcal{D}} : \mathcal{S}_{\mathcal{D}} \cdot \Pi \to \mathcal{S}_{\mathcal{D}} \cup \{\mathsf{NULL}\}$ *of* $\mathsf{SCERA}(\mathcal{D})$ *is defined by*

$$\delta_{\mathcal{D}}(S, c) = \begin{cases} S \cdot c & \text{if } (S \cdot c) \in \mathcal{S}_{\mathcal{D}}, \\ \mathsf{NULL} & \text{otherwise.} \end{cases}$$

Intuitively, $\delta_{\mathcal{D}}(\langle P_k \rangle[: j], \langle P_k \rangle[j+1]) = \langle P_k \rangle[: j+1]$, for any $P_k \in \mathcal{D}$ and $0 \leq j < |P_k|$. The states and goto function form a trie of all encoded patterns in $\langle \mathcal{D} \rangle$. For two states S and S' such that $S' = \delta_{\mathcal{D}}(S, c)$ for some $c \in \Pi$, we call S *the parent* of S' and S' *a child* of S. For convenience, for a state S and a string $X \in \Pi^*$, let $\delta_{\mathcal{D}}(S, X) = \delta_{\mathcal{D}}(\delta_{\mathcal{D}}(S, X[1]), X[2 :])$ and $\delta_{\mathcal{D}}(S, \varepsilon) = S$. Here we denote by π, the maximum number of possible outgoing transitions from any state.

Next, the failure function $\mathsf{fail}_{\mathcal{D}}$ of $\mathsf{SCERA}(\mathcal{D})$ is defined as follows.

Definition 6 (Failure function). *The failure function* $\mathsf{fail}_{\mathcal{D}} : \mathcal{S}_{\mathcal{D}} \to \mathcal{S}_{\mathcal{D}} \cup \{\mathsf{NULL}\}$ *of* $\mathsf{SCERA}(\mathcal{D})$ *is defined by* $\mathsf{fail}_{\mathcal{D}}(\varepsilon) = \mathsf{NULL}$ *and* $\mathsf{fail}_{\mathcal{D}}(\langle P_k \rangle[: j]) = \langle P_k[i : j] \rangle$ *for* $1 \leq j \leq |P_k|$, *where* $i = \min\{l \mid l > 1, \langle P_k[l : j] \rangle \in \mathsf{Pref}(\langle \mathcal{D} \rangle)\}$.

In other words, for any state $\langle P_k \rangle[: j]$, $\mathsf{fail}_{\mathcal{D}}(\langle P_k \rangle[: j]) = \langle P_k[i : j] \rangle$ iff $P_k[i : j]$ is the longest suffix of $P_k[: j]$ such that $\langle P_k[i : j] \rangle \in \mathsf{Pref}(\langle \mathcal{D} \rangle)$. Moreover, by the definition of prefix encoding, $P_k[i : j]$ is also the longest suffix of $P_k[: j]$ that $\approx$-matches a prefix in $\mathsf{Pref}(\mathcal{D})$. For convenience, assume that $\mathsf{fail}_{\mathcal{D}}(\mathsf{NULL}) = \mathsf{NULL}$, we define $\mathsf{fail}_{\mathcal{D}}$ recursively, namely $\mathsf{fail}_{\mathcal{D}}^k(S) = \mathsf{fail}_{\mathcal{D}}^{k-1}(\mathsf{fail}(S))$ and $\mathsf{fail}_{\mathcal{D}}^0(S) = S$.

The failure function has the following properties.

Lemma 2. *For any state* $\langle P_k \rangle[: j]$, *if* $\langle P_k[i : j] \rangle$ *is a state, there is* $q \geq 0$ *such that* $\mathsf{fail}_{\mathcal{D}}^q(\langle P_k \rangle[: j]) = \langle P_k[i : j] \rangle$.

Proof. Straightforward by induction on j. ☐

Lemma 3. *Consider two states* $\langle P_k \rangle[: j]$ *and* $\langle P_k \rangle[: j+1]$. *If* $\mathsf{fail}_{\mathcal{D}}(\langle P_k \rangle[: j+1]) \neq \varepsilon$, *there is* $q \geq 0$ *such that* $\mathsf{fail}_{\mathcal{D}}^q(\langle P_k \rangle[: j])$ *is the parent of* $\mathsf{fail}_{\mathcal{D}}(\langle P_k \rangle[: j+1])$.

Proof. Let $\langle P_k[i : j+1] \rangle = \mathsf{fail}_{\mathcal{D}}(\langle P_k \rangle[: j+1])$. From the condition $\mathsf{fail}_{\mathcal{D}}(\langle P_k \rangle[: j+1]) \neq \varepsilon$, we have $i \leq j+1$. Since $\langle P_k[i : j+1] \rangle \in \mathsf{Pref}(\langle \mathcal{D} \rangle)$, clearly $\langle P_k[i : j] \rangle \in \mathsf{Pref}(\langle \mathcal{D} \rangle)$ and $\langle P_k[i : j] \rangle$ is the parent of $\langle P_k[i : j+1] \rangle$. By Lemma 2, $\langle P_k[i : j] \rangle = \mathsf{fail}_{\mathcal{D}}^q(\langle P_k \rangle[: j])$ for some $q \geq 0$. ☐

Lemma 2 implies that any re-encoded suffix $\langle P_k[i : j] \rangle$ of $\langle P_k \rangle[: j]$ such that $\langle P_k[i : j] \rangle \in \mathsf{Pref}(\langle \mathcal{D} \rangle)$ can be found by executing failure function recursively starting from $\mathsf{fail}_{\mathcal{D}}(\langle P_k \rangle[: j])$. Moreover, Lemma 3 implies that for any state $\langle P_k \rangle[: j+1]$ such that $\mathsf{fail}_{\mathcal{D}}(\langle P_k \rangle[: j+1]) \neq \varepsilon$, the parent of $\mathsf{fail}_{\mathcal{D}}(\langle P_k \rangle[: j+1])$ can be found by executing failure function recursively starting from $\mathsf{fail}_{\mathcal{D}}(\langle P_k \rangle[: j])$.

Last, the output function $\mathsf{out}_{\mathcal{D}}$ of $\mathsf{SCERA}(\mathcal{D})$ is defined as follows.

Definition 7 (Output function). *The output function* $\mathsf{out}_{\mathcal{D}}$ *of* $\mathsf{SCERA}(D)$ *is defined by* $\mathsf{out}_{\mathcal{D}}(\langle P_k \rangle[: j]) = \{P \in \mathcal{D} \mid P \approx P_k[i : j] \text{ for some } 1 \leq i \leq j\}$.

For a state $\langle P_k \rangle[: j]$, $\mathsf{out}_\mathcal{D}(\langle P_k \rangle[: j])$ is the set patterns that $\approx$-match some suffix of $P_k[: j]$.

The output function has the following properties.

Lemma 4. *For any* $P \in \mathsf{out}_\mathcal{D}(\langle P_k \rangle[: j])$, $\langle P \rangle = \mathsf{fail}_\mathcal{D}^q(\langle P_k \rangle[: j])$ *for some* $q \geq 0$.

Proof. From the definition of $\mathsf{out}_\mathcal{D}$ and prefix encoding, $P \approx P_k[i : j]$ and $\langle P \rangle = \langle P_k[i : j] \rangle$ for some i. By Lemma 2, $\langle P_k[i : j] \rangle = \mathsf{fail}_\mathcal{D}^q(\langle P_k \rangle[: j])$ for some $q \geq 0$. □

Lemma 5. *For any state* $S \neq \varepsilon$, *if* $S \in \langle \mathcal{D} \rangle$, $\mathsf{out}_\mathcal{D}(S) = \{S\} \cup \mathsf{out}_\mathcal{D}(\mathsf{fail}_\mathcal{D}(S))$. *Otherwise,* $\mathsf{out}_\mathcal{D}(S) = \mathsf{out}_\mathcal{D}(\mathsf{fail}_\mathcal{D}(S))$.

Proof. Assume there is a pattern P such that $P \in \mathsf{out}_\mathcal{D}(\mathsf{fail}_\mathcal{D}(S))$ but $P \notin \mathsf{out}_\mathcal{D}(S)$. By Lemma 4, there exists q such that $\langle P \rangle = \mathsf{fail}_\mathcal{D}^q(\mathsf{fail}_\mathcal{D}(S))$. Since $\mathsf{fail}_\mathcal{D}^q((\mathsf{fail}_\mathcal{D}(S)) = \mathsf{fail}_\mathcal{D}^{q+1}(S)$, we have $P \in \mathsf{out}_\mathcal{D}(S)$ which contradicts the assumption.

Next, assume there exists a pattern P such that $P \in \mathsf{out}_\mathcal{D}(S)$ but $P \notin \mathsf{out}_\mathcal{D}(\mathsf{fail}_\mathcal{D}(S))$. By Lemma 4, there is q such that $\langle P \rangle = \mathsf{fail}_\mathcal{D}^q(S)$. If $q > 0$, $\langle P \rangle = \mathsf{fail}_\mathcal{D}^{q-1}(\mathsf{fail}(S))$ implies $P \in \mathsf{out}_\mathcal{D}(\mathsf{fail}_\mathcal{D}(S))$ which contradicts the assumption. Therefore, the remaining possibility is $q = 0$ which implies $S = \langle P \rangle \in \langle \mathcal{D} \rangle$. □

Lemma 5 implies that we can compute $\mathsf{out}_\mathcal{D}(S)$ by copying $\mathsf{out}_\mathcal{D}(\mathsf{fail}_\mathcal{D}(S))$ and adding S if $S \in \mathcal{D}$. We will utilize Lemma 5 to construct the output function efficiently.

Implementation and Space Complexity

We will describe how to implement SCER automata and show its space complexity. First, The goto function $\delta_\mathcal{D}$ can be implemented by using an associative array on each state. We have the following lemma for the required space and time to implement the goto function of $\mathsf{SCERA}(\mathcal{D})$.

Lemma 6. *Assume that the size of any symbol in* Π *is constant. The goto function* $\delta_\mathcal{D}$ *of* $\mathsf{SCERA}(\mathcal{D})$ *can be implemented in* $O(m)$ *space.*

Proof. The number of associative arrays used to implement $\delta_\mathcal{D}$ is $O(|\mathcal{S}|)$. Since for each S' there only exists one pair $(S.c) \in \mathcal{S} \cdot \Pi$ such that $\delta(S, c) = S'$, the total size of associative arrays is $O(|\mathcal{S}|)$. Therefore, $\delta_\mathcal{D}$ can be implemented in $O(m)$ space by Lemma 1. □

Next, the failure function can be implemented by using a state pointer on each state.

Lemma 7. *For a dictionary* $\mathcal{D}$, $\mathsf{fail}_\mathcal{D}$ *can be implemented in* $O(m)$ *space.*

Proof. Since $\mathsf{fail}_\mathcal{D}$ is defined for each state, $\mathsf{fail}_\mathcal{D}$ can be implemented using $O(|\mathcal{S}|)$ space. Therefore, $\mathsf{fail}_\mathcal{D}$ can be implemented in $O(m)$ space by Lemma 1. □

Last, similarly to the original AC-automata, the output function $\mathsf{out}_\mathcal{D}$ can be implemented in linear space by using a list.

Algorithm 1: A $\approx$-dictionary matching algorithm using SCERA($\mathcal{D}$)

---

1 **Function** Matching(T)
2   compute $\langle T \rangle$;
3     $v \leftarrow root$;
4   **for** $i \leftarrow 1$ **to** $|T|$ **do**
5   **while** $\delta(v, \langle T[i - \mathsf{dep}(v) : i]\rangle[\mathsf{dep}(v) + 1]) = $ NULL **do** $v \leftarrow \mathsf{fail}(v)$;
6         $v \leftarrow \delta(v, \langle T[i - \mathsf{dep}(v) : i]\rangle[\mathsf{dep}(v) + 1])$;
7   **for** $j \in \mathsf{out}(v)$ **do** **output** $(j, i - |P_j| + 1)$;

Lemma 8. *For a dictionary $\mathcal{D}$, $\mathsf{out}_{\mathcal{D}}$ can be implemented in $O(m)$ space.*

Proof. Each state S stores a pair (i, p), where i is the pattern number if $S \in \langle \mathcal{D} \rangle$ or NULL otherwise and p is a pointer to a state $S' = \mathsf{fail}_{\mathcal{D}}^q(\langle P_k \rangle[: j])$ for the smallest $q > 0$ such that $S' \in \langle \mathcal{D} \rangle$ if it exists or NULL otherwise. Since $|\mathcal{S}|$ is $O(m)$ by Lemma 1, $\mathsf{out}_{\mathcal{D}}$ can be implemented in $O(m)$ space. □

From Lemmas 1, 6, 7, and 8, we get the following theorem.

Theorem 1. *Assume that the size of any symbol in Π is constant. For a dictionary $\mathcal{D} = \{P_1, P_2, \ldots, P_d\}$ of total size $\|\mathcal{D}\| = m$, SCERA($\mathcal{D}$) can be implemented in $O(m)$ space.*

3.2 Dictionary Matching Using SCERA

In this section, we describe how to use SCER automata for dictionary matching. The proposed algorithm for $\approx$-dictionary matching is shown in Algorithm 1. For any state S, let $\mathsf{dep}(S)$ be the depth of S i.e. the length of the shortest path from $root$ to S. In order to simplify the algorithm we use an auxiliary state $\perp$ where $\delta(\perp, c) = root$ for any $c \in \Pi$, $\mathsf{fail}(root) = \perp$, and $\mathsf{dep}(\perp) = 0$.

The algorithm starts with $root$ as the active state and 1 as the active position. The algorithm reads the encoded text from the left to the right, while updating the active state and the active position. Let $v = \langle T[i - \mathsf{dep}(v) : i - 1]\rangle$ be the active state and i be the active position. The algorithm finds $\langle T[i - \mathsf{dep}(v) : i]\rangle[\mathsf{dep}(v)]$ transition from v. If $\langle T[i - \mathsf{dep}(v) : i]\rangle[\mathsf{dep}(v)]$ transition exists, the algorithm updates the active state to $\delta(v, \langle T[i - \mathsf{dep}(v) : i]\rangle[\mathsf{dep}(v)])$ and increments i. After the algorithm updates the active state, it outputs all patterns in $\mathsf{out}(v)$. Otherwise if $\langle T[i - \mathsf{dep}(v) : i]\rangle[\mathsf{dep}(v)]$ transition does not exist, the algorithm updates the active state to $\mathsf{fail}(v)$ without updating i. The algorithm repeats these operations until it reads all the text.

Lemma 9. *Let $v = \langle T[i - \mathsf{dep}(v) : i - 1]\rangle$ be the active state and i be the active position. $T[i - \mathsf{dep}(v) : i - 1]$ is the longest suffix of $T[: i - 1]$ such that $v \in \mathsf{Pref}(\langle \mathcal{D} \rangle)$.*

Proof. We will prove by induction. Initially, $v = root$ and $i = 1$, thus $v = \varepsilon$ is the longest suffix of $\langle T[: 0] \rangle = \varepsilon$. Assume that $v = T[i - \mathsf{dep}(v) : i - 1]$ is the longest suffix of $T[: i - 1]$ such that $v \in \mathsf{Pref}(\langle \mathcal{D} \rangle)$. Let u be the next active state and i be the next active position. From the algorithm, $u = \delta(\mathsf{fail}^q(v), \langle T[i - \mathsf{dep}(\mathsf{fail}^q(v)) : i] \rangle[\mathsf{dep}(\mathsf{fail}^q(v)) + 1])$, where q is the smallest integer such that $\delta(\mathsf{fail}^q(v), \langle T[i - \mathsf{dep}(\mathsf{fail}^q(v)) : i] \rangle[\mathsf{dep}(\mathsf{fail}^q(v)) + 1]) \neq \mathsf{NULL}$. Let $T[i - j : i]$ be the longest suffix of $T[: i]$ such that $\langle T[i - j : i] \rangle \in \mathsf{Pref}(\langle \mathcal{D} \rangle)$. If $\langle T[i - j : i] \rangle = \varepsilon$, we have $\delta(\langle T[i - l : i - 1] \rangle, \langle T[i - l : i] \rangle[l + 1]) = \mathsf{NULL}$ for any l, $0 \leq l \leq \mathsf{dep}(v)$, such that $\langle T[i - l : i - 1] \rangle \in \mathsf{Pref}(\langle \mathcal{D} \rangle)$. Thus, we have $\mathsf{fail}^q(v) = \bot$ and $u = root$. Otherwise if $\langle T[i - j : i] \rangle \neq \varepsilon$, we have $\langle T[i - j : i - 1] \rangle \in \mathsf{Pref}(\langle \mathcal{D} \rangle)$. Moreover, $\langle T[i-j : i-1] \rangle = \mathsf{fail}^{q'}(v)$ for some q' by Lemma 2 and $\delta(\langle T[i-j : i-1] \rangle, \langle T[i-j : i] \rangle[j + 1]) \neq \mathsf{NULL}$. Since $T[i - j : i]$ is the longest suffix of $T[: i]$ such that $\langle T[i - j : i] \rangle \in \mathsf{Pref}(\langle \mathcal{D} \rangle)$, $\delta(\langle T[i - l : i - 1] \rangle, \langle T[i - l : i] \rangle[l + 1]) = \mathsf{NULL}$, for any l, $j < l \leq \mathsf{dep}(v)$, such that $\langle T[i - l : i - 1] \rangle \in \mathsf{Pref}(\langle \mathcal{D} \rangle)$. Therefore, $q' = q$ which implies the correctness of Lemma 9. □

Theorem 2. *Given* $\mathsf{SCERA}(\mathcal{D})$ *and a text* T *of length* n, *Algorithm 1 outputs all occurrence positions of all patterns in* T *correctly in* $O(\xi(n) + n \cdot (\phi(\ell) + \log \pi) + occ)$ *time, where* ℓ *is the length of the longest pattern in* $\mathcal{D}$, $\xi(n)$ *is the time required to encode* T, $\phi(\ell)$ *is the time required to re-encode a symbol of a substring of* T *whose length is* ℓ *or less,* π *is the maximum number of possible outgoing transitions from any state, and* occ *is the number of occurrences of the patterns in* T.

Proof. First, we show the correctness of the algorithm. Assume there is an occurrence position o of P_k in T that has not been output by the algorithm. Let $o + |P_k|$ be the active position. By Lemma 9, the active state $v = \langle T[j : o + |P_k| - 1] \rangle$ is the longest suffix of $T[: o + |P_k| - 1]$ such that $\langle T[j : o + |P_k| - 1] \rangle \in \mathsf{Pref}(\langle D \rangle)$. Since $\langle T[o : o + |P_k| - 1] \rangle \in \mathsf{Pref}(\langle D \rangle)$, we have $j \leq o$. By the definition of the output function, $P_k \in \mathsf{out}(\langle T[j : o + |P_k| - 1] \rangle)$. Therefore, the algorithm outputs $P_k \in \mathsf{out}(\langle T[j : o + |P_k| - 1] \rangle)$ which contradicts the assumption. Moreover, for any active state $\langle T[i : j] \rangle$ the algorithm only outputs an occurrence position of P_k iff $P_k \approx$-matches a suffix of $T[i : j]$ by Lemma 5.

Next, we show the time complexity of the algorithm. The encoding $\langle T \rangle$ can be computed in $\xi(n)$. For each position i, the depth of the active position is increased by one, thus the depth of the active position is increased by n in total. Since the depth of the active position is decreased by at least one each time fail is executed, fail is executed at most n times. Next, each time δ is executed, either the depth of the active position is increased by one or fail is executed, thus δ is executed $O(n)$ times. Since we need to re-encode a symbol each time δ is executed and δ can be executed in $O(\log \pi)$ time by binary search, the algorithm takes $O(n \cdot (\phi(\ell) + \log \pi))$ time to execute δ in total. In order to output the occurrence positions, the algorithm takes $O(n)$ time to check whether there is any occurrence and $O(occ)$ time to output the occurrence positions. □

Algorithm 2: Computing goto function of SCERA($\mathcal{D}$)

```
1  Function constructGoto(𝒟)
2  |   compute ⟨𝒟⟩;
3  |   create states root and ⊥;
4  |   dep(root) ← 0; dep(⊥) ← 0;
5  |   δ(⊥, c) ← root for any symbol c;
6  |   for k ← 1 to d do
7  |   |   v ← root;
8  |   |   for j ← 1 to |Pₖ| do
9  |   |   |   if δ(v, ⟨Pₖ⟩[j]) ≠ NULL then  v ← δ(v, ⟨Pₖ⟩[j]);
10 |   |   |   else
11 |   |   |   |   create a state u;
12 |   |   |   |   δ(v, ⟨Pₖ⟩[j]) ← u;
13 |   |   |   |   label(u) ← i; dep(u) ← dep(v) + 1;
14 |   |   |   |   v ← u;
```

Algorithm 3: Computing failure function of SCERA($\mathcal{D}$)

```
1  Function constructFailure(𝒟)
2  |   compute ⟨𝒟⟩; fail(root) ← ⊥; push root to queue;
3  |   while queue ≠ ∅ do
4  |   |   pop v from queue;
5  |   |   for c such that δ(v, c) ≠ NULL do
6  |   |   |   u ← δ(v, c); push u to queue;
7  |   |   |   s ← fail(v); k ← label(u);
8  |   |   |   while δ(s, ⟨Pₖ[dep(u) − dep(s) : dep(u)]⟩[dep(s) + 1]) = NULL do
9  |   |   |   |   s ← fail(s);
10 |   |   |   fail(u) ← δ(s, ⟨Pₖ[dep(u) − dep(s) : dep(u)]⟩[dep(s) + 1)]));
```

3.3 Constructing SCERA

In this section, we describe an algorithm to construct SCERA($\mathcal{D}$). We divide the algorithm into three parts: goto function, failure function, and output function construction algorithms.

First, the goto function construction algorithm is shown in Algorithm 2. Initially, the algorithm computes $\langle P_k \rangle$ for all $P_k \in \mathcal{D}$; then it constructs the root state $root$ and the auxiliary state $\perp$. Next, for each pattern $\langle P_k \rangle \in \langle \mathcal{D} \rangle$, the algorithm finds the longest prefix of $\langle P_k \rangle$ that exists in the current automaton. After that, the algorithm creates states corresponding to the remaining prefixes from the shortest to the longest. After creating each state, the algorithm updates the goto function, adds a label i to the state, and compute the depth of the state.

Lemma 10. *Given a dictionary $\mathcal{D}$, Algorithm 2 constructs the goto function of* SCERA($\mathcal{D}$) *in* $O(\xi(m) + m \log \pi)$ *time.*

Algorithm 4: Computing output function of SCERA($\mathcal{D}$)

---

1 **Function** constructOutput($\mathcal{D}$)
2   compute $\langle \mathcal{D} \rangle$;
3   **for** $k \leftarrow 1$ **to** d **do**
4         $v \leftarrow root$;
5   **for** $j \leftarrow 1$ **to** $|P_k|$ **do**
6             $v \leftarrow \delta(v, \langle P_k \rangle[j])$;
7   **if** $j = |P_k|$ **then** out(v) $\leftarrow$ out(v) $\cup \{k\}$;
8   push $root$ to $queue$;
9   **while** $queue \neq \emptyset$ **do**
10   pop v from $queue$; out(v) $\leftarrow$ out(v) $\cup$ out(fail(v));
11   **for** c such that $\delta(v, c) \neq$ NULL **do**
12             $u \leftarrow \delta(v, c)$; push u to $queue$;

Proof. By assumption, each pattern $|P_k|$ in the dictionary can be encoded in $\xi(|P_k|)$ time. The operations in the inner loop are executed $O(m)$ times. $\delta(v, c)$ can be computed in $O(\log \pi)$ by binary search. $\qquad \square$

Next, we describe how to compute the failure function of SCERA($\mathcal{D}$). Algorithm 3 shows the algorithm for computing the failure function. The algorithm computes the failure function recursively by the breadth-first search. The algorithm uses the property in Lemma 3 to compute the failure function.

Consider computing fail($\langle P_k \rangle[: j]$). Since the algorithm computing fail by the breadth-first search, fail($\langle P_k \rangle[: j - 1]$) has been computed. By Lemma 3, there is $q \geq 1$ such that $\mathsf{fail}_{\mathcal{D}}^q(\langle P_k \rangle[: j - 1])$ is the parent of $\mathsf{fail}_{\mathcal{D}}(\langle P_k \rangle[: j])$ or $\mathsf{fail}_{\mathcal{D}}(\langle P_k \rangle[: j]) = \varepsilon$. We can find q by executing the failure function recursively from $\langle P_k \rangle[: j - 1]$ and checking whether $\delta(\langle P_k[i : j] \rangle, \langle P_k[i :] \rangle[j - i + 1]) = $ NULL.

Lemma 11. *Given a dictionary $\mathcal{D}$ and goto function of SCERA($\mathcal{D}$), Algorithm 3 constructs the failure function of SCERA($\mathcal{D}$) in $O(\xi(m) + m \cdot (\phi(\ell) + \log \pi))$ time.*

Proof. The dictionary can be encoded in $\xi(m)$ time. The running time of Algorithm 3 is bounded by the number of executions of fail. Let $x_{k,j}$ be the number of executions of fail when finding fail($\langle P_k[: j] \rangle$). Since $x_{k,j} \leq \mathsf{dep}(\mathsf{fail}(\langle P_k[: j - 1] \rangle)) - \mathsf{dep}(\mathsf{fail}(\langle P_k[: j] \rangle)) + 2$, we can compute $\Sigma_{k=1}^d \Sigma_{j=1}^{|P_k|} x_{k,j} \leq \Sigma_{k=1}^d 2|P_k| = 2m$. The goto function δ is executed each time fail be executed. Since we need to re-encode a substring each time δ be executed and δ can be executed in $O(\log \pi)$ time, the algorithm takes $O(\xi(m) + m \cdot (\phi(\ell) + \log \pi))$ time in total. $\qquad \square$

Last, Algorithm 4 shows an algorithm to compute the output function of SCERA($\mathcal{D}$). The algorithm first adds k to out($\langle P_k \rangle$) for each $P_k \in \mathcal{D}$. Next, the algorithm updates the output function recursively by the breadth-first search. The algorithm uses the property in Lemma 5 to compute the output function.

Consider computing $\mathsf{out}(\langle P_k \rangle[: j])$. Since the algorithm computes out by the breadth-first search, $\mathsf{out}(\mathsf{fail}(\langle P_k \rangle[: j]))$ has been computed. By Lemma 5, we can compute $\mathsf{out}(\langle P_k \rangle[: j])$ by just adding k to $\mathsf{out}(\mathsf{fail}(\langle P_k \rangle[: j]))$ if $j = |P_k|$ or by copying $\mathsf{out}(\mathsf{fail}(\langle P_k \rangle[: j]))$ if $j \neq |P_k|$. Note that it can be done efficiently by using pointers as described in Sect. 3.1 instead of copying the set.

Lemma 12. *Given a dictionary $\mathcal{D}$, the goto function, and the failure function of $\mathsf{SCERA}(\mathcal{D})$, Algorithm 4 constructs the output function of $\mathsf{SCERA}(\mathcal{D})$ in $O(\xi(m) + m \cdot \log \pi)$ time.*

Proof. The dictionary can be encoded in $\xi(m)$ time. Clearly the loops are executed $O(|\mathcal{S}|)$ times in total. The goto function can be executed in $O(\log \pi)$ time. Therefore, Algorithm 4 runs in $O(\xi(m) + m \cdot \log \pi)$ time. $\qquad \square$

From Lemmas 10, 11, and 12, we get the following theorem.

Theorem 3. *Given a dictionary $\mathcal{D}$, $\mathsf{SCERA}(\mathcal{D})$ can be constructed in $O(\xi(m) + m \cdot (\phi(\ell) + \log \pi))$ time, where ℓ is the length of the longest pattern in $\mathcal{D}$, $\xi(m)$ is the time required to encode $\mathcal{D}$, $\phi(\ell)$ is the time required to re-encode the last symbol of a substring of $P_k \in \mathcal{D}$, and π is the maximum number of possible outgoing transitions from any state.*

References

1. Aho, A.V., Corasick, M.J.: Efficient string matching: an aid to bibliographic search. Commun. ACM **18**(6), 333–340 (1975)
2. Amir, A., Aumann, Y., Lewenstein, M., Porat, E.: Function matching. Soc. Ind. Appl. Math. **35**(5), 1007–1022 (2006)
3. Amir, A., Farach, M., Muthukrishnan, S.: Alphabet dependence in parameterized matching. Inf. Process. Lett. **49**(3), 111–115 (1994)
4. Amir, A., Kondratovsky, E.: Sufficient conditions for efficient indexing under different matchings. In: 30th Annual Symposium on Combinatorial Pattern Matching (CPM 2019), pp. 6:1–6:12 (2019)
5. Antoniou, P., Crochemore, M., Iliopoulos, C., Jayasekera, I., Landau, G.: Conservative string covering of indeterminate strings. In: Prague Stringology Conference, vol. 2008, pp. 108–115 (2008)
6. Baker, B.S.: Parameterized pattern matching: algorithms and applications. J. Comput. Syst. Sci. **52**(1), 28–42 (1996)
7. Crochemore, M., Rytter, W.: Text Algorithm. Oxford University Press, Oxford (1994)
8. Crochemore, M., Rytter, W.: Jewels of Stringology. World Scientific Publishing Co. Pte. Ltd., Singapore (2002)
9. Diptarama, U.Y., Narisawa, K., Shinohara, A.: KMP based pattern matching algorithms for multi-track strings. In: Proceedings of Student Research Forum Papers and Posters at SOFSEM 2016, pp. 100–107 (2016)
10. Diptarama, Y.U., Yoshinaka, R., Shinohara, A.: Fast full permuted pattern matching algorithms on multi-track strings. In: Prague Stringology Conference (PSC) 2016, pp. 7–21 (2016)

11. Hendrian, D., Ueki, Y., Narisawa, K., Yoshinaka, R., Shinohara, A.: Permuted pattern matching algorithms on multi-track strings. Algorithms **12**(4), 73:1–73:20 (2019)
12. Idury, R.M., Schäffer, A.A.: Multiple matching of parameterized patterns. Theor. Comput. Sci. **154**(2), 203–224 (1996)
13. Katsura, T., Narisawa, K., Shinohara, A., Bannai, H., Inenaga, S.: Permuted pattern matching on multi-track strings. In: van Emde Boas, P., Groen, F.C.A., Italiano, G.F., Nawrocki, J., Sack, H. (eds.) SOFSEM 2013. LNCS, vol. 7741, pp. 280–291. Springer, Heidelberg (2013). https://doi.org/10.1007/978-3-642-35843-2_25
14. Kim, J., et al.: Order-preserving matching. Theor. Comput. Sci. **525**, 68–79 (2014)
15. Knuth, D.E., Morris Jr., J.H., Pratt, V.R.: Fast pattern matching in strings. SIAM J. Comput. **6**(2), 323–350 (1977)
16. Kubica, M., Kulczyński, T., Radoszewski, J., Rytter, W., Waleń, T.: A linear time algorithm for consecutive permutation pattern matching. Inf. Process. Lett. **113**(12), 430–433 (2013)
17. Matsuoka, Y., Aoki, T., Inenaga, S., Bannai, H., Takeda, M.: Generalized pattern matching and periodicity under substring consistent equivalence relations. Theor. Comput. Sci. **656**, 225–233 (2016)
18. Park, S.G., Amir, A., Landau, G.M., Park, K.: Cartesian tree matching and indexing. In: 30th Annual Symposium on Combinatorial Pattern Matching (CPM 2019), pp. 16:1–16:14 (2019)

Reconfiguring k-path Vertex Covers

Duc A. Hoang[1] , Akira Suzuki[2] , and Tsuyoshi Yagita[1](✉)

[1] Kyushu Institute of Technology, Fukuoka, Japan
hoanganhduc@ces.kyutech.ac.jp, yagita.tsuyoshi307@mail.kyutech.jp
[2] Tohoku University, Sendai, Japan
a.suzuki@ecei.tohoku.ac.jp

Abstract. A vertex subset I of a graph G is called a k-path vertex cover if every path on k vertices in G contains at least one vertex from I. The k-PATH VERTEX COVER RECONFIGURATION (k-PVCR) problem asks if one can transform one k-path vertex cover into another via a sequence of k-path vertex covers where each intermediate member is obtained from its predecessor by applying a given reconfiguration rule exactly once. We investigate the computational complexity of k-PVCR from the viewpoint of graph classes under the well-known reconfiguration rules: TS, TJ, and TAR. The problem for $k = 2$, known as the VERTEX COVER RECONFIGURATION (VCR) problem, has been well-studied in the literature. We show that certain known hardness results for VCR on different graph classes including planar graphs, bounded bandwidth graphs, chordal graphs, and bipartite graphs, can be extended for k-PVCR. In particular, we prove a complexity dichotomy for k-PVCR on general graphs: on those whose maximum degree is 3 (and even planar), the problem is PSPACE-complete, while on those whose maximum degree is 2 (i.e., paths and cycles), the problem can be solved in polynomial time. Additionally, we also design polynomial-time algorithms for k-PVCR on trees under each of TJ and TAR. Moreover, on paths, cycles, and trees, we describe how one can construct a reconfiguration sequence between two given k-path vertex covers in a yes-instance. In particular, on paths, our constructed reconfiguration sequence is shortest.

Keywords: Combinatorial Reconfiguration · Computational complexity · k-path vertex cover · PSPACE-completeness · Polynomial-time algorithms

1 Introduction

For the last decade, a collection of problems called *Combinatorial Reconfiguration* has been extensively studied. Work in this research area specifically aims to model dynamic situations where one needs to transform one feasible solution of a computational problem into another by locally changing a solution while

This work is partially supported by JSPS KAKENHI Grant Numbers JP17K12636, JP18H04091, and JP19K24349, and JST CREST Grant Number JPMJCR1402.

© Springer Nature Switzerland AG 2020
M. S. Rahman et al. (Eds.): WALCOM 2020, LNCS 12049, pp. 133–145, 2020.
https://doi.org/10.1007/978-3-030-39881-1_12

keeping its feasibility along the way. In a reconfiguration setting, two feasible solutions of a computational problem (e.g., SATISFIABILITY, INDEPENDENT SET, VERTEX COVER, DOMINATING SET, etc.) are given, along with a *reconfiguration rule* that describes an *adjacency relation* between solutions. A *reconfiguration problem* asks whether one feasible solution can be transformed into the other via a sequence of adjacent feasible solutions where each intermediate member is obtained from its predecessor by applying the given reconfiguration rule exactly once. Such a sequence, if exists, is called a *reconfiguration sequence*. One may recall the classic Rubik's cube puzzle as an example of a reconfiguration problem, where each configuration of the Rubik's cube corresponds to a feasible solution, and two configurations (solutions) are adjacent if one can be obtained from the other by rotating a face of the cube by either 90, 180, or 270°. The question is whether one can transform an arbitrary configuration to the one where each face of the cube has only one color. For an overview of this research area, readers are referred to the recent surveys by van den Heuvel [16] and Nishimura [22].

The k-Path Vertex Cover Reconfiguration Problem. Let $G = (V, E)$ be a simple graph. A *vertex cover* of G is a subset I of V where each edge contains at least one vertex from I. The VERTEX COVER (VC) problem, which asks whether there is a vertex cover of G whose size is at most some positive integer s, is one of the classic NP-complete problems in the computational complexity theory [14].

Let $k \geq 2$ be a fixed positive integer. A subset I of V is called a k-*path vertex cover* if every path on k vertices in G contains at least one vertex from I. The k-PATH VERTEX COVER (k-PVC) problem asks if there is a k-path vertex cover of G whose size is at most some positive integer s. Motivated by the importance of a problem related to secure communication in wireless sensor networks, Brešar et al. initiated the study of k-PVC in [8] (as a generalized concept of *vertex cover*). It is known that k-PVC is NP-complete for every $k \geq 2$ [1,8]. Subsequent work regarding the *maximum* variant [21] and *weighted* variant [9] of k-PVC has also been considered in the literature. Recently, the study of k-PVC and related problems has gained a lot of attraction from both theoretical aspect [19,23,24] and practical application [3,13].

In this paper, we initiate the study of k-PVC from the viewpoint of *reconfiguration*. Given two distinct k-path vertex covers I and J of a graph G and a *single* reconfiguration rule, the k-PATH VERTEX COVER RECONFIGURATION (k-PVCR) problem asks whether there is a reconfiguration sequence between I and J. We study the computational complexity of k-PVCR with respect to different graph classes under the well-known reconfiguration rules: *Token Sliding*, *Token Jumping*, and *Token Addition or Removal*. They are informally defined as follows. Imagine that a token is placed at each vertex of a k-path vertex cover in G. For each of the following rules, a common requirement is that the resulting token-set forms a k-path vertex cover of G.

- Token Sliding (TS): A TS-step involves moving a token on some vertex v to one of its unoccupied neighbors.

– Token Jumping (TJ): A TJ-step involves moving a token on v to any unoccupied vertex.
– Token Addition or Removal (TAR): A TAR-step involves either adding or removing a single token such that the resulting token-set is of size at most given positive integer u. We sometimes write "TAR(u)" instead of "TAR" to emphasize the upper bound u on the size of each token-set in a reconfiguration sequence under TAR.

Related Work. The *reoptimization* framework is closely related to *reconfiguration*. Roughly speaking, given an optimal solution of a problem instance $\mathcal{I}$, and some *perturbations* that change $\mathcal{I}$ into a new instance $\mathcal{I}'$, a reoptimization problem aims to find an optimal solution for the changed instance $\mathcal{I}'$. Recently, Kumar et al. [19] initiated the study of reoptimization problems for (both weighted and unweighted) k-PVC with $k \geq 3$, extending some known reoptimization paradigms for the well-known VC problem [2]. The perturbation they considered in [19] is changing the input graph of the current instance by inserting new vertices.

The VERTEX COVER RECONFIGURATION (VCR) problem is one of the most well-studied reconfiguration problems, from both classical and parameterized complexity viewpoints (e.g., see [22] for a quick summary of known results). It is well-known that if I is a vertex cover of a graph $G = (V, E)$ then $V \setminus I$ is an *independent set* of G, i.e., a vertex-subset whose members are pairwise non-adjacent. Consequently, from classical complexity viewpoint, results of INDEPENDENT SET RECONFIGURATION (ISR) and VERTEX COVER RECONFIGURATION are interchangeable.

We now mention some known complexity results of VCR (which are mostly interchanged with ISR) for some graph classes. It is well-known that VCR is PSPACE-complete under each of TS, TJ, and TAR for general graphs [17], planar graphs of maximum degree 3 [15], perfect graphs [18], and bounded bandwidth graphs [25]. Even on bipartite graphs, VCR remains PSPACE-complete under TS, and NP-complete under each of TJ and TAR [20]. On chordal graphs, VCR is known to be PSPACE-complete under TS [4]. On the positive side, polynomial-time algorithms have been designed for VCR on even-hole-free graphs (and therefore chordal graphs) under each of TJ and TAR [18], on bipartite permutation graphs and bipartite distance-hereditary graphs [12] under TS, on cographs [6,18], claw-free graphs [7], interval graphs [5,18], and trees [10,18] under each of TS, TJ, and TAR.

Our Results. In this paper, we investigate the complexity of k-PVCR with respect to different input graphs. More precisely, we show that:

– Several hardness results for VCR remain true for k-PVCR. More precisely, we show the PSPACE-completeness of k-PVCR on general graphs under each rule TS, TJ, and TAR using a reduction from a variant of VCR. As our reduction preserves some nice graph properties, we claim (as a consequence of our reduction) that the hardness results for VCR on several graphs (namely planar graphs, bounded bandwidth graphs, chordal graphs, bipartite graphs) can be converted into those for k-PVCR. Using a reduction from the NONDETER-MINISTIC CONSTRAINT LOGIC [15,26], we also show that k-PVCR remains PSPACE-complete even on planar graphs of bounded bandwidth and maximum degree 3. (Our reduction from VCR does not preserve the maximum degree.)

– On the positive side, we design polynomial-time algorithms for k-PVCR on some restricted graph classes: trees (under each of TJ and TAR), paths and cycles (under each of TS, TJ, and TAR). Our algorithms are constructive, i.e., we explicitly show how a reconfiguration sequence can be constructed in a yes-instance. On paths, we claim that our algorithm constructs a *shortest* reconfiguration sequence. As a result, we obtain a complexity dichotomy for k-PVCR on (planar) graphs with respect to their maximum degree.

Due to the page limitation, we omit almost all the proofs from this paper.

2   Preliminaries

In this section, we define some useful notation and terminology. For standard concepts on graphs, readers are referred to [11]. Let G be a simple graph with vertex-set $V(G)$ and edge-set $E(G)$. For two vertices u, v, we denote by $\text{dist}_G(u, v)$ the *distance* between u and v in G, i.e., the length of a shortest path between them. For a vertex $v \in V(G)$, we denote by $G - v$ the graph obtained from G by removing v. For two vertex-subsets I and J, we denote by $G[I \triangle J]$ the subgraph of G induced by their *symmetric difference* $I \triangle J = (I \setminus J) \cup (J \setminus I)$. For a fixed integer $k \geq 2$, we say that a vertex v *covers* a k-path (i.e., a path on k vertices) P_k in G if $v \in V(P_k)$. A vertex-subset I is called a k-path vertex cover if every k-path in G contains at least one vertex from I. In other words, vertices of I cover all k-paths in G. We denote by $\psi_k(G)$ the size of a minimum k-path vertex cover of G. Trivially, for $n \geq k \geq 2$, $\psi_k(P_n) = \lfloor n/k \rfloor$ and $\psi_k(C_n) = \lceil n/k \rceil$ for a path P_n and a cycle C_n on n vertices.

Throughout this paper, we denote by (G, I, J, R) an instance of k-PVCR under a reconfiguration rule $\mathsf{R} \in \{\mathsf{TJ}, \mathsf{TS}, \mathsf{TAR}\}$, where I and J are two k-path vertex covers of G. We shall respectively call a reconfiguration sequence under each of TS, TJ, and TAR by a TS-*sequence*, TJ-*sequence*, and TAR(u)-*sequence*. Formally, let $S = \langle I_0, I_1, \ldots, I_\ell \rangle$ be an *ordered* sequence of k-path vertex covers of G. The *length* of S is defined as ℓ, i.e., if S is a reconfiguration sequence then its length is exactly the number of steps it performs under the given reconfiguration rule. Imagine that a token is placed at each vertex of a k-path vertex cover of G. We may sometimes identify a token with the vertex where it is placed on and say "a token in a k-path vertex cover". We say that S is a TS-*sequence* between two k-path vertex covers I_0 and I_ℓ if for each $i \in \{0, \ldots, \ell - 1\}$, there exist two vertices x_i and y_i such that $I_i \setminus I_{i+1} = \{x_i\}$, $I_{i+1} \setminus I_i = \{y_i\}$, and $x_i y_i \in E(G)$. Roughly speaking, I_{i+1} is obtained from I_i by sliding the token placed on x_i to y_i along an edge $x_i y_i$. Similarly, we say that S is a TJ-*sequence* between I_0 and I_ℓ if for each $i \in \{0, \ldots, \ell - 1\}$, there exist two vertices x_i and y_i such that $I_i \setminus I_{i+1} = \{x_i\}$, $I_{i+1} \setminus I_i = \{y_i\}$. Intuitively, I_{i+1} is obtained from I_i by jumping the token placed on x_i to y_i. Now, if $\max\{|I_i| : 0 \leq i \leq \ell\} \leq u$

for some positive integer u, and for each $i \in \{0, \ldots, \ell - 1\}$, there exists a vertex x_i such that $I_i \Delta I_{i+1} = \{x_i\}$ then we say that S is a TAR(u)-sequence between I_0 and I_ℓ. Roughly speaking, I_{i+1} is obtained from I_i by either adding a token to x_i or removing a token from x_i. If a TS-, TJ-, or TAR(u)-sequence between two k-path vertex covers I and J exists, we say that I and J are *reconfigurable* under TS, TJ, or TAR, respectively.

Using a similar argument as in [18, Theorem 1], we can show that

Lemma 1. *There exists a* TJ-*sequence of length ℓ between two k-path vertex covers I, J of a graph G with $|I| = |J| = s$ if and only if there exists a* TAR($s+1$)-*sequence of length 2ℓ between them.*

A reconfiguration sequence of minimum length is called a *shortest* reconfiguration sequence. For a reconfiguration sequence $S = \langle I_0, I_1, \ldots, I_p \rangle$, we denote by $revS$ the *reverse* of S, i.e., the reconfiguration sequence $\langle I_p, \ldots, I_1, I_0 \rangle$. For two reconfiguration sequences $S = \langle I_0, I_1, \ldots, I_p \rangle$ and $S' = \langle I'_0, I'_1, \ldots, I'_q \rangle$ under the same reconfiguration rule, if $I_p = I'_0$ then we say that they can be *concatenated* and define their *concatenation* $S \oplus S'$ as the reconfiguration sequence $\langle I_0, I_1, \ldots, I_p, I'_1, \ldots, I'_q \rangle$. We assume for convenience that if S' is empty then $S \oplus S' = S' \oplus S = S$.

3 Hardness Results

In this section, we show the hardness of k-PVCR on some well-known graph classes. First of all, we show that

Theorem 2. k-PVCR *is* PSPACE-*complete under each of* TS, TJ, *and* TAR *even when the input graph is a planar graph of maximum degree 4, or a bounded bandwidth graph. Additionally, k-PVCR is* PSPACE-*complete under* TS *on chordal graphs and bipartite graphs. Under each of* TJ *and* TAR, *k-PVCR is* NP-*hard on bipartite graphs.*

Proof (sketch). Using a similar reduction as in [8], we can show the PSPACE-completeness of k-PVCR under TJ. Combining the above result, the known results for VCR, and Lemma 1, we can also show the hardness results on several graphs under each of TJ and TAR as mentioned in Theorem 2. Finally, we show that the hardness results under TS hold via the same reduction.   □

In the above discussion, we show the PSPACE-completeness for planar graphs of maximum degree 4. Furthermore, using a reduction from the NONDETERMINISTIC CONSTRAINT LOGIC [15,26], we can improve this result as follows.

Theorem 3. k-PVCR *remains* PSPACE-*complete under each of* TS, TJ, *and* TAR *even on planar graphs of bounded bandwidth and maximum degree 3.*

4 Polynomial-Time Algorithms

4.1 Trees

In this section, we show polynomial-time algorithms for k-PVCR on trees under each of TJ and TAR. We first show a polynomial-time algorithm for the problem under TJ. Then, using Lemma 1 and the above result, we show a polynomial-time algorithm for the problem under TAR.

First, in order to solve the problem under TJ, we claim that for an instance (T, I, J, TJ) of k-PVCR on a tree T, if $|I| = |J|$, one can construct in polynomial time a TJ-sequence between I and J. The idea is to construct a canonical k-path vertex cover I^* such that both I and J can be reconfigured to I^* under TJ.

Before constructing I^*, we prove the following lemma, which describes an useful algorithm for partitioning a tree into subtrees satisfying certain conditions.

Lemma 4. *Let T be a tree on n vertices rooted at a vertex r. Assume that $\psi_k(T) \geq 1$. Then, in $O(n)$ time, one can partition T into $\psi_k(T)$ subtrees $T_1(r), \ldots, T_{\psi_k(T)}(r)$ such that for each $i \in \{1, \ldots, \psi_k(T)\}$,*

(i) Each k-path vertex cover I satisfies $I \cap V(T_i(r)) \neq \emptyset$.
(ii) There is a vertex that covers all k-paths in $T_i(r)$.

Proof. To construct a partition $P(T) = \{T_1(r), \ldots, T_{\psi_k(T)}(r)\}$ of T satisfying the described conditions, we slightly modify the algorithm $\mathtt{PVCPTree}(T, k)$ in [8] as follows. A *properly rooted subtree* T_v of T is a subtree of T induced by the vertex v and all its descendants (with respect to the root r) satisfying the following conditions

1. T_v contains a path on k vertices;
2. $T_v - v$ does not contain a path on k vertices.

The modified algorithm $\mathtt{Partition}(T, k, r)$ systematically searches for a properly rooted tree T_v, decides whether T_v belongs to a solution $P(T)$, and if so, add T_v to $P(T)$, and remove T_v from the input tree T.

From [8], it follows that $\mathtt{Partition}(T, k, r)$ runs in $O(n)$ time. From the construction of $P(T)$, it is clear that (i) always holds. We show (ii) by induction on $\psi_k(T)$.

For a tree T with $\psi_k(T) = 1$, let T_v be a properly rooted subtree of T. Since any k-path vertex cover of T contains a vertex from T_v, it follows that $\psi_k(T - T_v) = \psi_k(T) - 1 = 0$, which implies that $T - T_v$ does not contain any properly rooted subtree, and therefore $P(T) = \{T\}$. To see that (ii) holds, note that v must cover all k-paths in T_v, and therefore it also covers all k-paths in T; otherwise, $T - T_v$ contains a k-path that is not covered by v, and then must contain a properly rooted subtree, which is a contradiction.

Assume that (ii) holds for any tree T with $\psi_k(T) < c$, for some constant $c > 1$. For a tree T rooted at some vertex r with $\psi_k(T) = c$, let T_v be a properly rooted subtree of T, where v is some vertex of T. From the algorithm

---

Algorithm 1: Partition(T, k, r).

Input: A tree T on n vertices rooted at r and a positive integer k;
Output: A partition of T into $\psi_k(T)$ subtrees;

1 $i := 1$;
2 **while** T contains a properly rooted subtree T_v **do**
3   **if** $T - T_v$ contains a properly rooted subtree **then**
4        $T_i(r) := T_v$;
5        $i := i + 1$;
6   **else**
7        $T_i(r) := T$;
8     $T := T - T_v$;
9 $P(T) = \{T_1(r), \ldots, T_i(r)\}$;
10 **return** $P(T)$;

Partition, it follows that v must cover all k-paths in $T_v = T_1(r)$. Since $c > 1$, the tree $T - T_v$ contains a properly rooted subtree. By inductive hypothesis, for each $i \in \{2, 3, \ldots, \psi_k(T)\}$, there is a vertex that covers all k-paths in $T_i(r)$. Therefore, (ii) holds for any tree T with $\psi_k(T) \geq 1$. □

We are now ready to show that

Theorem 5. *For any instance (T, I, J, TJ) of k-PVCR on a tree T, I and J are reconfigurable if and only if $|I| = |J|$. Moreover, a reconfiguration sequence between them, if exists, can be constructed in $O(n)$ time. Consequently, k-PVCR under TJ can be solved in linear time on trees.*

Proof. Clearly, if I and J are reconfigurable under TJ, they must be of the same size. To prove this theorem, it suffices to show that for an instance (T, I, J, TJ) of k-PVCR on a tree T, one can construct in polynomial time a TJ-sequence between I and J.

A minimum k-path vertex cover I^r can be easily constructed in linear time by modifying Partition as follows: Initially, $I^r = \emptyset$. In each iteration of the **while** loop, add to I^r the vertex v of the properly rooted subtree T_v that is currently considering. Let $I^\star$ be any k-path vertex cover of size $|I| = |J|$ such that $I^r \subseteq I^\star$. We claim that both I and J can be reconfigured to $I^\star$ under TJ. As a result, a TJ-sequence between I and J can be constructed by reconfiguring I to $I^\star$, and then $I^\star$ to J.

We now show how to construct a TJ-sequence between I and $I^\star$. Let $P(T) = \{T_1(r), \ldots, T_{\psi_k(T)}(r)\}$ be a partition of T resulting from the algorithm Partition and let $I_0 = I$.

- **Step 1:** For each $i \in \{1, \ldots, \psi_k(T)\}$, let $v_i \in I^r \cap V(T_i(r))$. If v_i does not contain a token in I_{i-1}, we jump a token from some vertex $x_i \in I_{i-1} \cap V(T_i(r))$ to v_i. Otherwise, we do nothing. Let I_i be the resulting set. Note that any k-path in T covered by x_i must also be covered by some v_j with $j \leq i$. A simple induction shows that $I_i = I_{i-1} \setminus \{x_i\} \cup \{v_i\}$ forms a k-path vertex cover of T.

– **Step 2:** For $x \in I_{\psi_k(T)} \setminus I^\star$ and $y \in I^\star \setminus I_{\psi_k(T)}$, we simply jump the token on x to y, and repeat the process with $I_{\psi_k(T)} \setminus \{x\}$ and $I^\star \setminus \{y\}$ instead of $I_{\psi_k(T)}$ and $I^\star$, respectively. Since $I^r \subseteq I_{\psi_k(T)} \cap I^\star$ is already a minimum k-path vertex cover, any TJ-step described above results a k-path vertex cover of T.

Since each token in I is jumped at most once, the above construction can be done in linear time. We have described how to construct a TJ-sequence from J to $I^\star$. In a similar manner, a TJ-sequence between J and $I^\star$ can be constructed. Our proof of Theorem 5 is complete.   □

Consequently, combining Theorem 5 and Lemma 1, we have

Theorem 6. *For any instance $(T, I, J, \mathsf{TAR}(u))$ of k-PVCR on a tree T, one can decide if I and J are reconfigurable in polynomial time.*

4.2   Paths and Cycles

Here, we describe polynomial-time algorithms for k-PVCR on paths and cycles. As paths and cycles are the only (planar) graphs of maximum degree 2, by combining Theorem 3 and our results, we have a complexity dichotomy of k-PVCR on (planar) graphs. Additionally, on paths, we claim that one can construct a *shortest* reconfiguration sequence between any two given k-path vertex covers (if exists) under each reconfiguration rule TS, TJ, and TAR. Due to the page limitation, we omit several technical details.

k-**PVCR on Paths.** By Theorems 5 and 6, clearly k-PVCR on paths can be solved in polynomial time under each of TJ and TAR. In this section, we slightly improve this result by showing that one can construct a *shortest* reconfiguration sequence between two k-path vertex covers on a path not only under each of TJ and TAR but also under TS.

Given an instance (P, I, J, TJ) of k-PVCR where $|I| = |J| = s$, one can construct a shortest TJ-sequence between I and J. Suppose that vertices in $I = \{v_{i_1}, \ldots, v_{i_s}\}$ and $J = \{v_{j_1}, \ldots, v_{j_s}\}$ are ordered such that $1 \leq i_1 < \cdots < i_s \leq n$ and $1 \leq j_1 < \cdots < j_s \leq n$. In each step of the algorithm, we move a token on the "rightmost" vertex $v_{i_p} \in I \setminus J$ to the "rightmost" vertex $v_{j_p} \in J \setminus I$ if $i_p > j_p$ or vice-versa otherwise, for $p \in \{1, \ldots, s\}$. As a reconfiguration sequence is reversible, one can easily form a TJ-sequence between I and J. Note that each step of the algorithm reduces $|I \triangle J|/2$ by exactly 1. Finally, we obtain a shortest TJ-sequence between I and J of length exactly $|I \triangle J|/2$.

Theorem 7. *Given an instance (P, I, J, TJ) of k-PVCR on a path P, the k-path vertex covers I and J are reconfigurable if and only if $|I| = |J|$. Moreover, we can compute a shortest reconfiguration sequence in $O(n)$ time.*

By Theorem 7 and Lemma 1, we obtain the following result on k-PVCR on a path P under TAR.

Theorem 8. *For any instance $(P, I, J, \mathsf{TAR}(u))$ of k-PVCR on a path P on n vertices, one can decide if I and J are reconfigurable in linear time.*

Now we sketch the idea for solving the problem under TS in polynomial time. Given an instance (P, I, J, TS) of k-PVCR where $|I| = |J| = s$, one can construct a shortest TS-sequence between I and J. Suppose that vertices in $I = \{v_{i_1}, \ldots, v_{i_s}\}$ and $J = \{v_{j_1}, \ldots, v_{j_s}\}$ are ordered such that $1 \leq i_1 < \cdots < i_s \leq n$ and $1 \leq j_1 < \cdots < j_s \leq n$. Our goal is to construct a shortest TS-sequence (of length $\sum_{p=1}^{s} \mathrm{dist}_P(v_{i_p}, v_{j_p})$) that repeatedly slides the token on the "leftmost" vertex $v_{i_p} \in I$ to the "leftmost" vertex $v_{j_p} \in J$ if $i_p < j_p$ or vice-versa otherwise, for $p \in \{1, \ldots, s\}$.

The key point is, in certain conditions, one can construct in polynomial time a function $\mathsf{Push}(P, I, i, j)$ whose task is to output a TS-sequence that moves the token placed at some vertex v_i of the k-path vertex cover I to vertex v_j in a given path $P = v_1 v_2 \ldots v_n$, where $1 \leq i < j \leq i + k \leq n$. Roughly speaking, $\mathsf{Push}(P, I, i, j)$ slides the token t on v_i toward v_j along the path $P_{ij} = v_i v_{i+1} \ldots v_j$ until either t ends up at v_j or there is some index $p \in \{i, \ldots, j - 1\}$ where t is already placed at v_p but cannot immediately move to v_{p+1} because there is already some token t' placed there. In the latter case, one can recursively call Push to slide t' from v_{p+1} to v_{p+2} and therefore enabling t (which is currently placed at v_p) to slide to v_{p+1}. Now, the same situation happens again with t and t', and the resolving procedure can be done in the same manner as before. This process stops when t is finally placed at v_j.

The following lemma says that if certain conditions are satisfied, the output of $\mathsf{Push}(P, I, i, j)$ is indeed a TS-sequence that reconfigures the k-path vertex cover I to some other k-path vertex cover of P.

Lemma 9. *Let $P = v_1 v_2 \ldots v_n$ be a path on n vertices, and let I be a k-path vertex cover of P. Let $i \in \{1, \ldots, n\}$ be such that either $i \leq k + 1$ or $\{v_{i-1}, \ldots, v_{i-k}\} \cap I \neq \emptyset$. If $\{v_i, v_{i+1}, \ldots, v_{i+p}\} \subseteq I$ and $v_{i+p+1} \notin I$ for some integer p satisfying $0 \leq p \leq n - i - 1$, then there exists a TS-sequence in P that reconfigures I to $I \setminus \{v_i, v_{i+1}, \ldots, v_{i+p}\} \cup \{v_{i+1}, \ldots, v_{i+p+1}\}$. Consequently, if the assumption is satisfied, the output of $\mathsf{Push}(P, I, i, j)$ is indeed a TS-sequence in P that reconfigures I to some k-path vertex cover of P.*

Clearly, the function $\mathsf{Push}(P, I, i_p, j_p)$ can be used to slide a token on v_{i_p} to v_{j_p} for $p \in \{1, \ldots, s\}$ and $i_p < j_p$. Thus, we have the following theorem.

Theorem 10. *Given an instance (P, I, J, TS) of k-PVCR on a path P, the k-path vertex covers I and J are reconfigurable if and only if $|I| = |J|$. Moreover, we can compute a shortest reconfiguration sequence in $O(n^2)$ time.*

k-PVCR on Cycles. Let $C = v_0 v_1 \ldots v_{n-1} v_0$ be a given n-vertex cycle, and let (C, I, J, R) be a k-PVCR instance on C under a reconfiguration rule $\mathsf{R} \in \{\mathsf{TJ}, \mathsf{TS}, \mathsf{TAR}(u)\}$. We remark that if $|I| = |J| = \lceil n/k \rceil$ and $n = c \cdot k$ for some c, then (C, I, J, R) where $\mathsf{R} \in \{\mathsf{TS}, \mathsf{TJ}\}$ is a no-instance. This is because no tokens can be moved in such instances.

Here we assume that the indices of vertices on the cycle increase in the clockwise manner. We claim that it is possible to apply the algorithms for paths to cycles, by cutting a cycle into a path with a vertex in $I \cap J$. Our algorithms do not always achieve the shortest reconfiguration sequence. However, we later show that achieving the shortest sequence even on cycles under TJ might not be trivially easy, since we can systematically create the instances such that the length of the shortest reconfiguration sequence is not equal to $|I \Delta J|/2$.

Now, we describe the sketch how to cut C under TJ, TS, and TAR. In the TS case, without loss of generality, we can assume that either $|I| \neq \lceil n/k \rceil$ or $n \neq c \cdot k$ holds. If v is already in $I \cap J$, we cut C by removing v. The following lemma ensures that if I and J are reconfigurable in $C - v$, then $I \cup \{v\}$ and $J \cup \{v\}$ are reconfigurable in C.

Lemma 11. *Let C be an n-vertex cycle and v be a token in $I \cap J$ of C. Then, for any k-path vertex cover I' of $C - v$, $I' \cup \{v\}$ is a k-path vertex cover of C.*

If $I \cap J = \emptyset$, there exists at least one token movable in the clockwise or counterclockwise direction. Here, we say a token u is *movable* if and only if (i) there exists a neighbor v of u such that no token is placed on v, and (ii) moving a token on u to v results a k-path vertex cover.

Lemma 12. *If either $|I| \neq \lceil n/k \rceil$ or $n \neq c \cdot k$ holds, then there exists at least one token movable by at least one step in the clockwise or counterclockwise direction. Furthermore, we can find such a token in linear time.*

After finding such a movable token, we can use *rotate* operation repeatedly until obtaining at least one vertex in $I \cap J$. Here, the rotate operation takes a token-set, a movable token which can be slid at least one step towards direction $d \in \{\text{clockwise}, \text{counterclockwise}\}$ as input, and outputs a TS-sequence that slides all tokens one step towards d. After obtaining at least one vertex in $I \cap J$, we can perform the cutting operation as before.

Next, we consider the TJ case. Since any TS-sequence is also a TJ-sequence, we can perform the same cutting operation as in the TS case. Then, using this cutting operation, we can show that

Theorem 13. *Given an instance (C, I, J, R) of k-PVCR on a cycle C where $R \in \{TS, TJ\}$, if $|I| = |J| = \lceil n/k \rceil$ and $n = c \cdot k$ for some c, then (C, I, J, R) is a no-instance. Otherwise, the k-path vertex covers I and J are reconfigurable if and only if $|I| = |J|$. Moreover, we can compute a reconfiguration sequence for TJ rule in $O(n)$ time, and for TS rule in $O(n^2)$ time.*

For the TAR case, we can use the result for the TJ case and Lemma 1 to show that

Theorem 14. *For any instance $(C, I, J, TAR(u))$ of k-PVCR on a cycle C, one can decide if I and J are reconfigurable in linear time.*

Consequently, we have

Theorem 15. *k-PVCR on cycles under each of TJ and TAR(u) can be solved in $O(n)$ time, and under TS can be solved in $O(n^2)$ time.*

To conclude this section, we give an example showing that even in a yes-instance (C, I, J, TJ) of k-PVCR ($k \geq 3$) under TJ on a cycle C, one may need to use more than $|I \Delta J|/2$ TJ-steps even in a shortest TJ-sequence. Intuitively, the lower bound $|I \Delta J|/2$ seems to be easy to achieve under TJ, simply by jumping tokens one by one from $I \setminus J$ to $J \setminus I$. However, as we show in the following lemma, to keep the k-path vertex cover property, sometimes a token in I may need to jump to some vertex not in $J \setminus I$ beforehand. This implies the non-triviality of finding a *shortest* reconfiguration sequence even under TJ.

Lemma 16. *For k-PVCR ($k \geq 3$) yes-instances (C, I, J, TJ) on cycles where $C = v_0 v_1 \ldots v_{3k-2} v_0$, $I = \{v_0, v_k, v_{2k}\}$ and $J = \{v_{3k-2}, v_{2k-2}, v_{k-1}\}$, the length of a shortest reconfiguration sequence from I to J is greater than $|I \Delta J|/2$.*

5   Concluding Remarks

In this paper, we have investigated the complexity of k-PVCR under each of TS, TJ, and TAR for several graph classes. In particular, several known hardness results for VCR (i.e., $k = 2$) can be generalized for k-PVCR when $k \geq 3$. Additionally, we proved a complexity dichotomy for k-PVCR by showing that it remains PSPACE-complete even if the input (planar) graph is of maximum degree 3 (using a reduction from NCL) and can be solved in polynomial time when the input (planar) graph is of maximum degree 2 (i.e., it is either a path or a cycle). On the positive side, we designed polynomial-time algorithms for k-PVCR on trees under each of TJ and TAR. We also showed how to construct a *shortest* reconfiguration sequence on paths, and presented an example showing the nontriviality of finding shortest reconfiguration sequences on cycles even under TJ. The question of whether one can solve k-PVCR on trees under TS in polynomial time remains open. Another target graphs may be chordal graphs (under each of TJ and TAR), cographs, and graphs of treewidth at most 2. Even on graphs of treewidth at most 2, the complexity of VCR remains open.

References

1. Acharya, H.B., Choi, T., Bazzi, R.A., Gouda, M.G.: The k-observer problem in computer networks. Networking Sci. **1**(1–4), 15–22 (2012)
2. Ausiello, G., Bonifaci, V., Escoffier, B.: Complexity and approximation in reoptimization. In: Computability in Context: Computation and Logic in the Real World, pp. 101–129. World Scientific, Singapore (2011)
3. Beck, M., Lam, K.Y., Ng, J.K.Y., Storandt, S., Zhu, C.J.: Concatenated k-path covers. In: Proceedings of ALENEX 2019, pp. 81–91 (2019)

4. Belmonte, R., Kim, E.J., Lampis, M., Mitsou, V., Otachi, Y., Sikora, F.: Token sliding on split graphs. In: Niedermeier, R., Paul, C. (eds.) Proceedings of STACS 2019. LIPIcs, vol. 126, pp. 13:1–13:7 (2019). Schloss Dagstuhl-Leibniz-Zentrum fuer Informatik

5. Bonamy, M., Bousquet, N.: Token sliding on chordal graphs. In: Bodlaender, H.L., Woeginger, G.J. (eds.) WG 2017. LNCS, vol. 10520, pp. 127–139. Springer, Cham (2017). https://doi.org/10.1007/978-3-319-68705-6_10

6. Bonsma, P.S.: Independent set reconfiguration in cographs and their generalizations. J. Graph Theor. **83**(2), 164–195 (2016)

7. Bonsma, P., Kamiński, M., Wrochna, M.: Reconfiguring independent sets in claw-free graphs. In: Ravi, R., Gørtz, I.L. (eds.) SWAT 2014. LNCS, vol. 8503, pp. 86–97. Springer, Cham (2014). https://doi.org/10.1007/978-3-319-08404-6_8

8. Brešar, B., Kardoš, F., Katrenič, J., Semanišin, G.: Minimum k-path vertex cover. Discrete Appl. Math. **159**(12), 1189–1195 (2011)

9. Brešar, B., Krivoš-Belluš, R., Semanišin, G., Šparl, P.: On the weighted k-path vertex cover problem. Discrete Appl. Math. **177**, 14–18 (2014)

10. Demaine, E.D., et al.: Linear-time algorithm for sliding tokens on trees. Theor. Comput. Sci. **600**, 132–142 (2015)

11. Diestel, R.: Graph Theory, Graduate Texts in Mathematics, vol. 173, 4th edn. Springer, Heidelberg (2010)

12. Fox-Epstein, E., Hoang, D.A., Otachi, Y., Uehara, R.: Sliding token on bipartite permutation graphs. In: Elbassioni, K., Makino, K. (eds.) ISAAC 2015. LNCS, vol. 9472, pp. 237–247. Springer, Heidelberg (2015). https://doi.org/10.1007/978-3-662-48971-0_21

13. Funke, S., Nusser, A., Storandt, S.: On k-path covers and their applications. In: Proceedings VLDB Endowment, vol. 7, no. 10, pp. 893–902 (2014)

14. Garey, M.R., Johnson, D.S.: Computers and Intractability; A Guide to the Theory of NP-Completeness. W.H. Freeman & Co., New York (1990)

15. Hearn, R.A., Demaine, E.D.: PSPACE-completeness of sliding-block puzzles and other problems through the nondeterministic constraint logic model of computation. Theor. Comput. Sci. **343**(1–2), 72–96 (2005)

16. van den Heuvel, J.: The complexity of change. In: Surveys in Combinatorics. London Mathematical Society Lecture Note Series, vol. 409, pp. 127–160. Cambridge University Press, Cambridge (2013)

17. Ito, T., et al.: On the complexity of reconfiguration problems. Theor. Comput. Sci. **412**(12–14), 1054–1065 (2011)

18. Kamiński, M., Medvedev, P., Milanič, M.: Complexity of independent set reconfigurability problems. Theor. Comput. Sci. **439**, 9–15 (2012)

19. Kumar, M., Kumar, A., Pandu Rangan, C.: Reoptimization of path vertex cover problem. In: Du, D.-Z., Duan, Z., Tian, C. (eds.) COCOON 2019. LNCS, vol. 11653, pp. 363–374. Springer, Cham (2019). https://doi.org/10.1007/978-3-030-26176-4_30

20. Lokshtanov, D., Mouawad, A.E.: The complexity of independent set reconfiguration on bipartite graphs. ACM Trans. Algorithms **15**(1), 7:1–7:19 (2019)

21. Miyano, E., Saitoh, T., Uehara, R., Yagita, T., van der Zanden, T.C.: Complexity of the maximum k-path vertex cover problem. In: Rahman, M.S., Sung, W.-K., Uehara, R. (eds.) WALCOM 2018. LNCS, vol. 10755, pp. 240–251. Springer, Cham (2018). https://doi.org/10.1007/978-3-319-75172-6_21

22. Nishimura, N.: Introduction to reconfiguration. Algorithms **11**(4), 52 (2018). (article 52)
23. Ran, Y., Zhang, Z., Huang, X., Li, X., Du, D.Z.: Approximation algorithms for minimum weight connected 3-path vertex cover. Appl. Math. Comput. **347**, 723–733 (2019)
24. Tsur, D.: Parameterized algorithm for 3-path vertex cover. Theor. Comput. Sci. **783**, 1–8 (2019)
25. Wrochna, M.: Reconfiguration in bounded bandwidth and treedepth. J. Comput. Syst. Sci. **93**, 1–10 (2018)
26. Zandenvan der Zanden, T.C.: Parameterized complexity of graph constraint logic. In: Proceedings of IPEC 2015. LIPIcs, vol. 9076, pp. 282–293 (2015)

Computational Complexity
of the Chromatic Art Gallery Problem
for Orthogonal Polygons

Chuzo Iwamoto$^{(\boxtimes)}$ (iD) and Tatsuaki Ibusuki

Hiroshima University, Higashi-Hiroshima 739-8521, Japan
chuzo@hiroshima-u.ac.jp
https://home.hiroshima-u.ac.jp/chuzo/

Abstract. The art gallery problem is to find a set of guards who together can observe every point of the interior of a polygon P. We study a chromatic variant of the problem, where each guard is assigned one of k distinct colors. The *chromatic art gallery problem* is to find a guard set for P such that no two guards with the same color have overlapping visibility regions. We study the decision version of this problem for orthogonal polygons with r-visibility when the number of colors is $k = 2$. Here, two points are r-*visible* if the smallest axis-aligned rectangle containing them lies entirely within the polygon. In this paper, it is shown that determining whether there is an r-visibility guard set for an orthogonal polygon with holes such that no two guards with the same color have overlapping visibility regions is NP-hard when the number of colors is $k = 2$.

Keywords: Chromatic art gallery problem · Orthogonal polygons · r-visibility · NP-hard

1 Introduction

The art gallery problem is to determine the minimum number of guards who can observe the interior of a gallery. Chvátal [3] proved that $\lfloor n/3 \rfloor$ guards are the lower and upper bounds for this problem; namely, $\lfloor n/3 \rfloor$ guards are always sufficient and sometimes necessary for observing the interior of an n-vertex simple polygon. This $\lfloor n/3 \rfloor$-bound is replaced by $\lfloor n/4 \rfloor$ if the instance is restricted to a simple orthogonal polygon [8].

Another perspective to the art gallery problem is to study the complexity of locating the minimum number of guards in a polygon. The NP-hardness and APX-hardness of this problem were shown by Lee and Lin [12] and by Eidenbenz et al. [4], respectively. Furthermore, Schuchardt and Hecker [16] proved that this problem remains NP-hard if we restrict our attention to simple orthogonal polygons. Even guarding the vertices of a simple orthogonal polygon was shown to be NP-hard [11].

This work was supported by JSPS KAKENHI Grant Number 16K00020.

M. S. Rahman et al. (Eds.): WALCOM 2020, LNCS 12049, pp. 146–157, 2020.
https://doi.org/10.1007/978-3-030-39881-1_13

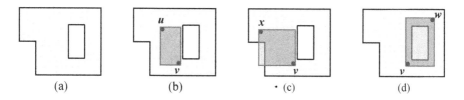

Fig. 1. (a) Orthogonal polygon with a hole. (b) Two points v and u are r-visible. (c) v and x are not r-visible. (d) v and w are not r-visible.

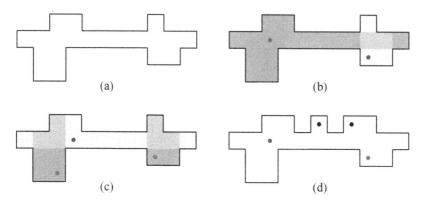

Fig. 2. (a) Orthogonal polygon P. (b, c) Every point in P can be observed by a 2-color guard set. Every point in P is observed by either a single guard or two guards having different colors. (d) Orthogonal polygon which can be observed by a 3-color guard set, but not by any 2-color guard set.

We study a chromatic variant of the art gallery problem, where each guard is assigned one of k distinct colors. The chromatic art gallery problem is to find a guard set for a polygon such that no two guards with the same color have overlapping visibility regions. We study the decision version of this problem for orthogonal polygons with r-visibility when the number of colors is $k = 2$. Here, two points u and v are r-*visible* if the smallest axis-aligned rectangle containing them lies entirely within the polygon (see Figs. 1 and 2). In this paper, it is shown that determining whether there is an r-visibility guard set for an orthogonal polygon with holes such that no two guards with the same color have overlapping visibility regions is NP-hard when the number of colors is $k = 2$.

The computational complexity of the chromatic art gallery problem was firstly investigated in [6]; the NP-hardness was proved for deciding whether there is a chromatic guard set for a given polygon with holes when the number of colors is $k = 2$. This result is for general (non-orthogonal) polygons under standard visibility (and not r-visibility). Recently, the NP-hardness of the chromatic art gallery problem was proved for finding a *conflict-free* guard set under the vertex-to-vertex guarding condition [2]. Here, a colored guard set is called *conflict-free* if each point of the polygon is seen by some guard whose color appears exactly once

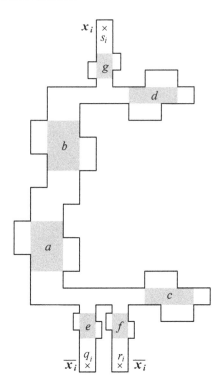

Fig. 3. Variable gadget for x_i.(Color figure online)

among the guards visible to that point. Several results on the lower and upper bounds of the minimum number of colors can be found in [1,5,9] for general and orthogonal polygons under standard and orthogonal visibility conditions. The r-visibility guard set problem for orthogonal polygons with holes was shown to be NP-hard [10]; this result is on the standard (non-chromatic) art gallery problem with r-visibility.

2 Definitions and Results

The definition of visibility is mostly from [17]. Two points v and u in a polygon P are said to be r-*visible* if the smallest axis-aligned rectangle containing them lies entirely within P (see Fig. 1). Here, the rectangle must not contain any hole of the polygon. A point v is said to *observe* a region if every point of the region is r-visible from v.

The definitions of a polygon and a polygon with holes are mostly from [13,15]. A *polygon* is defined by a finite set of segments such that every segment extreme is shared by exactly two edges and no subset of edges has the same property. The segments are the *edges* and their endpoints are the *vertices* of the polygon.

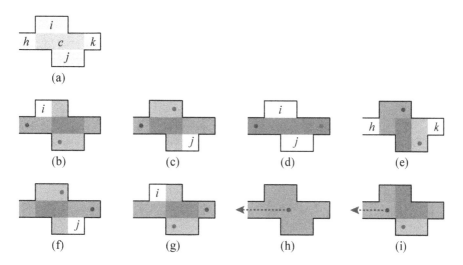

Fig. 4. (a) Area c and its surrounding regions h, i, j, and k. (b)–(g) If no guard is place on area c, then there remain white regions which are observed by no guards. (h, i) If a guard is placed on area c, every point on regions c and h, i, j, k is observed by either a single guard or two guards having different colors. (Color figure online)

If each edge of a polygon is perpendicular to one of the coordinate axes, then the polygon is called *orthogonal*. If no non-consecutive pair of edges overlap, then the polygon is said to be *simple*.

A *polygon with holes* is a polygon P enclosing several other polygons $H_1, H_2, \ldots, H_h$, the holes. None of the boundaries of P, $H_1, H_2, \ldots, H_h$ may intersect, and each of the holes is empty. P is said to bound a *multiply-connected* region with h holes: the region of the plane interior to or on the boundary of P, but exterior to or on the boundary of $H_1, H_2, \ldots, H_h$. Similarly, we define an *orthogonal polygon with holes* to be an orthogonal polygon with orthogonal holes, with all edges aligned with the same pair of orthogonal axes.

An instance of the *chromatic art gallery problem for orthogonal polygons with holes* is $(P, H_1, H_2, \ldots, H_h; k)$, where P is an orthogonal polygon with holes $H_1, H_2, \ldots, H_h$, and k is the number of colors. The problem asks whether there is a set of colored guards with r-visibility for P with holes such that no two guards with the same color have overlapping visibility regions when the number of colors is k.

Theorem 1. *The chromatic art gallery problem for orthogonal polygons with holes is NP-hard when the number of colors is two.*

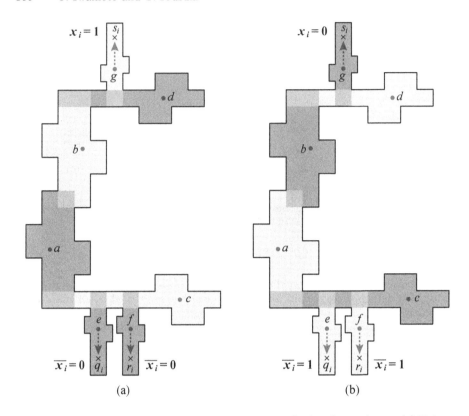

Fig. 5. Variable gadget for x_i. There are two types of colored guard sets: (a) Point s_i is observed by a blue guard, which corresponds to the assignment $x_i = 1$. (b) Point s_i is observed by a red guard, which corresponds to $x_i = 0$. (Color figure online)

3   NP-Completeness

3.1   3SAT Problem

The definition of 3SAT is mostly from [7] and [14]. Let $U = \{x_1, x_2, \ldots, x_n\}$ be a set of Boolean *variables*. Boolean variables take on values 0 (false) and 1 (true). If x is a variable in U, then x and $\overline{x}$ are *literals* over U. The value of $\overline{x}$ is 1 (true) if and only if x is 0 (false). A *clause* over U is a set of literals over U, such as $\{\overline{x_1}, x_3, x_4\}$. A clause is *satisfied* by a truth assignment if and only if at least one of its members is true under that assignment.

An instance of PLANAR 3SAT is a collection $C = \{c_1, c_2, \ldots, c_m\}$ of clauses over U such that (i) $|c_j| = 3$ for each $c_j \in C$ and (ii) the graph $G = (V, E)$, defined by $V = U \cup C$ and $E = \{(x_i, c_j) \mid x_i \in c_j \in C \text{ or } \overline{x_i} \in c_j \in C\}$, is planar. PLANAR 3SAT asks whether there exists some truth assignment for U that simultaneously satisfies all the clauses in C. If E is replaced with

$$E_1 = E \cup \{(c_j, c_{j+1}) \mid 1 \leq j \leq m - 1\},$$

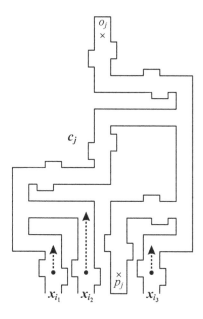

Fig. 6. Clause gadget transformed from $c_j = \{x_{i_1}, x_{i_2}, x_{i_3}\}$.

then the problem is called CLAUSE-LINKED PLANAR 3SAT. This problem is NP-complete, since VARIABLE-CLAUSE-LINKED PLANAR 3SAT was shown to be NP-complete in [14], where the edge set E_2 is defined as

$$E_2 = E \cup \{(x_i, x_{i+1}) \mid 1 \le i \le n - 1\} \cup \{(x_n, c_1)\}$$
$$\cup \{(c_j, c_{j+1}) \mid 1 \le j \le m - 1\} \cup \{(c_m, x_1)\}.$$

For example, $U = \{x_1, x_2, x_3, x_4\}$, $C = \{c_1, c_2, c_3\}$, and $c_1 = \{x_1, \overline{x_2}, x_3\}$, $c_2 = \{x_1, x_2, \overline{x_4}\}$, $c_3 = \{\overline{x_2}, \overline{x_3}, x_4\}$ provide an instance of CLAUSE-LINKED PLANAR 3SAT. For this instance, the answer is "yes," since there is a truth assignment $(x_1, x_2, x_3, x_4) = (1, 1, 0, 1)$ satisfying all clauses.

3.2   Transformation from an Instance of CLAUSE-LINKED PLANAR 3SAT to an Orthogonal Polygon

We present a polynomial-time transformation from an arbitrary instance C of CLAUSE-LINKED PLANAR 3SAT to an orthogonal polygon P with holes such that C is satisfiable if and only if there is a set of colored guards with r-visibility for P such that no two guards with the same color have overlapping visibility regions, where each guard is assigned one of two distinct colors (see Fig. 11).

Each variable $x_i \in \{x_1, x_2, \ldots, x_n\}$ is transformed into the variable gadget as illustrated in Fig. 3. (In this figure, we suppose that variable x_i appears once positively and twice negatively in C.) There are seven yellow rectangle

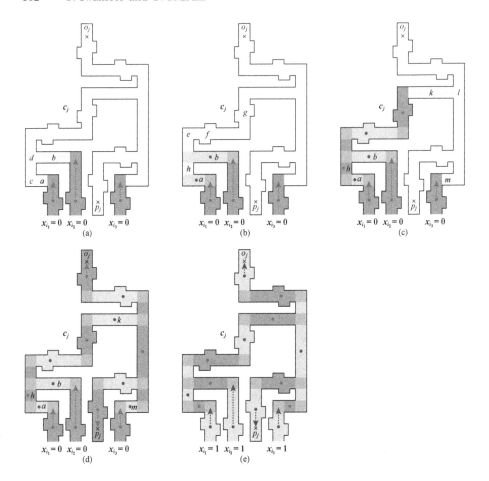

Fig. 7. (a)–(d) If $x_{i_1} = x_{i_2} = x_{i_3} = 0$, then points o_j and p_j will be observed by red guards. (e) If $x_{i_1} = x_{i_2} = x_{i_3} = 1$, then points o_j and p_j will be observed by blue guards. (Color figure online)

areas $a, b, \ldots, g$ and three points q_i, r_i, and s_i. At least one guard must be placed in each of the seven yellow areas; the reason is given in the next paragraph.

Consider area c (see also Fig. 4(a)). Assume that there is no guard in area c. In this case, even if a blue guard and a red guard are placed on two of the four grey areas h, i, j, and k (see Figs. 4(b)–(g)), there exists a point (see white areas in the figure) which can be observed by neither a blue guard nor a red guard. Therefore, at least one guard must be placed on area c (see Figs. 4(h) and (i)).

By a similar observation, one can see that at least one guard must be placed in each of the seven yellow areas $a, b, \ldots, f$ (see Fig. 5). The visibility regions of the guards on areas a, d, e, f overlap the visibility regions of the guards on areas b, c, g. There are two types of guard sets, which correspond to assignment $x_i = 1$ and $x_i = 0$ (see Fig. 5(a) and (b), respectively). If point s_i is observed

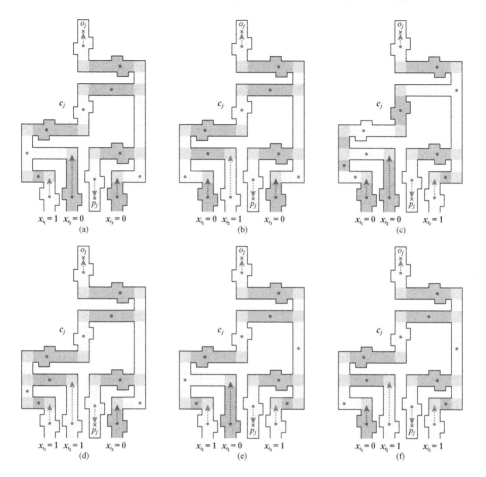

Fig. 8. If at least one of the variables x_{i_1}, x_{i_2}, and x_{i_3} is 1, then points o_j and p_j can be observed by blue guards. For the case $x_{i_1} = x_{i_2} = x_{i_3} = 1$, see Fig. 7(e). (Color figure online)

by a blue guard (resp. a red guard), then points q_i and r_i are observed by red guards (resp. blue guards). This corresponds to the assignment $x_i = 1$ (resp. $x_i = 0$).

Clause $c_j \in \{c_1, c_2, \ldots, c_m\}$ is transformed into the clause gadget as illustrated in Fig. 6. Let $c_j = \{x_{i_1}, x_{i_2}, x_{i_3}\}$. If $x_{i_1} = x_{i_2} = x_{i_3} = 0$ (resp. $x_{i_1} = x_{i_2} = x_{i_3} = 1$) as illustrated in Fig. 7(d) (resp. Fig. 7(e)), then points o_j and p_j must be observed by red guards (resp. blue guards) because of the following reason.

In Fig. 7(a), we need two blue guards in order to observe grey areas a and b. If a blue guard is placed at c (resp. d), then area b (resp. a) can be observed by neither a blue guard nor a red guard. Thus, two blue guards must be placed in areas a and b (see (b)). (b) If a red guard is placed at e in order to observe

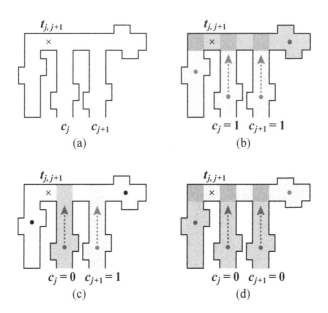

Fig. 9. XNOR gadget. (Color figure online)

area h, then a blue guard must be placed at f. In this case, neither a red guard nor a blue guard can be placed at g. Thus, the red guard is placed at h (see the red guard h in (c)). (c) If a blue guard is placed at l in order to observe area k, then area m can be observed by neither a blue guard nor a red guard. Thus, the blue guard is placed in area k (see (d)). Hence, points o_j and p_j are observed by red guards when $x_{i_1} = x_{i_2} = x_{i_3} = 0$. By the same observation, if $x_{i_1} = x_{i_2} = x_{i_3} = 1$, then points o_j and p_j must be observed by blue guards (see Fig. 7(e)).

On the other hand, if at least one of the variables x_{i_1}, x_{i_2}, and x_{i_3} is 1 (see Fig. 8), then points o_j and p_j can be observed by blue guards.

Figure 9 is an XNOR gadget, which connects clause gadgets c_j and c_{j+1} for every $j \in \{0, 1, \ldots, m-1\}$ (see Fig. 11). Clauses c_j and c_{j+1} have two distinct colors (see Fig. 9(c)) if and only if point $t_{j,j+1}$ can be observed by neither a blue guard nor a red guard. Figures 10(a, c) and (b, d) are Z-turn and U-turn gadgets, respectively. They connect clause and variable gadgets or clause and XNOR gadgets.

Figure 11 is an orthogonal polygon P with holes transformed from $U = \{x_1, x_2, x_3, x_4\}$ and $C = \{c_1, c_2, c_3\}$, where $c_1 = \{x_1, \overline{x_2}, x_3\}$, $c_2 = \{x_1, x_2, \overline{x_4}\}$, and $c_3 = \{\overline{x_2}, \overline{x_3}, x_4\}$. In this figure, $c_0 = \{x_0, x_0, x_0\}$ is a dummy clause, where x_0 is a dummy variable.

Lemma 1. *The instance C of 3SAT is satisfiable if and only if there is a set of 2-colored guards with r-visibility for P with holes such that no two guards with the same color have overlapping visibility regions.*

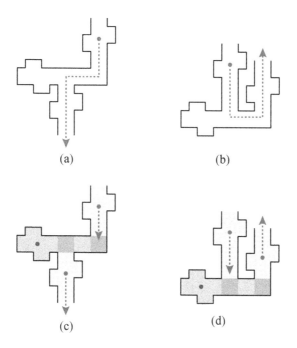

(a)   (b)

(c)   (d)

Fig. 10. (a, c) Z-turn gadget. (b, d) U-turn gadgets.

Proof. ($\Rightarrow$) Suppose that the instance C of 3SAT is satisfiable. In Fig. 11, *start point* s_0 can be observed by a blue guard, since the dummy clause $c_0 = \{x_0, x_0, x_0\}$ is satisfied if the dummy variable $x_0 = 1$. Then, *target point* $t_{0,1}$ can be observed by a red guard if c_1 is satisfied. Suppose that c_0 and c_1 are satisfied. Then, target point $t_{1,2}$ can be observed by a red guard if c_2 is satisfied. By continuing this observation, one can see that all target points $t_{0,1}, t_{1,2}, \ldots, t_{m-1,m}$ can be observed by red guards if all of $c_1, c_2, \ldots, c_m$ are satisfied.

($\Leftarrow$) Suppose that the instance C of 3SAT is not satisfiable. Consider an arbitrary assignment $(b_1, b_2, \ldots, b_n) \in \{0, 1\}^n$ for $(x_1, x_2 \ldots, x_n)$. Since C is not satisfiable, there exists at least one clause $c_j = \{x_{i_{j_1}}, x_{i_{j_2}}, x_{i_{j_3}}\}$ such that $x_{i_{j_1}} = x_{i_{j_2}} = x_{i_{j_3}} = 0$ when the assignment is $(b_1, b_2, \ldots, b_n)$.

Furthermore, for the assignment $(b_1, b_2, \ldots, b_n)$, there exists at least one clause $c_k = \{x_{i_{k_1}}, x_{i_{k_2}}, x_{i_{k_3}}\}$ such that $x_{i_{k_1}} = x_{i_{k_2}} = x_{i_{k_3}} = 1$ because of the following reason: Assume for contradiction that there is no $c_k = \{x_{i_{k_1}}, x_{i_{k_2}}, x_{i_{k_3}}\}$ such that $x_{i_{k_1}} = x_{i_{k_2}} = x_{i_{k_3}} = 1$. Then, every clause contains at least one literal x whose value is 0. Now, consider the "inverted" assignment $(\overline{b_1}, \overline{b_2}, \ldots, \overline{b_n})$. For the inverted assignment, every clause contains at least one literal of value $\overline{x} = 1$. This implies that C is satisfiable, a contradiction.

Therefore, in any unsatisfiable instance C of 3SAT, there are two clauses $c_j = \{x_{i_{j_1}}, x_{i_{j_2}}, x_{i_{j_3}}\}$ and $c_k = \{x_{i_{k_1}}, x_{i_{k_2}}, x_{i_{k_3}}\}$ such that $x_{i_{j_1}} = x_{i_{j_2}} = x_{i_{j_3}} = 0$ and $x_{i_{k_1}} = x_{i_{k_2}} = x_{i_{k_3}} = 1$ for every assignment $(b_1, b_2, \ldots, b_n)$. The 2-color

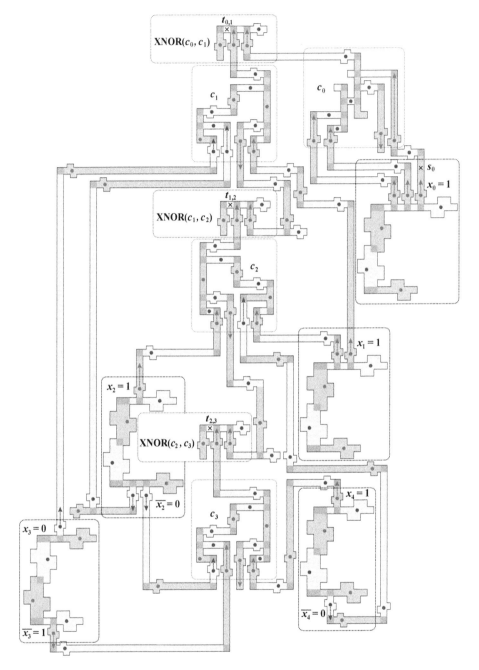

Fig. 11. Orthogonal polygon with holes transformed from $C = \{c_1, c_2, c_3\}$, where $c_1 = \{x_1, \overline{x_2}, x_3\}$, $c_2 = \{x_1, x_2, \overline{x_4}\}$, and $c_3 = \{\overline{x_2}, \overline{x_3}, x_4\}$. In this figure, $c_0 = \{x_0, x_0, x_0\}$ is a dummy clause, where x_0 is a dummy variable. From the positions of red and blue guards, one can see that $(x_1, x_2, x_3, x_4) = (1, 1, 0, 1)$ satisfies all the clauses. (Color figure online)

guard sets of c_j and c_k are given in Figs. 7(d) and (e), respectively. If $j < k$, there exists an integer $j' \in \{j, j+1, \ldots, k-1\}$ such that the point $t_{j',j'+1}$ is observed by neither a blue guard nor a red guard (see Fig. 9(c)). The case $k < j$ is similar. This completes the proof of Lemma 1.

References

1. Bärtschi, A., Suri, S.: Conflict-free chromatic art gallery coverage. In: 29th International Symposium on Theoretical Aspects of Computer Science, pp. 160–171 (2012)
2. Çağırıcı, O., Ghosh, S.K., Hliněný, P., Roy, R.: On conflict-free chromatic guarding of simple polygons. arXiv:1904.08624, 26 pages (2019)
3. Chvátal, V.: A combinatorial theorem in plane geometry. J. Comb. Theory B **18**, 39–41 (1975)
4. Eidenbenz, S.J., Stamm, C., Widmayer, P.: Inapproximability results for guarding polygons and terrains. Algorithmica **31**, 79–113 (2001)
5. Erickson, L.H., LaValle, S.M.: A chromatic art gallery problem. Technical report, Department of Computer Science, University of Illinois at Urbana-Champaign (2011)
6. Fekete, S.P., Friedrichs, S., Hemmer, M., Mitchell, J.B.M., Schmidt, C.: On the chromatic art gallery problem. In: 30th Canadian Conference on Computational Geometry, Halifax, pp. 73–79 (2014)
7. Garey, M.R., Johnson, D.S.: Computers and Intractability: A Guide to the Theory of NP-Completeness. W.H. Freeman, New York (1979)
8. Hoffmann, F.: On the rectilinear art gallery problem. In: Paterson, M.S. (ed.) ICALP 1990. LNCS, vol. 443, pp. 717–728. Springer, Heidelberg (1990). https://doi.org/10.1007/BFb0032069
9. Hoffman, F., Kriegel, K., Suri, S., Verbeek, K., Willert, M.: Tight bounds for conflict-free chromatic guarding of orthogonal art galleries. Comput. Geom. Theory Appl. **73**, 24–34 (2018)
10. Iwamoto, C., Kume, T.: Computational complexity of the r-visibility guard set problem for polyominoes. In: Akiyama, J., Ito, H., Sakai, T. (eds.) JCDCGG 2013. LNCS, vol. 8845, pp. 87–95. Springer, Cham (2014). https://doi.org/10.1007/978-3-319-13287-7_8
11. Katz, M.J., Roisman, G.S.: On guarding the vertices of rectilinear domains. Comput. Geom. Theory Appl. **39**(3), 219–228 (2008)
12. Lee, D.T., Lin, A.K.: Computational complexity of art gallery problems. IEEE Trans. Inf. Theory **32**(2), 276–282 (1986)
13. O'Rourke, J.: Art Gallery Theorems and Algorithms. Oxford University Press, New York (1987)
14. Pilz, A: Planar 3-SAT with a clause-variable cycle. In: 16th Scandinavian Symposium and Workshops on Algorithm Theory, pp. 31:1–31:13 (2018)
15. Preparata, F.P., Shamos, M.I.: Computational Geometry: An Introduction. MCS. Springer, New York (1985). https://doi.org/10.1007/978-1-4612-1098-6
16. Schuchardt, D., Hecker, H.-D.: Two NP-hard art-gallery problems for orthopolygons. Math. Logic Q. **41**(2), 261–267 (1995)
17. Worman, C., Keil, J.M.: Polygon decomposition and the orthogonal art gallery problem. Int. J. Comput. Geom. Appl. **17**(2), 105–138 (2007)

Maximum Bipartite Subgraph
of Geometric Intersection Graphs

Satyabrata Jana[1], Anil Maheshwari[2], Saeed Mehrabi[2(✉)], and Sasanka Roy[1]

[1] Indian Statistical Institute, Kolkata, India
satyamtma@gmail.com, sasanka.ro@gmail.com
[2] School of Computer Science, Carleton University, Ottawa, Canada
anil@scs.carleton.ca, saeed.mehrabi@carleton.ca

Abstract. We study the *Maximum Bipartite Subgraph* (MBS) problem, which is defined as follows. Given a set S of n geometric objects in the plane, we want to compute a maximum-size subset $S' \subseteq S$ such that the intersection graph of the objects in S' is bipartite. We first show that the MBS problem is NP-hard on geometric graphs for which the maximum independent set is NP-hard (hence, it is NP-hard even on unit squares and unit disks). On the algorithmic side, we first give a simple $\mathcal{O}(n)$-time algorithm that solves the MBS problem on a set of n intervals. Then, we give an $\mathcal{O}(n^2)$-time algorithm that computes a near-optimal solution for the problem on circular-arc graphs. Moreover, for the approximability of the problem, we first present a PTAS for the problem on unit squares and unit disks. Then, we present efficient approximation algorithms with small-constant factors for the problem on unit squares, unit disks and unit-height rectangles. Finally, we study a closely related geometric problem, called *Maximum Triangle-free Subgraph (MTFS)*, where the objective is the same as that of MBS except the intersection graph induced by the set S' needs to be triangle-free only (instead of being bipartite).

Keywords: Bipartite subgraph · Geometric intersection graphs ·
NP-hardness · Approximation schemes · Triangle-free subgraph

1 Introduction

In this paper, we study the following geometric problem. Given a set S of n geometric objects in the plane, we are interested in computing a maximum-size subset $S' \subseteq S$ such that the intersection graph induced by the objects in S' is bipartite. We refer to this problem as the *Maximum Bipartite Subgraph (MBS)* problem. The MBS problem is closely related to the Odd Cycle Transversal (OCT) problem: given a graph G, the objective of the OCT problem is to compute a minimum-cardinality subset of $S \subseteq V(G)$ such that the intersection of S and the vertices of every odd cycle of the graph is non-empty. Notice that MBS and OCT

This work is supported in part by Natural Sciences and Engineering Research Council of Canada (NSERC).

© Springer Nature Switzerland AG 2020
M. S. Rahman et al. (Eds.): WALCOM 2020, LNCS 12049, pp. 158–169, 2020.
https://doi.org/10.1007/978-3-030-39881-1_14

are equivalent for the class of graphs on which OCT is polynomial-time solvable: an exact solution S for OCT gives $V(G) \setminus S$ as an exact solution for MBS within the same time bound (see below for a summary of the main known results on the OCT problem). However, on classes of graphs for which OCT is NP-hard, an α-approximation algorithm for OCT might not provide any information on the approximability of MBS on the same classes of graphs.

Another problem that is related to MBS is the Feedback Vertex Set (FVS) problem. The objective of FVS is the same as that of OCT except the set S has a non-empty intersection with every cycle of the graph (not only the odd ones). The FVS problem has been extensively studied in graph theory from both hardness [11,29] and algorithmic [4,9,14,17] points of view.

We also study a simpler variant of MBS, called the *Maximum Triangle-free Subgraph (MTFS)* problem. Let S be a set of n geometric objects in the plane. Then, the objective of the MTFS problem is to compute a maximum-size subset $S' \subseteq S$ such that the intersection graph induced by the objects in S' is triangle free (as opposed to being bipartite).

Related Work. The MBS problem is NP-complete for planar graphs with maximum degree four [6]. For graphs with maximum degree three, Choi et al. [6] showed that for a given constant k there is a vertex set of size k or less whose removal leaves an induced bipartite subgraph if and only if there is an edge set of size k or less whose removal leaves a bipartite spanning subgraph. As edge deletion graph bipartization problem is NP-complete for cubic graphs [28], the MBS problem is NP-complete for cubic graphs. Moreover, since the maximum edge deletion graph bipartization problem is solvable in $\mathcal{O}(n^3)$ time for planar graphs [1,12] where n is the number of vertices of the input graph, this immediately implies that MBS is $\mathcal{O}(n^3)$-time solvable for planar graphs with maximum degree three. For the vertex-weighted version of the MBS problem, Baiou et al. [2] showed that the MBS problem can be solved in $\mathcal{O}(n^{3/2} \log n)$ time for planar graphs with maximum degree three. Finally, Cornaz et al. [8] considered the maximum induced bipartite subgraph problem: given a graph with non-negative weights on the edges, the goal is to find a maximum-weight bipartite subgraph. An edge subset $F \subseteq E$ is called independent if the subgraph induced by the edges in F (incident vertices) is bipartite; otherwise, it is called dependent. They showed that the minimum dependent set problem with non-negative weights can be solved in polynomial time.

The OCT problem is known to be NP-complete on planar graphs with degree at most 6 [6]. For planar graphs with degree at most 3, OCT can be solved in $\mathcal{O}(n^3)$ time [6] (even the weighted version of the problem). There are several results known concerning the parameterized complexity of OCT (i.e., given a graph G on n vertices and an integer k, is there a vertex set U in G of size at most k such that $G \setminus U$ is bipartite). Reed et al. [26] first gave an algorithm with running time $\mathcal{O}(4^k kmn)$. Lokshtanov et al. [19] improved this running time to $\mathcal{O}(3^k kmn)$. Lokshtanov et al. [20] provide an algorithm with running time $\mathcal{O}(2^{\mathcal{O}(k \log k)} n)$ for planar graphs. Moreover, assuming the exponential time hypothesis, the running time cannot be improved to $2^{\mathcal{O}(k)} n^{\mathcal{O}(1)}$.

The MBS problem is also closely related to the Maximum Independent Set (MIS) problem. Observe that any feasible solution for MIS is also a feasible solution for MBS and, moreover, a feasible solution $S \subseteq V(G)$ for the MBS problem provides a feasible solution of size at least $|S|/2$ for the MIS problem. Hence, $\mathrm{OPT}(\mathtt{MIS}) \leq \mathrm{OPT}(\mathtt{MBS}) \leq 2\mathrm{OPT}(\mathtt{MIS})$. This implies that an α-approximation algorithm for MIS on a class of graphs is a 2α-approximation for MBS on the same class. In particular, the PTASes for MIS on unit disks and unit squares [13] imply polynomial-time $(2 + \epsilon)$-approximation algorithms for MBS on unit disks and unit squares. Moreover, we obtain a polynomial-time $\mathcal{O}(\log \log n)$-approximation [5] (or, $\mathcal{O}(\log \mathrm{OPT})$-approximation [3]) algorithm for MBS on rectangles.

Our Results. In this paper, we present the following results.

- On the hardness side, we show that the MBS problem is NP-hard on the classes of geometric intersection graphs for which MIS is NP-hard (Sect. 2); this in particular includes unit disks and unit squares. We also extend this result to a corresponding W[1]-hardness result.
- On the algorithmic side, we give a linear-time algorithm for MBS on interval graphs, and an $\mathcal{O}(n^2)$-time algorithm that computes a near-optimal solution for MBS on any circular-arc graph with n vertices (Sect. 3).
- On the approximation side, we obtain a PTAS for the MBS problem on unit disks and unit squares. For a set of n unit-height rectangles in the plane, we give an $\mathcal{O}(n \log n)$-time 2-approximation algorithm for the problem. Moreover, we design an $\mathcal{O}(n^2)$-time 4-approximation algorithm for the same problem on unit disks (Sect. 4).
- Finally, we show that the MTFS problem is NP-hard on the intersection graph of axis-parallel rectangles in the plane (Sect. 5).

2   NP-Hardness

In this section, we show that the MBS problem is NP-complete on the classes of geometric intersection graphs for which MIS is NP-complete. The MIS problem is known to be NP-complete on a wide range of geometric intersection graphs, even restricted to unit disks and unit squares [7], 1-string graphs [16], and B_1-VPG graphs [18]. Let $G = (V, E)$ be an intersection graph induced by a set S of n geometric objects in the plane. We construct a new graph G' from the disjoint union of two copies of G by adding edges as follows. For each vertex in V, we add an edge from each vertex in one copy of G to the corresponding vertex in the other copy. For each edge $(u, v) \in E$, we add four edges $(u, v), (u', v'), (u, v')$, and (v, u') to G', where u' and v' are the corresponding vertices of u and v, respectively in the other copy. Graph G' is the intersection graph of $2n$ geometric objects S, where each object has occurred twice in the same position. Clearly, the number of vertices and edges in G' are polynomial in the number of vertices of G; hence, the construction can be done in polynomial time.

Lemma 1. *G has an independent set of size at least k if and only if G' has a bipartite subgraph of size at least $2k$.*

Proof. Let U be an independent set of G with $|U| \geq k$. Let H be the subgraph of G' induced by U along with all the corresponding vertices of U in the other copy. Then, H is a bipartite subgraph with size at least $2k$. Conversely, if G' has a bipartite subgraph of size at least $2k$, then G' must have an independent set of size at least k. By the construction of G', if G' has an independent set of size at least k, then G must have an independent set of size at least k.   □

By Lemma 1, we have the following theorem.

Theorem 1. *The MBS problem is NP-complete on the classes of geometric intersection graphs for which MIS is NP-complete.*

Remark. By the definition of parameterized reduction [10], one can verify that the above reduction is in fact a parameterized reduction and so we have the following result.

Corollary 1. *The MBS problem is W[1]-complete on the classes of geometric intersection graphs for which MIS is W[1]-complete.*

We note that Marx [22,23] proved that MIS is W[1]-complete on unit squares, unit disks, and even unit line segments. As such, by Corollary 1, the MBS problem is W[1]-complete on all these geometric intersection graphs.

3   Algorithmic Results

In this section, we present our algorithms for the MBS problem on interval graphs and circular-arc graphs. We start with interval graphs.

3.1   Interval Graphs

In this section, we consider the MBS problem on a set S of n intervals and give a linear-time algorithm for the problem. Notice that for interval graphs, the MBS problem is the same as FVS; to the best of our knowledge, the best-known algorithm for solving FVS on interval graphs takes $\mathcal{O}(|V| + |E|))$ time [21]. Since interval graphs are a subclass of chordal graphs, the MBS problem on interval graphs reduces to the problem of computing a maximum-size subset of intervals in S whose induced graph is triangle free. Consequently, a point can "stab" at most two intervals in any feasible solution for the MBS problem on intervals. Algorithm 1 exploits this property to solve the problem exactly.

In the following, we assume that (i) the endpoints of intervals in S are $2n$ distinct points on the real line, and (ii) the intervals are sorted from left to right by the increasing order of their right endpoint; we denote them as $I_1, I_2, \ldots, I_n$. Moreover, the variable x (resp., y) denotes the x-coordinate of the rightmost point on the real line such that it is contained in two intervals (resp., one interval) of the current solution computed by the algorithm. For an interval I, we denote the left and right endpoints of I by $\texttt{left}(I)$ and $\texttt{right}(I)$, respectively.

Algorithm 1. BIPARTITEINTERVAL(S)

1: let initially $M = \emptyset$;
2: $x := -\infty$ and $y := -\infty$;
3: **for** $i := 1$ to n **do**
4:      **if** left(I_i) $> y$ **then**
5:          $M := M \cup I_i$;
6:          $y := $ right(I_i);
7:      **else if** $x < $ left(I_i) $< y$ **then**
8:          $M := M \cup I_i$;
9:          $x := y$;
10:         $y := $ right(I_i);
11:     **end if**
12: **end for**
13: **return** M;

Correctness. Let M_i, for all $1 \leq i \leq n$, denote the set M at the end of iteration i of the for-loop. Consider the following invariant.

Invariant I. For all $i = 1, \ldots, n$, at the end of iteration i of the for-loop, the set M_i is an optimal solution for the set of intervals $\{I_1, I_2, \ldots, I_i\}$.

We prove Invariant I by induction on $|S|$. If $|S| = 1$, then $M = \{I_1\}$ by line 5 of the algorithm and we are done. Moreover, if $|S| = 2$, then there are two cases depending on whether $\{I_1, I_2\}$ form a clique or an independent set. In either case, $M = \{I_1, I_2\}$ and we are done. Now, suppose that Invariant I is true for all $|S| = 1, 2, \ldots, n - 1$. Let S be a set of n intervals and consider the set $S \setminus I_n$ (where I_n is the interval with rightmost right endpoint in S). By induction hypothesis, let M_{n-1} be the optimal solution for $S \setminus I_n$ computed by the algorithm and consider the values of x and y before returning M_{n-1} in line 13. We must have that either (i) left(I_n) $> y$, (ii) $x < $ left(I_n) $< y$, or (iii) left(I_n) $< x$. In cases (i) and (ii), the algorithm adds I_n to M_{n-1} resulting in an optimal solution. In case (iii), the algorithm returns M_{n-1} without adding I_n to the solution. Observe that this is optimal as no feasible solution can add I_n.

The algorithm clearly runs in time linear in n and so we have the following theorem.

Theorem 2. *The MBS problem on a set of n intervals can be solved in $\mathcal{O}(n)$ time, assuming that the intervals are already sorted on their right endpoint.*

3.2 Circular-Arc Graphs

We now give a near-optimal solution for the MBS problem on circular-arc graphs. For an optimization problem, a *near-optimal* solution is a feasible solution whose objective function value is within a specified range from the optimal objective function value. A circular-arc graph is the intersection graph of arcs on a circle. That is, every vertex is represented by an arc, and there is an edge between two vertices if and only if the corresponding arcs intersect. Observe that interval graphs are a proper subclass of circular-arc graphs. For the rest of this section, let $G = (V, E)$ be a circular-arc graph and assume that a geometric representation of G (i.e., a set of $|V(G)|$ arcs on a circle $\mathcal{C}$) is given as part of the input. First, we prove the following lemmas.

Lemma 2. *If G is triangle-free, then it can have at most one cycle.*

Proof. Suppose for the sake of contradiction that G has more than one cycle. Let A_1 and A_2 be two cycles of G. Now, since G is a triangle-free circular-arc graph, the corresponding arcs of the vertices of any cycle in G together cover the circle $\mathcal{C}$. So, there must exist three distinct vertices $v \in A_1, u \in A_1$ and $w \in A_2$ such that v, u, w are pairwise adjacent. Which is a contradiction to the fact that G is triangle-free. □

Lemma 3. *If B and T are optimal solutions for the MBS and MTFS problems on G, respectively, then $|T| - 1 \le |B| \le |T|$.*

Proof. Since a bipartite subgraph contains no triangle, $|B| \le |T|$. Now, if $G[T]$ (i.e., the subgraph of G induced by T) is odd-cycle free, then it induces a bipartite subgraph. Otherwise, $G[T]$ can have at most one cycle by Lemma 2. If this cycle is odd, then by removing any single vertex form the cycle, we obtain a bipartite subgraph of G with size at least $|T| - 1$. □

Since $G[T]$ contains at most one cycle, following lemma trivially holds.

Lemma 4. *If H is a maximum-size induced forest in G, then $|V(H)| \ge |T| - 1$.*

By the above lemmas, our goal now is to find a maximum acyclic subgraph H of G. Notice that there must be a clique K ($|K| \ge 1$) in G that is not in H. Now, for each arc u in the circular-arc representation of G, let $l(u)$ and $r(u)$ denote the two endpoints of u in the clockwise order of the endpoints u. Then, we consider two vertex sets $S_u^1 = \{w\colon w \in V, l(u) \notin [l(w), r(w)]\}$ and $S_u^2 = \{z\colon z \in V, r(u) \notin [l(z), r(z)]\}$. Both S_u^1 and S_u^2 are interval graphs. Since there are n vertices in G, we compute $2n$ interval graphs in total. Then, for each of these interval graphs, we apply Algorithm 1 to compute an optimal solution for MBS, and will return the one with maximum size as the final solution. Since Algorithm 1 runs in $\mathcal{O}(n)$ time, the total time to find H is $\mathcal{O}(n^2)$; so we have the following theorem.

Theorem 3. *Let OPT be a maximum-size induced bipartite subgraph of a circular-arc graph G with n vertices. Then, there is an algorithm that computes an induced bipartite subgraph H of G such that $|V(H)| \ge |OPT| - 1$. The algorithm runs in $\mathcal{O}(n^2)$ time.*

4   Approximation Algorithms

Recall that since MIS is NP-complete on unit disks and unit squares, the MBS problem is NP-complete on these graphs by Theorem 1. In this section, we first give PTASes for MBS on both unit squares and unit disks, and will then consider the problem on unit-height rectangles.

4.1   Unit Disks and Unit Squares

We first show the PTAS for unit disks, and will then discuss it for unit squares as well as the *weighted* MBS problem.

Let S be a set of n unit disks in the plane, and let $k > 1$ be a fixed integer. A PTAS running in $\mathcal{O}(n^{\mathcal{O}(1)} \cdot n^{\mathcal{O}(1/\epsilon^2)})$ time, for any $\epsilon > 0$, is straightforward using the shifting technique of Hochbaum and Maass [13] and the following packing argument: for an instance of the MBS problem, where the unit disks lie inside a $k \times k$ square, an optimal solution cannot have more than k^2 unit disks. Hence, we can compute an exact solution for such an instance of the problem in $\mathcal{O}(n^{\mathcal{O}(1)} \cdot n^{\mathcal{O}(k^2)})$ time. Consequently, by setting $k = 1/\epsilon$, we obtain a PTAS running in time $\mathcal{O}(n^{\mathcal{O}(1)} \cdot n^{\mathcal{O}(1/\epsilon^2)})$.

To improve the running time to $\mathcal{O}(n^{\mathcal{O}(1)} \cdot n^{\mathcal{O}(1/\epsilon)})$, we rely on the shifting technique again, but instead of applying the plane partitioning twice, we only partition the plane into horizontal slabs and solve the MBS problem for each of them exactly. This gives us the desired running time for our PTAS. We next describe the details of how to solve MBS exactly for a slab.

Algorithm for a Slab. Let H be a horizontal slab of height k and let $D \subseteq S$ be the set of disks that lie entirely inside H. The idea is to build a vertex-weighted directed acyclic graph G such that finding a maximum-weight path from the source vertex to the target vertex corresponds to an exact solution for the MBS problem [24]. To this end, let a and b ($a < b$) be two integers such that every disk in D lies inside the rectangle R bounded by H and the vertical lines $x = a$ and $x = b$. Partition R vertically into unit-width boxes B_i, where the left side of B_i has the x-coordinate $a + i$, for all integers $0 \leq i < b - a$; let $D_i \subseteq D$ denote the set of disks whose centers lie inside B_i. Since B_i has height k and width 1, we can compute all feasible (not necessarily exact) solutions for the MBS problem on D_i in $\mathcal{O}(n^{\mathcal{O}(1)} \cdot n^{\mathcal{O}(k)})$ time; let $\mathcal{M}_i$ be the set of all such feasible solutions. We now build a directed vertex-weighted acyclic graph G as follows. The vertex set of $V(G)$ is $V \cup \{s, t\}$, where V has one vertex for each solution in $\mathcal{M}_i$, for all i. Moreover, the weight of each vertex is the number of disks in the corresponding bipartite graph. For every pair i, j, where $1 \leq i < j < n$, consider two solutions $M \in \mathcal{M}_i$ and $M' \in \mathcal{M}_j$. Then, there exists an edge from the vertex of M to that of M' in G if the intersection graph induced by the disks in $M \cup M'$ is bipartite. Finally, for all i and for all $M \in \mathcal{M}_i$: there exists an edge from s to M, and there exists an edge from M to t. The weights of vertices s and t are zero.

Lemma 5. *The MBS problem has a feasible solution of size k on G if and only if there exists a directed path from s to t with the total weight k.*

Proof. For a given directed st-path with total weight k, let X be the union of all the disks corresponding to the interval vertices of this path. Then, the intersection graph of X is bipartite because the disks in $X \cap \mathcal{M}_i$ are disjoint from the disks in $S \cap \mathcal{M}_j$ when $j > i + 1$. Moreover, when $j = i + 1$, the disks in $X \cap (\mathcal{M}_i \cup \mathcal{M}_j)$ must form an induced bipartite graph by the definition of an edge in G. Since the total weight of the vertices on the path is k, we have

$|X| = k$. On the other hand, let Y be a feasible solution of size k for the MBS problem on G. Then, the intersection graph of disks in $Y \cap D_i$ is bipartite, for all i. Hence, selecting the vertices corresponding to $Y \cap D_i$ for all i gives us a path with total weight k from s to t.   □

By Lemma 5, the MBS problem for H reduces to the problem of finding the maximum-weighted path from s to t on G. The number of vertices of G that correspond to feasible solutions for the MBS problem on disks in $S \cap D_i$ is bounded by $\mathcal{O}(n^{\mathcal{O}(k)})$, which is the bound on the number of vertices of G that correspond to these feasible solutions. Hence, we can compute the edge set of G in $\mathcal{O}(n^{\mathcal{O}(1)} \cdot n^{\mathcal{O}(k)})$ time (we can check whether a subset of disks form a bipartite graph in $\mathcal{O}(n^{\mathcal{O}(1)})$ time). Since G is directed and acyclic, the maximum-weighted st-path problem can be solved in linear time. Therefore, by setting $k = 1/\epsilon$, we have the following theorem.

Theorem 4. *There exists a PTAS for MBS on unit disks that runs in $\mathcal{O}(n^{\mathcal{O}(1)} \cdot n^{\mathcal{O}(1/\epsilon)})$ time, for any $\epsilon > 0$.*

PTAS on Unit Squares. One can verify that the above algorithm can be applied to obtain a PTAS for MBS on a set of n unit squares, as well. Moreover, the algorithm extends to the weighted MBS problem on unit disks and unit squares. The only modification is, instead of assigning the number of disks (resp., squares) in a solution as the weight of the corresponding vertex, we assign the total weight of the disks (resp., squares) in the solution as the vertex weight.

Theorem 5. *There exists a PTAS for the MBS problem on unit squares running in $\mathcal{O}(n^{\mathcal{O}(1)} \cdot n^{\mathcal{O}(1/\epsilon)})$ time, for any $\epsilon > 0$. Moreover, the weighted MBS problem also admits a PTAS running within the same time bound on unit disks and unit squares.*

A 4-Approximation on Unit Disks. Recall from Sect. 1 that OPT(MIS) $\leq$ OPT(MBS) $\leq$ 2OPT(MIS). Nandy et al. [25] designed a factor-2 approximation algorithm for the MIS problem on unit disks, which runs in $\mathcal{O}(n^2)$ time. Consequently, we obtain an $\mathcal{O}(n^2)$-time 4-approximation algorithm for the MBS problem on unit disks.

4.2   Unit-Height Rectangles

Here, we give an $\mathcal{O}(n \log n)$-time 2-approximation algorithm for MBS on a set of n unit-height rectangles. To this end, suppose that the bottom side of the bottommost rectangle has y-coordinate a and the top side of the topmost rectangle has y-coordinate b. Consider the set of horizontal lines $y := a + i + \epsilon$ for all $i = 0, \ldots, b$, where $\epsilon > 0$ is a small constant; we may assume w.l.o.g. that each rectangle intersects exactly one line. Ordering the lines from bottom to top, let S_i be the set of rectangles that intersect the horizontal line i. We now run BIPARTITEINTERVAL(S), once for when $S = S_1 \cup S_3 \cup S_5 \ldots$ and once for when $S = S_2 \cup S_4 \cup S_6 \ldots$, and will then return the largest of these two solutions. We perform an initial sorting that takes $\mathcal{O}(n \log n)$ time, and BIPARTITEINTERVAL(S) runs in $\mathcal{O}(n)$ time. This gives us the following theorem.

Theorem 6. *There exists an $\mathcal{O}(n \log n)$-time 2-approximation algorithm for the* MBS *problem on a set of n unit-height rectangles in the plane.*

5   NP-**Hardness of** MTFS

Here, we show that MTFS problem is NP-hard when geometric objects are axis-parallel rectangles. We give a reduction from the independent set problem on 3-regular planar graphs, which is known to be NP-complete [11].

Rim et al. [27] proved that MIS is NP-hard for planar rectangle intersection graphs with degree at most 3. They also gave a reduction from the independent set problem on 3-regular planar graphs. Given a 3-regular planar graph $G = (V, E)$, they construct an instance $H = (V', E')$ of the MIS problem on rectangle intersection graphs. First we outline their construction of H from G. For any cubic planar graph G, it is always possible to get a rectilinear planar embedding of G such that each vertex $v \in V$ is drawn as a point p_v, and each edge $e = (u, v) \in E$ is drawn as a rectilinear path, connecting the points p_u and p_v, having at most four bends, and thus consisting of at most five straight line segments. They [27] construct a family of rectangles B in the following way. For each point p_{v_i} where $v_i \in V$, a rectangle b_i is placed surrounding the point p_{v_i}. In each rectilinear path connecting p_{v_i} and p_{v_j}, they place six rectangles $b_{ij}^1, b_{ij}^2, \ldots, b_{ij}^6$ such that (i) b_i intersects b_{ij}^1, (ii) b_j intersects b_{ij}^6, (iii) b_{ij}^k intersects b_{ij}^{k+1} for $k = 1, 2, \ldots, 5$ (iv) $b_{ij}^1, b_{ij}^2, \ldots, b_{ij}^6$ do not intersect any other rectangles in B. For an illustration see Fig. 1.

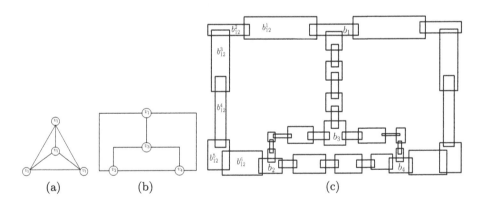

Fig. 1. (a) A cubic planar graph G. (b) A rectilinear embedding of G. (c) Family of rectangle B.

Clearly, $H(V', E')$ is an axis-parallel rectangle intersection graph with degree at most 3 where $|V'| = |V| + 6|E|$ and $|E'| = 7|E|$. In their reduction, the following lemma holds.

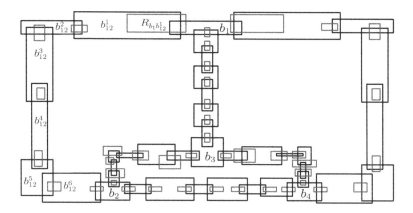

Fig. 2. Construction of an instance of MTFS problem from B. (Color figure online)

Lemma 6. *[27] $G = (V, E)$ has an independent set of size $\geq m$ if and only if H has an independent set of size $m + 3|E|$.*

Given H, we construct an instance G_H of MTFS problem in axis-parallel rectangles intersection graphs. For the sake of understanding, let all rectangles corresponding to vertices in H, i.e., all rectangles in B, be colored black. To get G_H, we insert a family R of red rectangles in the following way. For each pair of adjacent rectangles b and b' in B, we place a red rectangle $R_{b,b'}$ such that (i) $R_{b,b'}$ intersects both b and b', (ii) $R_{b,b'}$ does not intersect any other rectangles in $B \cup R$. As per construction of H, it is always possible to place such a rectangle for each pair of adjacent rectangles in B. See Fig. 2 for an illustration of this transformation. This completes the construction of our instance of the MTFS problem on axis-parallel rectangles intersection graphs. Since R contains $7|E|$ rectangles, the above transformation can be done in polynomial time. Now G_H is an axis-parallel rectangle intersection graph with underlying geometric objects $B \cup R$. Clearly the number of vertices in G_H is $(|V| + 13|E|)$. We now prove the following lemma whose proof is given in the full version of the paper [15] due to space constraints.

Lemma 7. *H has an independent set of size $\geq k$ if and only if G_H has a triangle-free subgraph on $\geq k + 7|E|$ vertices.*

By Lemmas 6 and 7, we have the following.

Lemma 8. *$G = (V, E)$ has an independent set of size $\geq m$ if and only if G_H has a triangle-free subgraph on $m + 10|E|$ vertices.*

We can now conclude the following theorem.

Theorem 7. *The MTFS problem is NP-complete on axis-parallel rectangle intersection graphs.*

6   Conclusion

In this paper, we studied the problem of computing a maximum-size bipartite subgraph on geometric intersection graphs. We showed that the problem is NP-hard on the geometric graphs for which maximum independent set is NP-hard. On the positive side, we gave polynomial-time algorithms for solving the problem on interval graphs and circular-arc graphs. We furthermore obtained several approximation algorithms for the problem on unit squares, unit disks, and unit-height rectangles. Finally, we showed the NP-hardness of a simpler problem in which the goal is to compute a maximum-size induced triangle-free subgraph. We conclude by the following open questions:

- Does MBS admit a PTAS on unit-height rectangles, or is it APX-hard?
- Is there a polynomial-time algorithm for MBS on a set of n unit disks intersecting a common horizontal line?
- Can we improve the 4-approximation algorithm for unit disks with the same time bound $\mathcal{O}(n^2)$ (or, even better)?

Acknowledgment. We thank Michiel Smid for useful discussions on the problem.

References

1. Aoshima, K., Iri, M.: Comments on F. Hadlock's paper: finding a maximum cut of a planar graph in polynomial time. SIAM J. Comput. **6**(1), 86 (1977)
2. Baïou, M., Barahona, F.: Maximum weighted induced bipartite subgraphs and acyclic subgraphs of planar cubic graphs. SIAM J. Discrete Math. **30**(2), 1290–1301 (2016)
3. Bose, P., et al.: Computing maximum independent set on outerstring graphs and their relatives. In: Proceedings of the 16th International Symposium on Algorithms and Data Structures (WADS 2019), Edmonton, AB, Canada (2019)
4. Brandstädt, A.: On improved time bounds for permutation graph problems. In: Mayr, E.W. (ed.) WG 1992. LNCS, vol. 657, pp. 1–10. Springer, Heidelberg (1993). https://doi.org/10.1007/3-540-56402-0_30
5. Chalermsook, P., Chuzhoy, J.: Maximum independent set of rectangles. In: Proceedings of the 20th Annual ACM-SIAM Symposium on Discrete Algorithms (SODA 2009), New York, NY, USA, 4–6 January 2009, pp. 892–901 (2009)
6. Choi, H., Nakajima, K., Rim, C.S.: Graph bipartization and via minimization. SIAM J. Discrete Math. **2**(1), 38–47 (1989)
7. Clark, B.N., Colbourn, C.J., Johnson, D.S.: Unit disk graphs. Discrete Math. **86**(1–3), 165–177 (1990)
8. Cornaz, D., Mahjoub, A.R.: The maximum induced bipartite subgraph problem with edge weights. SIAM J. Discrete Math. **21**(3), 662–675 (2007)
9. Daniel Liang, Y., Chang, M.S.: Minimum feedback vertex sets in cocomparability graphs and convex bipartite graphs. Acta Informatica **34**(5), 337–346 (1997)
10. Downey, R.G., Fellows, M.R.: Parameterized Complexity. Monographs in Computer Science. Springer, New York (1999)
11. Garey, M.R., Johnson, D.S.: Computers and Intractability, vol. 29. W.H. Freeman, New York (2002)

12. Hadlock, F.: Finding a maximum cut of a planar graph in polynomial time. SIAM J. Comput. **4**(3), 221–225 (1975)
13. Hochbaum, D.S., Maass, W.: Approximation schemes for covering and packing problems in image processing and VLSI. J. ACM **32**(1), 130–136 (1985)
14. Honma, H., Nakajima, Y., Sasaki, A.: An algorithm for the feedback vertex set problem on a normal helly circular-arc graph. J. Comput. Commun. **4**(08), 23 (2016)
15. Jana, S., Maheshwari, A., Mehrabi, S., Roy, S.: Maximum bipartite subgraph of geometric intersection graphs. CoRR abs/1909.03896 (2019)
16. Kratochvíl, J., Nešetřil, J.: Independent set and clique problems in intersection-defined classes of graphs. Commentationes Mathematicae Universitatis Carolinae **31**(1), 85–93 (1990)
17. Kratsch, D., Müller, H., Todinca, I.: Feedback vertex set on AT-free graphs. Discrete Appl. Math. **156**(10), 1936–1947 (2008)
18. Lahiri, A., Mukherjee, J., Subramanian, C.R.: Maximum independent set on B_1-VPG graphs. In: Lu, Z., Kim, D., Wu, W., Li, W., Du, D.-Z. (eds.) COCOA 2015. LNCS, vol. 9486, pp. 633–646. Springer, Cham (2015). https://doi.org/10.1007/978-3-319-26626-8_46
19. Lokshtanov, D., Saurabh, S., Sikdar, S.: Simpler parameterized algorithm for OCT. In: Combinatorial Algorithms, 20th International Workshop, IWOCA 2009, Hradec nad Moravicí, Czech Republic, 28 June–2 July 2009, Revised Selected Papers. pp. 380–384 (2009)
20. Lokshtanov, D., Saurabh, S., Wahlström, M.: Subexponential parameterized odd cycle transversal on planar graphs. In: IARCS Annual Conference on Foundations of Software Technology and Theoretical Computer Science, FSTTCS 2012, Hyderabad, India, 15–17 December 2012, pp. 424–434 (2012)
21. Lu, C.L., Tang, C.Y.: A linear-time algorithm for the weighted feedback vertex problem on interval graphs. Inf. Process. Lett. **61**(2), 107–111 (1997)
22. Marx, D.: Efficient approximation schemes for geometric problems? In: 13th Annual European Symposium on Algorithms - ESA 2005, Palma de Mallorca, Spain, 3–6 October 2005, Proceedings, pp. 448–459 (2005)
23. Marx, D.: Parameterized complexity of independence and domination on geometric graphs. In: Bodlaender, H.L., Langston, M.A. (eds.) IWPEC 2006. LNCS, vol. 4169, pp. 154–165. Springer, Heidelberg (2006). https://doi.org/10.1007/11847250_14
24. Matsui, T.: Approximation algorithms for maximum independent set problems and fractional coloring problems on unit disk graphs. In: Akiyama, J., Kano, M., Urabe, M. (eds.) JCDCG 1998. LNCS, vol. 1763, pp. 194–200. Springer, Heidelberg (2000). https://doi.org/10.1007/978-3-540-46515-7_16
25. Nandy, S.C., Pandit, S., Roy, S.: Faster approximation for maximum independent set on unit disk graph. Inf. Process. Lett. **127**, 58–61 (2017)
26. Reed, B.A., Smith, K., Vetta, A.: Finding odd cycle transversals. Oper. Res. Lett. **32**(4), 299–301 (2004)
27. Rim, C.S., Nakajima, K.: On rectangle intersection and overlap graphs. IEEE Trans. Circ. Syst. I Fundam. Theory Appl. **42**(9), 549–553 (1995)
28. Yannakakis, M.: Node- and edge-deletion NP-complete problems. In: Proceedings of the 10th Annual ACM Symposium on Theory of Computing, San Diego, California, USA, 1–3 May 1978, pp. 253–264 (1978)
29. Yannakakis, M.: Node-deletion problems on bipartite graphs. SIAM J. Comput. **10**(2), 310–327 (1981)

The Stub Resolution of 1-Planar Graphs

Michael Kaufmann[1], Jan Kratochvil[2], Fabian Lipp[3],
Fabrizio Montecchiani[4(✉)], Chrysanthi Raftopoulou[5], and Pavel Valtr[2]

[1] Universität Tübingen, Tübingen, Germany
mk@informatik.uni-tuebingen.de
[2] Charles University, Prague, Czech Republic
{honza,valtr}@kam.mff.cuni.cz
[3] Universität Würzburg, Würzburg, Germany
fabian.lipp@uni-wuerzburg.de
[4] Università degli Studi di Perugia, Perugia, Italy
fabrizio.montecchiani@unipg.it
[5] National Technical University of Athens, Athens, Greece
crisraft@mail.ntua.gr

Abstract. The resolution of a drawing plays a crucial role when defining criteria for its quality. In the past, grid resolution, edge-length resolution, angular resolution and crossing resolution have been investigated. In this paper, we investigate the *stub resolution*, a recently introduced criterion for nonplanar drawings. A crossed edge is divided into parts, called stubs, which should not be too short for the sake of readability. Thus, the stub resolution of a drawing is defined as the minimum ratio between the length of a stub and the length of the entire edge, over all the edges of the drawing. We consider 1-planar graphs and we explore scenarios in which near optimal stub resolution, i.e. arbitrarily close to $\frac{1}{2}$, can be obtained in drawings with zero, one, or two bends per edge, as well as further resolution criteria, such as angular and crossing resolution. In particular, our main contributions are as follows: (i) Every 1-planar graph with independent crossing edges has a straight-line drawing with near optimal stub resolution; (ii) Every 1-planar graph has a 1-bend drawing with near optimal stub resolution.

1 Introduction

The question of drawing graphs with high resolution is one of the most prominent when it comes to better understanding of a diagram. We quote from an early graph drawing tutorial by Cruz and Tamassia (1994): "Display devices and the human eye have only finite resolution". This viewpoint inspired the convention

---

Research by J. Kratochvíl and P. Valtr was supported by the Czech Science Foundation (GAČR) grant no. 18-19158S. Research by F. Montecchiani partially supported by: (i) MIUR, under Grant 20174LF3T8 AHeAD: efficient Algorithms for HArnessing networked Data. (ii) Dipartimento di Ingegneria dell'Università degli Studi di Perugia, under grant RICBA18WD: "Algoritmi e sistemi di analisi visuale di reti complesse e di grandi dimensioni".

© Springer Nature Switzerland AG 2020
M. S. Rahman et al. (Eds.): WALCOM 2020, LNCS 12049, pp. 170–182, 2020.
https://doi.org/10.1007/978-3-030-39881-1_15

to use an integer underlying grid for the drawings, which guarantees a certain minimum distance between any two vertices, as well as criteria like the ratio between the shortest and the longest edge (known as *edge-length resolution*) [26].

The *angular resolution* of a drawing is the minimum angle that occurs at a vertex (often called *vertex angle*), and it is expressed as a function of the maximum vertex degree of the graph. This branch has been started by Formann et al. [19]. Important contributions on planar graphs have been made by Malitz and Papakostas [27], and by Duncan and Kobourov [18]. An early work by Di Battista and Vismara [12] characterized the realizability of planar straight-line drawings for a given set of vertex angles and lead the way for the minimization of the largest vertex angle. Special graph classes, e.g., trees, allow more direct approaches to get a good angular resolution, especially with respect to the used area (see, e.g., [17,20]). From there, the research line on the planar slope number developed, where only a fixed set of slopes can be used to draw the edges of a graph. While this approach does not lead to good angular resolution for planar straight-line drawings [24], Angelini et al. [2] showed how to compute planar drawings with one bend per edge using a set of slopes that guarantees asymptotically optimal angular resolution.

Huang et al. [22] experimentally showed the detrimental effect on readability when crossing angles are "sharp". This, together with the seminal paper by Didimo et al. [14], started a line of research on nonplanar graph drawings where sharp angles are forbidden, i.e., with good *crossing resolution*. The ultimate goal are *right-angle crossing (RAC)* drawings, where crossing edge segments always form right angles. Angelini et al. [3] studied the effect of drawing planar graphs with large or right crossing angles. Di Giacomo et al. [13] considered RAC drawings on 2 parallel lines. Most notably on the RAC model are the results on maximum edge density when allowing zero, one or two bends per edge [6,14], as well as the NP-hardness result by Argyriou et al. [4].

Vertex and crossing resolutions have been considered together only for very restricted types of graphs and drawings [5,15].

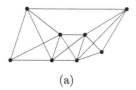   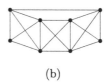

(a)   (b)

Fig. 1. Two RAC drawings of the same 1-planar graph. The drawing in (b) has better stub resolution (equal to $\frac{1}{2}$) than the one in (a).

In this paper we investigate a recent criterion for nonplanar drawings, called *stub resolution* [23]. A crossed edge is divided into parts, called *stubs*, which should not be too short to guarantee adequate readability. Hence, the stub resolution of a drawing is defined as the minimum ratio between the length of the shortest stub of an edge and the length of the entire edge. As we indicate in Fig. 1, not only the crossing resolution is helpful for the sake of readability, but good stub resolution is essential as well. An earlier research direction in the same spirit is *partial edge drawing* (see, e.g., [9]), which follows the idea that for the effective display of a crossed edge, only long enough end segments are important, while the crossings might lead to visual clutter and could be omitted.

Contribution and Paper Organization. After a formal introduction of the model and an overview of our approach (Sect. 2), we consider 1-planar graphs as a meaningful graph class where crossings are naturally involved. A graph is *1-planar* if it can be drawn with at most one crossing per edge (refer to [25] for a survey). This family of graphs is among the most investigated ones in the rapidly growing literature about graph drawing beyond planarity [16]. A natural question is whether 1-planar graphs admit 1-planar drawings with bounded stub resolution. As a preliminary result, we proved in [23] that a class of maximal 1-planar graphs allow straight-line 1-planar drawings with stub resolution $\frac{1}{5}$.

Our contribution is as follows. (1) We first study 1-planar straight-line drawings (Sect. 3), and we show that stub resolution equal to $\frac{1}{2}$ (i.e., optimal) cannot be always achieved, while stub resolution arbitrarily close to $\frac{1}{2}$ is possible for 1-planar graphs with independent crossings. (2) We then study 1-planar drawings with at most one bend per edge (Sect. 4), and we show that stub resolution arbitrarily close to $\frac{1}{2}$, or angular resolution that is lower bounded by a function of the maximum vertex degree of the graph (similar as the one in [27]) is always possible. Note that the study of 1-bend drawings is also motivated by the fact that there exist 1-planar graphs that do not admit a 1-planar straight-line drawing [21,28], while 1-planar 1-bend RAC drawings exist for all 1-planar graphs [7,10]. (3) Finally, we study 1-planar drawings with at most two bends per edge (Sect. 4), and we show that stub resolution arbitrarily close to $\frac{1}{2}$ and right-angle crossings can be achieved simultaneously.

Proofs whose statements are marked with (*) have been omitted.

2   Preliminaries and Proof Strategy

Drawings and Embeddings. We consider simple undirected graphs. A *drawing* Γ of a graph G maps the vertices of G to distinct points in the plane and the edges of G to simple Jordan arcs connecting their endpoints. Γ is *planar* if no edges cross, and *1-planar* if each edge is crossed at most once. Γ is *IC-planar* if it is 1-planar and there are no two crossed edges that share a vertex (i.e., the set of crossing edges is a matching in G). In the following, we shall not distinguish between a vertex (an edge) of G and its corresponding point (arc) in Γ.

A planar drawing Γ of a graph G induces an *embedding*, which is the class of topologically equivalent drawings. In particular, an embedding specifies the regions of the plane, called *faces*, whose boundary consists of a cyclic sequence of edges. The unbounded face is called the *outer face*. For a 1-planar drawing, we can still derive an embedding by allowing the boundary of a face to consist also of edge segments from a vertex to a crossing point. A graph with a given planar

(1-planar, IC-planar) embedding is called a *plane* (*1-plane, IC-plane*) graph. A *kite* $K = \{a, b, c, d\}$ is a graph isomorphic to K_4 with an embedding such that there is a crossing-free 4-cycle $\langle a, b, c, d \rangle$, and the two edges (a, c) and (b, d) cross inside this cycle; see Fig. 2(a). Let G be a 1-plane graph, and let $K = \{a, b, c, d\}$ be a kite such that $K \subseteq G$. K is an *empty kite*, if there is no vertex of G inside the 4-cycle $\langle a, b, c, d \rangle$. An *outer kite* $K = \{a, b, c, d\}$ is a graph isomorphic to K_4 with an embedding such that there is a crossing-free 4-cycle $\langle a, b, c, d \rangle$, and the two edges (a, c) and (b, d) cross outside this cycle; see Fig. 2(b).

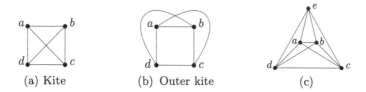

| (a) Kite | (b) Outer kite | (c) |

Fig. 2. (a)–(b) Crossing configurations. (c) Unique 1-planar embedding of K_5.

Drawing Resolutions. A drawing Γ of a graph G is *straight-line* if all edges are segments, while it is *b-bend* ($b > 0$) if each edge is a polyline with at most $b + 1$ segments. Drawing Γ is *right-angle crossing (RAC)* if the angles at any crossing point are right angles. The *angular resolution* of Γ is the minimum angle that any two incident edges form at a vertex. Note that for a graph with maximum vertex degree Δ, the angular resolution cannot be larger than $\frac{2\pi}{\Delta}$. We recall the following result concerning the angular resolution of planar drawings.

Lemma 1 (Theorem 2.2 in [27]). *Every triangulated planar graph with maximum vertex degree Δ admits a planar straight-line drawing with angular resolution $\Omega(0.15^\Delta)$.*

We shall assume (and ensure) that no more than two edges cross at any point of a drawing Γ. An edge e of Γ that is crossed k times is divided into $k + 1$ parts called *stubs*. Let l_e and s_e be the length of e and of its shortest stub, respectively. The *stub resolution* of e is $\mathrm{sr}_e = \frac{s_e}{l_e}$. The *stub resolution* of Γ is the minimum stub resolution over all edges of Γ, i.e., $\mathrm{sr}_\Gamma = \min_{e \in \Gamma} \mathrm{sr}_e$.

Observation 1. *A drawing in which the maximum number of crossings per edge is $k \geq 0$ has stub resolution at most $\frac{1}{k+1}$.*

3   Straight-Line Drawings

We first show that K_5 has no 1-planar straight-line drawing with optimal stub resolution, and so this is the case for any 1-planar graph having K_5 as a subgraph.

Observation 2. *Let Γ be a 1-planar straight-line drawing of a graph G with $\mathrm{sr}_\Gamma = \frac{1}{2}$, and let (a, c) and (b, d) be a pair of edges crossing in Γ. Then vertices a, b, c, d form a parallelogram in Γ.*

Lemma 2 (*). *Let Γ be a straight-line drawing of K_5. Then $\mathrm{sr}_\Gamma < \frac{1}{2}$.*

The next theorem proves that IC-planar graphs can be realized with 1-planar straight-line drawings with worst-case optimal stub resolution. We remark that IC-planar graphs also admit straight-line RAC drawings [8].

Theorem 1 (*). *Every IC-planar graph G has a 1-planar straight-line drawing Γ with stub resolution $\mathrm{sr}_\Gamma = \frac{1}{2} - \varepsilon$, for any fixed $\varepsilon > 0$.*

Proof sketch: If G is a subgraph of K_5 the statement immediately follows as we can use the embedding of Fig. 2(c) and draw $\langle a, b, c, d \rangle$ almost as a square (placing vertex e sufficiently far). Hence, we assume that G has at least six vertices. Start from an IC-planar embedding of $G = (V, E)$ and use the transformation by Brandenburg et al. [8, Lemma 1] to obtain a 3-connected IC-plane graph $G' = (V, E')$ with $E \subseteq E'$ such that each pair of crossing edges induces an empty kite (hence there is no outer kite) and all faces are triangles. We prove that G' admits a 1-planar straight-line drawing Γ' with stub resolution $\frac{1}{2} - \varepsilon$, with the additional property that the outer face of Γ' is a prescribed triangle T. Removing the edges in $E' \setminus E$ from Γ' cannot decrease the stub resolution of the drawing, and hence the drawing obtained by removing these edges is the desired representation of G.

The proof is by induction on the number of empty kites. In the base case G' has no empty kites, i.e., G' is a plane graph. Then we can apply the algorithm by Chiba et al. [11] to compute a planar straight-line drawing Γ' of G' such that the outer face of G' corresponds to the prescribed triangle T. By induction, if G' has $k \geq 0$ empty kites, then our claim holds. We shall prove that the claim still holds in the case where G' has $k + 1$ empty kites. We first distinguish two cases, based on whether G' contains a separating triangle or not.

CASE A: G' contains a separating triangle $C = \{a, b, c\}$. We claim that the three edges of C are all crossing-free or they can be redrawn (interpreting an embedding as a drawing) to be crossing-free. Suppose for a contradiction that one edge of C, say (b, c), is crossed and it cannot be redrawn without crossings. Observe that in this case no other edge of C is crossed, as otherwise G' would not be IC-plane. Let c_1 and c_2 be two components of $G' \setminus C$ such that c_1 contains the edge that crosses (b, c). Since the other two edges of C are not crossed, c_2 lies completely inside or outside the closed curve defined by C, say inside; in particular, vertices a, b, c all lie on the outer face of $c_2 \cup C$. Consider the inner face of $c_2 \cup C$ that contain edge (b, c). Since we cannot reroute edge (b, c) inside this face without creating new crossings, it means that an edge of c_2 is crossed by an edge of another component. Hence there exists a kite merging the two components; contradicting the fact that the two components are distinct. Denote by C_{in} (resp., C_{out}) the subgraph of G' that lies in the interior (resp., exterior) of C. Note that C is the outer face (resp., an empty face) of C_{in} (resp., C_{out}). If C_{out} has no empty kites, then we draw it with the algorithm of Chiba et al. [11] inside the prescribed triangle T. Note that C_{in} contains $k + 1$ kites, and it must be drawn inside the triangle defined by C, that is, we can assume that G' corresponds to C_{in}, and that the prescribed triangle T is C.

Similarly, if C_{in} has no empty kites, then we assume that G' corresponds to C_{out} and that it must be drawn inside the prescribed triangle T. Once we obtain a drawing of C_{out}, we can again use the algorithm of Chiba et al. [11] for C_{in} with the drawing of C as prescribed outer face. By the above discussion, we can assume that both C_{in} and C_{out} have at least one empty kite. Then by the induction hypothesis C_{out} and C_{in} can be drawn with stub resolution $\frac{1}{2} - \varepsilon$, as desired.

CASE B: G' has no separating triangles. We distinguish two further cases depending on whether there exists an empty kite K such that none of its edges is part of the outer face of G' or not. Note that if an edge of K belongs to the outer face of G', then this edge is not crossed.

CASE B.1: Suppose first that G' contains an empty kite $K = \{a, b, c, d\}$ such that none of its crossing-free edges belongs to the outer face of G'. Let f_1, f_2, f_3, and f_4 denote the faces incident to the crossing-free edges of K. Recall that these faces are triangles, and denote as v_i the vertex of f_i that does not belong to K, for $i = 1, 2, 3, 4$. Some of these vertices can coincide, but not all of them, otherwise K would be a K_5 in G' and there would be a separating triangle in G' (recall that no edge of K belongs to the outer face of G'). For the same reason, any crossing-free edge of K belongs to at most one triangle of G', and this triangle is a face of G' distinct from the outer face. The general idea is that we collapse K into a single vertex r. The derived graph G'' has fewer empty kites than G' and therefore we can obtain a drawing Γ'' of G'' with stub resolution $\frac{1}{2} - \varepsilon$ and straight-line edges, inside the prescribed triangle T. Then, we can reinsert the kite as a parallelogram and connect its vertices to their neighbours with crossing-free straight-line segments; to this aim, we distinguish three main cases, based on whether some of $v_1, \ldots, v_4$ coincide.

CASE B.2: Suppose now that every empty kite of G' has a non-crossing edge on the outer face of G'. Since every face of G' is a triangle and G' is IC-planar, it follows that in this case G' has only one empty kite $K = \{a, b, c, d\}$ so that an edge of K, say (c, d) belongs to the outer face of G'; see Fig. 3(a). Let e be the third vertex of the outer face of G'. Also, let u_1 be the common neighbor of vertices a and d, and u_2 be the common neighbor of vertices b and c. Then, we contract edge (a, d) to a single vertex w_1 and edge (b, c) to vertex w_2. After removing parallel edges, the derived graph G'' has no kites and remains fully triangulated. Hence, by the base case of our induction, we can compute a planar drawing Γ'' of G'' with vertices w_1, w_2 and e on its outer face.

Suppose that edge (w_1, w_2) is drawn as a horizontal segment (up to a rotation of the drawing); refer to Fig. 3(b). We want to draw kite $K = \{a, b, c, d\}$ as an isosceles trapezoid $\mathcal{P}$ with $|ab| < |cd|$ as its two parallel bases and so that a and b are drawn at the points of vertices w_1 and w_2 in Γ''. Consider vertex w_1. In a clockwise traversal of the edges incident at w_1 and starting from edge (w_1, e), first we encounter the neighbors of d in the same order as they appear in G', and once we encounter u_1 the neighbors of a follow in the same order as they appear in G'. We need to ensure that we can redraw the edges of d as straight lines and without introducing any crossing in the drawing. Let $C_1(w_1, r_1)$ be a circle with center w_1 and radius r_1, for a value of r_1 that is sufficiently small as explained below; refer to Fig. 3(b). Consider the sector of C_1 that is bounded above by

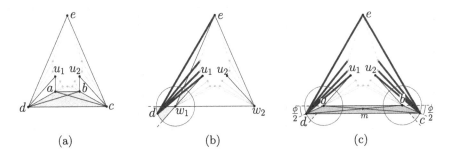

Fig. 3. Illustration for **CASE B.2**.

the line through w_1 and w_2 and the line through w_1 and u_1 (gray shaded in Fig. 3(b)). We choose r_1 to be sufficiently small such that by drawing vertex d on the arc of this sector, we have that d can be connected to all its neighbors in G' with straight lines without introducing any new crossings (blue thick edges in Fig. 3(b)). Observe that such a choice is always feasible because as r_1 decreases, the difference between the slopes of the edges incident to d, and the slopes of the edges incident to w_1 also decreases. A similar argument holds for vertex w_2. We draw a circle $C_2(w_2, r_2)$, where r_2 is again sufficiently small such that there exists an arc of C_2 where we can draw vertex c and connect c to its neighbors in G' without introducing any crossings. Let $r < \min\{r_1, r_2\}$ and let ϕ be the smaller angle of the two arcs (on C_1 and C_2). We draw an isosceles trapezoid $\mathcal{P}$ so that $|ab| < |cd|$ are its two parallel bases, and its base angles are equal to $\frac{\phi}{2}$; refer to Fig. 3(c). We draw edges (a, c) and (b, d) as straight lines and let m be their crossing point. Since $\mathcal{P}$ is isosceles, we have that $\frac{|am|}{|mc|} = \frac{|bm|}{|md|} = \frac{|ab|}{|cd|}$. We want to prove that by appropriately choosing the radius r, the stub-resolution of edges (a, c) and (b, d) is at least $\frac{1}{2} - \varepsilon$. Actually, we chose a value $\varepsilon' \leq \varepsilon$ such that $r = \frac{2\varepsilon' |ab|}{(1 - 2\varepsilon') \cos \frac{\phi}{2}}$. On the one hand, this choice guarantees that $|ab| = |cd| \frac{1 - 2\varepsilon'}{1 + 4\varepsilon'}$, and hence that the stub-resolution is equal to $\frac{1}{2} - \varepsilon' \geq \frac{1}{2} - \varepsilon$. On the other hand, by choosing ε' small enough we can ensure that $r < \min\{r_1, r_2\}$ (observe that r decreases as ε' decreases). This concludes the proof of Theorem 1. $\square$

4   Polyline Drawings

While there exist 1-planar graphs that do not admit a 1-planar straight-line drawing [21,28], every 1-planar graph has a 1-planar 1-bend RAC drawing [7,10]. This section shows that for 1-planar 1-bend drawings, it is possible to optimize also the stub-resolution (Theorem 2), and the angular resolution (Theorem 3). In particular, the main contribution of this section is the following theorem.

Theorem 2 (*). *Every 1-planar graph has a 1-planar 1-bend drawing Γ with stub resolution $\mathrm{sr}_\Gamma \geq \frac{1}{2} - \varepsilon$, for any fixed $0 < \varepsilon < \frac{1}{2}$. Furthermore all crossing-free edges are drawn straight-line.*

We begin by proving how to construct a drawing with near-optimal stub resolution. The proof is based on a constructive argument that uses the next two technical lemmas as building blocks.

Lemma 3. *Let $K = \{a, b, c, d\}$ be an empty kite and let $\mathcal{P}$ be a convex polygon with four corners. There exists an embedding-preserving drawing Γ of K such that: (i) $\mathrm{sr}_\Gamma = \frac{1}{2}$; (ii) Vertices $\{a, b, c, d\}$ are placed at the corners of $\mathcal{P}$; (iii) The two crossing edges (a, c) and (b, d) are drawn with at most one bend each, while the crossing-free edges are straight-line; (iv) The bend point of one of the two crossing edges is at the crossing point.*

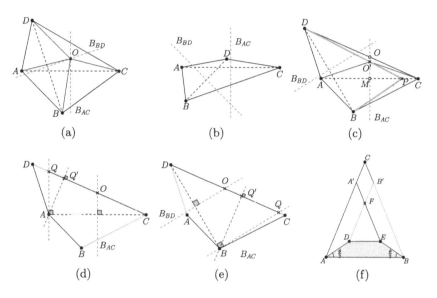

Fig. 4. (a)–(e) Configurations of Lemma 3. (f) Configuration of Lemma 4.

Proof. Let $\{A, B, C, D\}$ be the corners of $\mathcal{P}$. We start by placing vertices a, b, c, d at corners A, B, C, D respectively, and draw the crossing-free edges of K on the boundary of $\mathcal{P}$ as straight-line. Let B_{AC} and B_{BD} be the perpendicular bisectors of the diagonals AC and BD, respectively. We consider two cases depending on whether B_{AC} and B_{BD} cross in the interior of $\mathcal{P}$ or not.

If B_{AC} and B_{BD} cross in the interior of $\mathcal{P}$ at point O (refer to Fig. 4(a)), we draw edge (a, c) and edge (b, d) with a bend at point O. Since $O \in B_{AC}$ and $O \in B_{BD}$, we have that $|AO| = |OC|$ and $|BO| = |OD|$. Hence (a, c) and (b, d) cross at O with stub resolution equal to $\frac{1}{2}$. Furthermore, the bend of both crossing edges is at their crossing point as claimed.

Suppose now that B_{AC} and B_{BD} cross in the exterior of $\mathcal{P}$ (note that they cannot be parallel because P is convex); refer to Fig. 4(c). In the following we prove that we can draw edges (a, c) and (b, d) with one bend each so that: (i) the bend point, say O, of one of the two edges, say (a, c), is along its bisector B_{AC}, and (ii) edge (b, d) crosses (a, c) at O with stub resolution $\frac{1}{2}$. For the sake of contradiction, suppose that this is not true. W.l.o.g. we can assume that bisector B_{AC} separates vertices A and D from vertex C (note that we do not make any assumption about the relative position of vertex B and bisector B_{AC}). We claim that we can also assume that B_{BD} separates B and C from D as in Fig. 4(c). Suppose momentarily that this is not true, i.e. B_{BD} separates B from D and C; see Fig. 4(b). This case is symmetric to the previous one, if we rename points $\{A, B, C, D\}$ as $\{B, C, D, A\}$.

Hence we have the configuration of Fig. 4(c). This implies that for any point X of B_{AC} in the interior of $\mathcal{P}$ it is $|XB| < |XD|$, and for any point X' of B_{BD} in the interior of $\mathcal{P}$ it is $|X'A| < |X'C|$. Furthermore, B_{AC} and B_{BD} both cross the boundary edge AD of $\mathcal{P}$. Consider first edge (a, c); similar arguments hold also for edge (b, d). Let O be the crossing point of B_{AC} with CD and M its crossing point with AC. For any point $O' \in MO$ we draw edge (b, d) as follows: we start from point D with a straight line through point O' until we cross diagonal AC, we add a bend point, say P, on AC and continue with a straight line up to B (see the thick green edge in Fig. 4(c)). If $|DO'| < |O'P| + |PB|$, we have that the midpoint of edge (b, d) is not on DO'. As we move the bend point of (a, c) from P towards O', the midpoint of (b, d) also changes, and when the bend point coincides with O' the midpoint is on DO' (since $|DO'| > |O'B|$). Hence, one can find a bend-point on $O'P$, so that the midpoint of (b, d) is O' and the lemma holds. So, we can assume that $|DO'| > |O'P| + |PB|$ for any point $O' \in MO$. For $O' = O$ the bend point P coincides with C and we have $|DO| > |OC| + |CB|$. We draw the parallel of B_{AC} through point A that crosses line CD at point Q; refer to Fig. 4(d). Then $|QO| = |OC|$ and from the previous inequality $|DO| > |QO|$, i.e. $Q \in CD$. In particular we have that $|DO| > |QO| + |CB| \Rightarrow |DQ| > |CB|$. We also draw the perpendicular line of CD through point A crossing CD at point Q'. Looking at triangle $\{A, C, D\}$ we have that $|DQ| < |DQ'|$, and considering the orthogonal triangle $\{A, D, Q'\}$ it is $|AD| > |DQ'|$. Combining the above:

$$|AD| > |DQ'| > |DQ| > |CB| \Rightarrow |AD| > |CB|. \tag{1}$$

Arguing similarly for edge (b, d), we can either conclude that the lemma holds or that $|CB| > |AD|$ as shown in Fig. 4(e), contradicting Eq. 1. $\square$

Lemma 4. *Let $K = \{a, b, c, d\}$ be an outer kite, let $T = \{A, B, C\}$ be an isosceles triangle, and let $0 < \varepsilon < \frac{1}{2}$. There exists an embedding-preserving drawing Γ of K such that: (i) $\mathrm{sr}_\Gamma = \frac{1}{2} - \varepsilon$; (ii) Vertices $\{a, b, c, d\}$ are placed at the corners of a trapezoid $\mathcal{P}$ such that its larger base coincides with AB, and its smaller base is inside T; (iii) The two crossing edges (a, c) and (b, d) are drawn with at most one bend each, while the crossing-free edges are straight-line.*

Proof. Suppose that T is drawn so that its base AB is horizontal, and let ϕ be the value of its base angles. We draw an isosceles trapezoid $\mathcal{P} = \{A, B, D, E\}$ so that $DE < AB$ are its two parallel bases, and $\angle B, A, D = \angle A, B, E = \frac{\phi}{2}$;

refer to Fig. 4(f). Vertices a, b, c, d of K are placed on the corners A, B, E, D of $\mathcal{P}$ respectively, so that uncrossed edges of K are drawn on the boundary of $\mathcal{P}$ as straight lines. In order to draw edge (a, c) we start from point E parallel to BC until we cross AC at point A', we bend at A' and follow $A'A$ up to point A. Edge (b, d) is drawn symmetrically, and let F be the crossing point of (a, c) and (b, d). Since triangle $T' = \{D, E, F\}$ is similar to triangle T, the stub resolution of (a, c) and (b, d) is the same. Now edge (a, c) has two stubs: the first one consists of segments AA' and $A'F$, and the second one only of segment FE, where $|AA'| + |A'F| > |FE|$. We want to prove that by appropriately choosing the height h of trapezoid $\mathcal{P}$, the stub-resolution of (a, c) is equal to $\frac{1}{2} - \varepsilon$. Since $|A'F| = |A'C|$ and $\cos \frac{\phi}{2} = \frac{|DE|}{2|FE|} = \frac{|AB|}{2|AC|}$ the stub-resolution equals $\frac{|FE|}{|FE| + |AC|} = \frac{|DE|}{|DE| + |AB|}$. For $h = \frac{2\varepsilon |AB|}{1 + 2\varepsilon} \tan \frac{\phi}{2}$, we have that $|DE| = |AB| \frac{1 - 2\varepsilon}{1 + 2\varepsilon}$, and stub-resolution is equal to $\frac{1}{2} - \varepsilon$, therefore completing the proof. □

We first describe how to construct drawings for 3-connected 1-planar graphs, and then extend our technique to all 1-planar graphs. We assume an embedding is given in input, although our technique may need to change it.

Lemma 5. *Every 3-connected 1-plane graph G has a 1-planar 1-bend drawing Γ with $sr_\Gamma = \frac{1}{2}$, except for at most one pair of crossing edges whose stub resolution is $\frac{1}{2} - \varepsilon$, for any fixed $0 < \varepsilon < \frac{1}{2}$. All crossing-free edges are drawn straight-line.*

Proof. After possibly changing the embedding of G, we may assume that all pairs of crossing edges of G induce an empty kite, except for at most one pair of crossing edges that are part of the outer face and form an outer kite [1].

Let G' be the plane graph obtained from G by removing all pairs of crossing edges. We say that a quadrangular face $f = \{u, w, v, z\}$ of G' is *marked*, if (u, v) and (w, z) are two crossing edges of G. We first compute a straight-line drawing Γ' of G' by using the algorithm of Chiba et al. [11]. The algorithm in [11] has two main properties. First, it produces a drawing in which all faces are convex. Second, it allows to specify any convex polygon P to represent the outer face of the input graph. We now describe how to specify the outer polygon. If the outer face of G' is not marked, then we can use any convex polygon. Else, the outer face of G' is the 4-cycle of an outer kite K of G. In this case we let T be any isosceles triangle, and we apply Lemma 4 to obtain a drawing of K. This drawing fixes a trapezoid P for the four vertices of K, which we use as input polygon. It remains to show how to reinsert all edges in $G \setminus G'$ that belong to a marked inner face. For each such face $f = \{u, w, v, z\}$ of Γ', we reinsert the pair of crossing edges (u, v) and (w, z) by applying Lemma 3, where the convex polygon is the drawing of f in Γ'. This concludes the proof. □

Proof sketch for Theorem 2: We exploit a decomposition in 3-connected components for 1-planar graphs [7], and we apply a variant of Lemma 5 that allows to merge distinct components attached to a same separation pair of G. □

The next theorem states that angular resolution bounded from below by a function of the maximum vertex degree of the graph and independent of its size can be obtained. Also, if two bends per edge are allowed, right-angle crossings and stub resolution close to $\frac{1}{2}$ can be simultaneously achieved.

Theorem 3 (*). *Every 1-plane graph G with maximum degree Δ has a 1-planar 1-bend drawing with angular resolution $\Omega(0.15^{6\Delta})$. Also, all crossing-free edges are drawn straight-line.*

Theorem 4 (*). *Every 3-connected 1-plane graph G has a 1-planar 2-bend RAC drawing Γ with $\mathrm{sr}_\Gamma = \frac{1}{2}$, except for at most one pair of edges whose stub resolution is $\frac{1}{2} - \varepsilon$, for any fixed $0 < \varepsilon < \frac{1}{2}$. All crossing-free edges are straight-line.*

5   Open Problems

Interesting open problems arise from our research, among them: (i) Is there a constant $\delta > 2$ such that every straight-line drawable 1-planar graph has a 1-planar straight-line drawing with stub resolution at least $\frac{1}{\delta}$? (ii) Does every 1-planar graph with maximum vertex degree Δ admit a 1-planar 1-bend drawing with $\Omega(\frac{1}{\Delta})$ angular resolution? (iii) Can we generalize our results to k-planar graphs? In this direction, we have preliminary results showing that 2-planar drawings with bounded stub resolution are possible for optimal 2-planar graphs if we allow a constant number of bends for the crossing edges. (iv) Finally, it would be interesting to study stub resolution in combination with other aesthetics, such as compact area or few slopes for the edge segments.

Acknowledgments. Research started at the 2017 Bertinoro Workshop on Graph Drawing, we thank all participants for fruitful discussions.

References

1. Alam, M.J., Brandenburg, F.J., Kobourov, S.G.: Straight-line grid drawings of 3-connected 1-planar graphs. In: Wismath, S., Wolff, A. (eds.) GD 2013. LNCS, vol. 8242, pp. 83–94. Springer, Cham (2013). https://doi.org/10.1007/978-3-319-03841-4_8
2. Angelini, P., Bekos, M.A., Liotta, G., Montecchiani, F.: Universal slope sets for 1-bend planar drawings. Algorithmica **81**(6), 2527–2556 (2019)
3. Angelini, P., et al.: Large angle crossing drawings of planar graphs in subquadratic area. In: Márquez, A., Ramos, P., Urrutia, J. (eds.) EGC 2011. LNCS, vol. 7579, pp. 200–209. Springer, Heidelberg (2012). https://doi.org/10.1007/978-3-642-34191-5_19
4. Argyriou, E.N., Bekos, M.A., Symvonis, A.: The straight-line RAC drawing problem is NP-hard. J. Graph Algorithms Appl. **16**(2), 569–597 (2012)
5. Argyriou, E.N., Bekos, M.A., Symvonis, A.: Maximizing the total resolution of graphs. Comput. J. **56**(7), 887–900 (2013)
6. Arikushi, K., Fulek, R., Keszegh, B., Moric, F., Tóth, C.D.: Graphs that admit right angle crossing drawings. Comput. Geom. **45**(4), 169–177 (2012)
7. Bekos, M.A., Didimo, W., Liotta, G., Mehrabi, S., Montecchiani, F.: On RAC drawings of 1-planar graphs. Theor. Comput. Sci. **689**, 48–57 (2017)

8. Brandenburg, F.J., Didimo, W., Evans, W.S., Kindermann, P., Liotta, G., Montecchiani, F.: Recognizing and drawing IC-planar graphs. Theor. Comput. Sci. **636**, 1–16 (2016)
9. Bruckdorfer, T., Cornelsen, S., Gutwenger, C., Kaufmann, M., Montecchiani, F., Nöllenburg, M., Wolff, A.: Progress on partial edge drawings. J. Graph Algorithms Appl. **21**(4), 757–786 (2017)
10. Chaplick, S., Lipp, F., Wolff, A., Zink, J.: Compact drawings of 1-planar graphs with right-angle crossings and few bends. In: Biedl, T., Kerren, A. (eds.) GD 2018. LNCS, vol. 11282, pp. 137–151. Springer, Cham (2018). https://doi.org/10.1007/978-3-030-04414-5_10
11. Chiba, N., Yamanouchi, T., Nishizeki, T.: Linear algorithms for convex drawings of planar graphs. Prog. Graph Theory **173**, 153–173 (1984)
12. Di Battista, G., Vismara, L.: Angles of planar triangular graphs. SIAM J. Discrete Math. **9**(3), 349–359 (1996)
13. Di Giacomo, E., Didimo, W., Eades, P., Liotta, G.: 2-layer right angle crossing drawings. Algorithmica **68**(4), 954–997 (2014)
14. Didimo, W., Eades, P., Liotta, G.: Drawing graphs with right angle crossings. Theor. Comput. Sci. **412**(39), 5156–5166 (2011)
15. Didimo, W., Kaufmann, M., Liotta, G., Okamoto, Y., Spillner, A.: Vertex angle and crossing angle resolution of leveled tree drawings. Inf. Process. Lett. **112**(16), 630–635 (2012)
16. Didimo, W., Liotta, G., Montecchiani, F.: A survey on graph drawing beyond planarity. ACM Comput. Surv. **52**(1), 4:1–4:37 (2019)
17. Duncan, C.A., Eppstein, D., Goodrich, M.T., Kobourov, S.G., Nöllenburg, M.: Drawing trees with perfect angular resolution and polynomial area. Discrete Comput. Geom. **49**(2), 157–182 (2013)
18. Duncan, C.A., Kobourov, S.G.: Polar coordinate drawing of planar graphs with good angular resolution. J. Graph Algorithms Appl. **7**(4), 311–333 (2003)
19. Formann, M., et al.: Drawing graphs in the plane with high resolution. SIAM J. Comput. **22**(5), 1035–1052 (1993)
20. Garg, A., Tamassia, R.: Planar drawings and angular resolution: algorithms and bounds (extended abstract). In: van Leeuwen, J. (ed.) ESA 1994. LNCS, vol. 855, pp. 12–23. Springer, Heidelberg (1994). https://doi.org/10.1007/BFb0049393
21. Hong, S.-H., Eades, P., Liotta, G., Poon, S.-H.: Fáry's theorem for 1-planar graphs. In: Gudmundsson, J., Mestre, J., Viglas, T. (eds.) COCOON 2012. LNCS, vol. 7434, pp. 335–346. Springer, Heidelberg (2012). https://doi.org/10.1007/978-3-642-32241-9_29
22. Huang, W., Eades, P., Hong, S.: Larger crossing angles make graphs easier to read. J. Vis. Lang. Comput. **25**(4), 452–465 (2014)
23. Kaufmann, M., Kratochvíl, J., Lipp, F., Montecchiani, F., Raftopoulou, C., Valtr, P.: Bounded stub resolution for some maximal 1-planar graphs. In: Panda, B.S., Goswami, P.P. (eds.) CALDAM 2018. LNCS, vol. 10743, pp. 214–220. Springer, Cham (2018). https://doi.org/10.1007/978-3-319-74180-2_18
24. Keszegh, B., Pach, J., Pálvölgyi, D.: Drawing planar graphs of bounded degree with few slopes. SIAM J. Discrete Math. **27**(2), 1171–1183 (2013)
25. Kobourov, S.G., Liotta, G., Montecchiani, F.: An annotated bibliography on 1-planarity. Comput. Sci. Rev. **25**, 49–67 (2017)
26. Krug, M., Wagner, D.: Minimizing the area for planar straight-line grid drawings. In: Hong, S.-H., Nishizeki, T., Quan, W. (eds.) GD 2007. LNCS, vol. 4875, pp. 207–212. Springer, Heidelberg (2008). https://doi.org/10.1007/978-3-540-77537-9_21

27. Malitz, S.M., Papakostas, A.: On the angular resolution of planar graphs. SIAM J. Discrete Math. **7**(2), 172–183 (1994)
28. Thomassen, C.: Rectilinear drawings of graphs. J. Graph Theory **12**(3), 335–341 (1988)

Dispersion of Mobile Robots on Grids

Ajay D. Kshemkalyani[1], Anisur Rahaman Molla[2], and Gokarna Sharma[3(⊠)]

[1] University of Illinois at Chicago, Chicago, USA
`ajay@uic.edu`
[2] Indian Statistical Institute, Kolkata, India
`molla@isical.ac.in`
[3] Kent State University, Kent, USA
`sharma@cs.kent.edu`

Abstract. The dispersion problem on graphs asks $k \leq n$ robots initially placed arbitrarily on the nodes of an n-node anonymous graph to reposition autonomously to reach a configuration in which each robot is on a distinct node of the graph. This problem is of significant interest due to its relationship to many other fundamental robot coordination problems, such as exploration, scattering, load balancing, relocation of self-driven electric cars (robots) to recharge stations (nodes), etc. The objective in this problem is to simultaneously minimize (or provide trade-off between) two fundamental performance metrics: (i) time to achieve dispersion and (ii) memory requirement at each robot. The existing algorithms for trees and arbitrary graphs either minimize time or memory but not both. In this paper, we consider for the very first time the dispersion problem on a grid graph embedded in the Euclidean plane and present solutions that simultaneously minimize both the metrics. The grid graph is appealing as it naturally discretizes the 2-dimensional Euclidean plane and finds applications in many real-life robotic systems. Particularly, we provide two deterministic algorithms on an anonymous grid graph that achieve simultaneously optimal bounds for both the metrics.

1 Introduction

The dispersion of autonomous mobile robots to spread them out evenly in a region is a problem of significant interest in distributed robotics, e.g., see [4,5]. Recently, this problem has been formulated in the context of graphs as follows: Given any arbitrary initial configuration of $k \leq n$ robots positioned on the nodes of an n-node graph, the robots reposition autonomously to reach a configuration where each robot is positioned on a distinct node of the graph (which we call the DISPERSION problem) [1]. This problem has many practical applications, for example, in relocating self-driven electric cars (robots) to recharge stations (nodes), assuming that the cars have smart devices to communicate with each other to find a free/empty charging station [1,6]. This problem is also important due to its relationship to many other well-studied autonomous robot coordination problems, such as exploration, scattering, and load balancing [1,6].

© Springer Nature Switzerland AG 2020
M. S. Rahman et al. (Eds.): WALCOM 2020, LNCS 12049, pp. 183–197, 2020.
https://doi.org/10.1007/978-3-030-39881-1_16

Table 1. Results on DISPERSION for $k \leq n$ robots on n-node square grids (for grids, maximum degree $\Delta = 4$ and diameter $D = O(\sqrt{n})$). Time always remains sub-optimal on a grid applying the arbitrary graph algorithms for DISPERSION from the literature. Theorem 1 is simultaneously time-memory optimal for grid when $k = \Omega(n)$ and Theorem 2 is for any $k \leq n$.

Algorithm	Memory/robot (in bits)	Time (in rounds)	Communication model
Lower bound	$\Omega(\log k)$	$\Omega(\sqrt{k})$	Local/global
Applying first algorithm of [6]	$O(k)$	$O(n)$	Local
Applying second algorithm of [6]	$O(D) = O(\sqrt{n})$	$O(4^D) = O(4^{\sqrt{n}})$	Local
Applying third algorithm of [6]	$O(\log k)$	$O(nk)$	Local
Applying algorithm of [7]	$O(\log n)$	$O(k \log k)$	Local
Theorem 1 (this paper)	$O(\log k)$	$O(\min(k, \sqrt{n}))$	**Local**
Applying algorithm of [8]	$O(\log k)$	$O(k)$	Global
Theorem 2 (this paper)	$O(\log k)$	$O(\sqrt{k})$	**Global**

The objective in this problem is to simultaneously minimize (or provide trade-off between) two fundamental performance metrics: (i) time to achieve dispersion and (ii) memory requirement at each robot. Several papers studied this problem recently on trees and arbitrary graphs giving algorithms with increasingly better bounds on both the metrics [1,6–8]. However, the existing algorithms (for trees and arbitrary graphs) are only able to minimize either time or memory but not both (details in **related work**). Particularly, some of the existing algorithms obtained optimal memory bounds but no algorithm established optimal time bounds. Therefore, the following question naturally arises: *Is it possible to solve* DISPERSION *on some graph classes simultaneously minimizing both time and memory?* In this paper, we answer this question in the affirmative, considering for the very first time DISPERSION on a grid graph embedded in the Euclidean plane. Surprisingly, we are not only able to simultaneously minimize both the metrics but also able to achieve optimal bounds for both. Grid graph setting is simple yet appealing as it naturally discretizes the 2-dimensional Euclidean plane and finds applications in many real-life robotic systems, for example, see [9,12].

Specifically, we provide two novel deterministic algorithms for DISPERSION on an anonymous grid graph (Table 1). Our first algorithm works in the *local communication* model where a robot can only communicate with other robots that are present at the same node. Our second algorithm works in the *global communication* model where a robot can communicate with any other robot in the graph possibly at different nodes (but the graph structure is not known to robots). The global communication model seems stronger than the local model at first sight, however many challenges that occur in the local model carryover to the global model. For example, two robots in two neighboring nodes of G cannot figure out just by communicating which edge of the nodes leads to each other. Therefore, the robots still need to explore through the edges as in the local model. The global communication model has been of much interest in the past in distributed robotics, e.g., see [2,3,11]. Among the previous works [1,6–8] on DISPERSION, papers [1,6,7] considered the local communication model and [8] considered the global communication model. Applying the arbitrary graph results of [6–8] to a grid, memory bound obtained is optimal but time remains sub-optimal.

Overview of the Model and Results. We consider a system of $k \leq n$ robots are operating on an n-node graph G. G is assumed to be a connected, undirected graph with m edges, diameter D, and maximum degree Δ. In addition, G is *anonymous*, i.e., nodes have no unique IDs and hence are indistinguishable but the ports (leading to incident edges) at each node have unique labels from $[1, \delta]$, where δ is the degree of that node. The robots are *distinguishable*, i.e., they have unique IDs in the range $[1, k]$. The robot activation setting is *synchronous* – all robots are activated in a round and they perform their operations simultaneously in synchronized rounds. Runtime is measured in rounds (or steps). We establish the following result in the local model.

Theorem 1. *Given any initial configuration of* $k \leq n$ *mobile robots on an anonymous n-node square grid graph G,* DISPERSION *can be solved in* $O(\min(k, \sqrt{n}))$ *time with* $O(\log k)$ *bits memory at each robot in the local communication model.*

Theorem 1 is simultaneously time-memory optimal when $k = \Omega(n)$ since a time lower bound of $\Omega(D)$ trivially holds for DISPERSION of n robots on any graph [1] and the diameter of a n-node grid is $D = \Omega(\sqrt{n})$. Furthermore, $\Omega(\log k)$ bits are necessary at a robot to distinguish the robots from each other. We also extend Theorem 1 on a rectangular grid G. We establish the following result in the global model.

Theorem 2. *Given any initial configuration of* $k \leq n$ *mobile robots on an anonymous n-node square grid graph G,* DISPERSION *can be solved in* $O(\sqrt{k})$ *time with* $O(\log k)$ *bits memory at each robot in the global communication model.*

Theorem 2 is simultaneously time-memory optimal for DISPERSION for any $k \leq n$. We also extend Theorem 2 for DISPERSION on a rectangular grid graph G.

Challenges and Techniques. The solutions proposed in the literature in the local [1,6,7] and global [8] communication models for DISPERSION on arbitrary graphs and trees (grids were not considered before) heavily use a DFS (depth first search) traversal approach. The best bound so far for grid in the local model is obtained while applying the arbitrary graph algorithm of [7] (the bounds obtained in [7] for arbitrary graphs are given in related work): $O(k \log k)$ time with $O(\log k)$ bits at each robot (see Table 1). However, for a grid, the time lower bound is $\Omega(D) = \Omega(\sqrt{k})$ for any $k \leq n$ and memory lower bound is obviously $\Omega(\log k)$ at each robot to distinguish the robots from each other (consider the case of all k robots are at a single node of G initially). Therefore, we develop two techniques for grids, one specific to the local model and another to the global model. The technique for the local model uses the grid properties (instead of a DFS traversal). Particularly, the grid technique in the local model uses the idea of repositioning first all robots to the boundary nodes of G using the grid properties (despite the grid being anonymous), then collect them to a boundary corner node of G, and finally distribute them to the nodes of G leaving one robot on each node. The technique for the global model uses the grid properties with a limited use of the DFS traversal tailored specifically to the information available to robots during execution due to the global model. Particularly, the grid technique in the global model uses the DFS traversal for $O(\sqrt{k})$ time, then forms a square boundary of perimeter $4(\sqrt{k} - 1)$, and disperses all k robots within that boundary in $O(\sqrt{k})$ time. Although the techniques above seem rather straightforward in high level, we need to overcome many challenges for them to correctly work while putting together to solve DISPERSION.

Related Work. As discussed above, for DISPERSION, there are three previous studies [1,6,7] in the local communication model. For $k = n$, Augustine and Moses Jr. [1] proved a memory lower bound of $\Omega(\log n)$ bits at each robot and a time lower bound of $\Omega(D)$ ($\Omega(n)$ on arbitrary graphs) for any deterministic algorithm on any graph. They then provided two algorithms for arbitrary graphs for $k = n$: (i) The first algorithm with $O(mn)$ time using $O(\log n)$ bits at each robot and (ii) The second algorithm with $O(m)$ time using $O(n \log n)$ bits at each robot. Kshemkalyani and Ali [6] provided an $\Omega(k)$ time lower bound for arbitrary graphs for any $k \leq n$. They then provided three deterministic algorithms for arbitrary graphs: (i) The first algorithm with $O(m)$ time using $O(k \log \Delta)$ bits at each robot, (ii) The second algorithm with $O(\Delta^D)$ time using $O(D \log \Delta)$ bits at each robot, and (iii) The third algorithm with $O(mk)$ time using $O(\log(\max(k, \Delta)))$ bits at each robot. Recently, Kshemkalyani et al. [7] provided an algorithm for arbitrary graphs that runs in $O(\min(m, k\Delta) \cdot \log k)$ time using $O(\log n)$ bits at each robot. There is one previous study [8] for DISPERSION in the global communication model, which provides a deterministic algorithm for arbitrary graphs that runs in $O(\min(m, k\Delta))$ time using $O(\log(\max(k, \Delta)))$ bits at each robot, improving the time in [7] in the local model by an $O(\log k)$ factor. Randomized algorithms for DISPERSION are presented in [10] where the random bits are used to reduce the memory requirement. Other closely related works to DISPERSION are omitted due to space constraints.

Roadmap. We discuss model details in Sect. 2. We present an algorithm in the local model in Sect. 3. We present an algorithm in the global model in Sect. 4. Finally, we conclude in Sect. 5 with a short discussion on possible future work. Due to space constraints, many details and proofs are deferred to a full version.

2   Model Details and Preliminaries

Grid Graph. Let $G = (V, E)$ be an n-node grid graph embedded in the 2-dimensional Euclidean plane, i.e., $|V| = n$ and $|E| = m = O(n)$. G is assumed to be connected, unweighted, and undirected with no holes. G is *anonymous*, i.e., nodes do not have identifiers but, at any node, its incident edges are uniquely identified by a *label* (aka port number) in the range $[1, \delta]$, where $\delta \leq 4$ is the *degree* of that node. The *maximum degree* of G is $\Delta = 4$, which is the maximum among the degree δ of the nodes in G. Any edge e connecting two nodes $u, v \in G$ has two port numbers associated with it, one at the end of e towards u and another at the end of e towards v. We assume that there is no correlation between two port numbers of an edge. For a grid graph G, the nodes on 2 boundary rows and 2 boundary columns are called *boundary nodes* and the 4 corner nodes on boundary rows (or columns) are called *boundary corner nodes*. In an n-node square grid graph G, there are exactly $4\sqrt{n} - 4$ boundary nodes. Any number of robots are allowed to move along an edge at any time. The grid nodes do not have memory, i.e., they are not able to store any information.

Robots. Let $\mathcal{R} = \{r_1, r_2, \ldots, r_k\}$ be a set of $k \leq n$ robots residing on the nodes of G. For simplicity, we sometime use i to denote robot r_i. No robot resides on the edges of G, but one or more robots can occupy the same node of G. Each robot has a unique $\lceil \log k \rceil$-bit ID taken from $[1, k]$. When a robot moves from node u to node v in G, it is aware of the port of u it used to leave u and the port of v it used to enter v. Furthermore, it is assumed that each robot is equipped with memory to store information, which may also be read and modified by other robots present on the same node.

Communication Model. We have two communication models: local and global. In the local communication model, a robot can only communicate with other robots present on the same node. In contrast, in the global communication model, a robot is capable to communicate with any other robot in the graph G, irrespective of their positions in the graph. However, they will not have the position information as graph nodes are anonymous and there is no correlation between two port numbers of an edge.

Cycle. At any time a robot $r_i \in \mathcal{R}$ could be active or inactive. When a robot r_i becomes active, it performs the "Communicate-Compute-Move" (CCM) cycle as follows.

- *Communicate:* r_i can communicate with and observe the memory of some other robot $r_j \in \mathcal{R}$ (at the same node or a different node) depending on the communication model used. Robot r_i can also observe its own memory.
- *Compute:* r_i may perform an arbitrary computation using the information observed during the "communicate" portion of that cycle. This includes determination of a (possibly) port to use to exit v_i and the information to store in the robot r_j at v_i.
- *Move:* At the end of the cycle, r_i writes new information (if any) in the memory of r_j at v_i, and exits v_i using the computed port to reach to a neighbor v_i' of v_i. After entering v_i', it will keep track of the port at v_i' from which it entered v_i'.

Time and Memory Complexity. We consider the synchronous setting where every robot is active in every CCM cycle and they perform the cycle in a syn-

---

Algorithm 1: $Grid\_Disperse(k)$ in the local communication model

---

1 **Input:** An n-node square grid G with $k \leq n$ robots positioned arbitrarily on its
  nodes.
2 **if** $k \geq \sqrt{n}$ **then**
3   **Stage 1:** the robots not already on the boundary nodes move to the
      boundary nodes;
4   **Stage 2:** the robots not on the boundary corner nodes move to the corner
      nodes;
5   **Stage 3:** the robots not on the boundary corner nodes move to a corner node;
6   **Stage 4:** the robots distribute on the nodes of a grid side, at most $\sqrt{k}$ on
      each node;
7   **Stage 5:** the robots disperse with one robot on a node;
8 **else**
9   The robots move in a direction from their position to find a free node to
      settle;

---

chrony. Therefore, time is measured in *rounds* or *steps* (a cycle is a round or
step). Another parameter is memory which comes from a single source – the
number of bits stored at each robot.

Mobile Robot Dispersion. DISPERSION can be formally defined as follows.

Definition 1 (DISPERSION). *Given any n-node anonymous grid $G = (V, E)$
having $k \leq n$ mobile robots positioned initially arbitrarily on its nodes, the robots
reposition autonomously to reach a configuration where each robot is on a distinct
node of G.*

The goal is to optimize two performance metrics: (i) **Time** – the number of
rounds (steps), and (ii) **Memory** – the number of bits stored at each robot.

3   Algorithm in the Local Communication Model (Theorem 1)

We present and analyze an algorithm, $Grid\_Disperse(k)$, that solves DISPER-
SION for $k \leq n$ robots on n-node square grid graphs in $O(\min(k, \sqrt{n}))$ time using
$O(\log k)$ bits at each robot in the local communication model. The high level
pseudocode is given in Algorithm 1. We discuss algorithm $Grid\_Disperse(k), k = \Omega(n)$, here. $Grid\_Disperse(k), k \leq o(n)$, and the algorithm for rectangular grid
graphs for any $k \leq n$ are omitted here and discussed in a full version due to
space constraints.

High Level Overview of the Algorithm. $Grid\_Disperse(k), k = \Omega(n)$, for n-
node square grid graphs has five stages, Stages 1–5 (Algorithm 1), which execute
sequentially one after another. The goal in Stage 1 is to move all the robots (not
already on boundary nodes) to position them on the boundary nodes of G. The
goal in Stage 2 is to move the robots, now all on the boundary nodes, to the

four boundary corner nodes of G. The goal in Stage 3 is to collect all robots, now on four corner nodes, at one corner node of G. The goal in Stage 4 is to distribute robots equally on the nodes of a boundary row or column of G. The goal in Stage 5 is to distribute the robots, now on the nodes of a boundary row or column, so that each node of G has exactly one robot positioned on it. We will show that each stage can be performed correctly in $O(\sqrt{n})$ rounds, giving overall $O(\sqrt{n})$ time complexity for $Grid\_Disperse(k)$. Algorithm $Grid\_Disperse(k)$ for $k \leq o(n)$ differentiates the cases of $\sqrt{n} \leq k \leq o(n)$ and $k < \sqrt{n}$ and handles them through separate algorithms. For $\sqrt{n} \leq k \leq o(n)$, we provide a $O(\sqrt{n})$-time algorithm, and for $k < \sqrt{n}$, we provide a $O(k)$-time algorithm, giving overall $O(\min(k, \sqrt{n}))$ runtime for any $k \leq n$. We then extend all these ideas for DISPERSION on rectangular grid graphs.

Algorithm for Square Grid Graphs, $k = \Omega(n)$. We describe in detail how Stages 1–5 of $Grid\_Disperse(k)$ are executed for square grids. The high level pseducode is in Algorithm 1. Figure 1 illustrates the working principle of $Grid\_Disperse(k)$ for $n = k = 49$. Each robot $r_i \in \mathcal{R}$ stores five variables $r_i.round$ (initially 0), $r_i.stage$ (values 1 to 5, initially $null$), $r_i.port\_entered$ (values 1 to 4, initially $null$), $r_i.port\_exited$ (values 1 to 4, initially $null$), and $r_i.settled$ (values 0 and 1, initially 0). We assume that in each round r_i updates $r_i.round \leftarrow r_i.round + 1$. Moreover, we denote the rounds of each stage by $\alpha.\beta$, where $\alpha \in \{1, 2, 3, 4, 5\}$ denotes the stage and β denotes the round within the stage. Therefore, the first round $(\alpha + 1).1$ for Stage $\alpha + 1$ is the next round after the last round of Stage α.

Stage 1. The goal in Stage 1 is to reposition the robots that are not already on boundary nodes of G to the boundary nodes of G. The robots that are already on (or reach during Stage 1) the boundary nodes do nothing until Stage 1 finishes. The idea here is, for each robot independently, to choose a port of a non-boundary node to exit in the current round based on the port from which it entered that non-boundary node in the previous round. Pick a robot $r_i \in \mathcal{R}$ at some node $v \in G$ which is not the boundary node. In round 1.1, r_i does not have information on the port of v from which it entered v (r_i has not moved yet). Therefore, it randomly picks one of the four ports at v and exits v. Suppose, in the beginning of round 1.2, r_i reaches a neighbor of v, say w. If w is a boundary node, we are done as r_i can differentiate a boundary node from a non-boundary node (a boundary node has three ports instead of four ports at a non-boundary node). While reaching w, the model provides r_i with the information on the port p at w from which r_i entered w. That means, in round 1.2 and after, r_i has information on port from which it entered the node in the previous round where it is positioning in the current round. Therefore, in round 1.2, r_i moves as follows. Besides port p, w has three other ports. r_i orders (in clockwise or counterclockwise direction) the three remaining ports at w starting from p. It then picks the second port in either order and exits w using that port. The process is then repeated by r_i in round 1.3 and after until it reaches a boundary node.

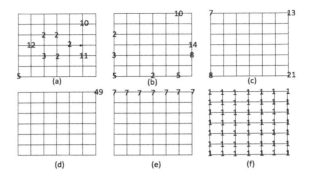

Fig. 1. An illustration of the five stages of algorithm $Grid\_Disperse(k)$ for $n = k = 49$: (a) An initial configuration, (b) Stage 1 that moves robots to boundary nodes of G, (c) Stage 2 that moves robots to four boundary corner nodes of G, (d) Stage 3 that moves the robots to one boundary corner node of G, (e) Stage 4 that distributes robots equally in a row (or a column) with each node having $\sqrt{n}$ robots, and (f) Stage 5 that distributes robots so that each node of G having exactly one robot each. The numbers denote the number of robots positioned at that node.

Lemma 1. *Pick a robot r_i at non-boundary node $v \in G$. Let r_i moves to a neighbor node w of v in round 1.1. Let $L_{\overrightarrow{vw}}$ be a line with one end v and passing through w. In round 1.2 and after (until r_i reaches a boundary node), r_i moves following the nodes of G that are on $L_{\overrightarrow{vw}}$ in the same direction in each round.*

Proof. Let the four ports of v be p_{v1}, p_{v2}, p_{v3}, and p_{v4}. Suppose r_i exits v using p_{v1} in round 1.1 and reaches node w in the beginning of round 1.2. If w is a boundary node, we are done. Otherwise, it remains to show that in round 1.2 and after, r_i always moves on the nodes of $L_{\overrightarrow{vw}}$ in the same direction. Let p_{w1} be the port at w from which r_i entered w in round 1.1. The three remaining ports at w are p_{w2}, p_{w3}, and p_{w4}. Since r_i picks second port in the clockwise (or counterclockwise) order in round 1.2 and after, the port r_i picks at w is always opposite port of port p_{w1} that it used to enter w from v in round 1.1. Notice that r_i is aware of p_{w1} while at w. Therefore, in the beginning of round 1.3, r_i reaches a neighbor node of w on $L_{\overrightarrow{vw}}$ (opposite of v). Continuing this way, the move decision in round 1.3 (and after) resembles the approach of round 1.2, takes r_i farther way from v on $L_{\overrightarrow{vw}}$ in each subsequent round.   □

Lemma 2. *At the end of Stage 1, all $k = \Omega(n)$ robots in any initial configuration are positioned on boundary nodes of G. Stage 1 finishes in $\sqrt{n} - 1$ rounds.*

Stage 2. The goal in Stage 2 is to collect all $k = \Omega(n)$ robots on boundary nodes of G to four boundary corners of G. Let L_{ab} be a boundary row or column of G passing through boundary corners a, b of G. For the rest three boundary row or columns the ideas are analogous and run in parallel. In Stage 2, $Grid\_Disperse(k)$ collects the robots on the nodes on L_{ab} to either node a or node b or some on a and some on b. The idea here is to move each robot independently in a direction on the boundary row or column of the grid until the robot reaches a boundary corner node. In this process, while selecting a port to leave from a current node to the next node in the boundary row or column, the robot may reach to a non-boundary node, in which case it will return back to the boundary node in the next round using the same port from which it entered the non-boundary node. Details are in a full version. We have the following lemma.

Lemma 3. *At the end of Stage 2, all $k = \Omega(n)$ robots in G are positioned on (at most) 4 boundary corner nodes of G. Stage 2 finishes in $3(\sqrt{n} - 1)$ rounds after Stage 1.*

Stage 3. The goal in Stage 3 is to collect all $k = \Omega(n)$ robots on a boundary corner node of G. Let a, b, c, d be the four boundary corner nodes of G. Suppose the smallest ID robot $r_1 \in \mathcal{R}$ is positioned on a. Pick any robot $r_i \neq r_1$. If r_i is already on a, it does nothing in Stage 3. Otherwise, it is on b, c, or d (say b) and it moves in Stage 3 to reach a. Since r_i needs to move on the boundary of G, the technique of Stage 2 can be modified for r_i so that it reaches a following the boundary nodes (details omitted).

Lemma 4. *At the end of Stage 3, all $k = \Omega(n)$ robots in G are positioned on a boundary corner nodes of G. Stage 3 finishes in $9(\sqrt{n} - 1)$ rounds after Stage 2.*

Stage 4. The goal is Stage 4 is to distribute $k = \Omega(n)$ robots (that are at a boundary corner node a after Stage 3) to a boundary row or column so that there will be no more than $\sqrt{n}$ robots on each node. In the first round, the smallest ID robot r_1 moves to pick a direction (i.e., row or a column). From the second round onwards, the idea similar to Stage 2 is used to move the robots on the boundary row or column leaving $\sqrt{n}$ robots on each node of that row or column. Details are omitted.

Lemma 5. *At the end of Stage 4, all $k = \Omega(n)$ robots in G are distributed on a boundary row or column of G so that there will be exactly $\sqrt{n}$ or less robots on a node. Stage 4 finishes in $3\sqrt{n} - 1$ rounds after Stage 3.*

Stage 5. The goal in Stage 5 is to distribute robots to nodes of G so that there will be at most one robot on each node. Let c be a boundary node with $\sqrt{n}$ or less robots on it and r_i is on c. In round 5.1, if r_i is the largest ID robot r_{max} among the robots on c, it settles at c assigning $r_i.settled \leftarrow 1$. Otherwise, in round 5.1, r_i moves as follows. While executing Stage 4, r_i stores the port of c it used to enter c (say $r_i.port\_entered = p_{c1}$) and the port of c used by the robot that left c exited through (say $r_i.port\_exited = p_{c2}$). Robot r_i then exits through port p_{c3}, which is not $r_i.port\_entered$ and $r_i.port\_exited$. This way r_i reaches a non-boundary node c'. All other robots except r_{max} also reach c' in the beginning of round 5.2. In round 5.2, the largest ID robot $r_{max'}$ settles at c'. The other at most $\sqrt{n} - 2$ robots exit c' using the port of c' selected through the port ordering technique described in Stage 1. This process then continues until a single robot remains at a node z, which settles at z.

Lemma 6. *At the end of Stage 5, all $k = \Omega(n)$ robots in G are distributed such that there is exactly one robot on a node of G. Stage 5 finishes in $\sqrt{n}$ rounds after Stage 4.*

Theorem 3. *Grid_Disperse(k) solves* DISPERSION *correctly for $k = \Omega(n)$ robots on an n-node square grid graph G in $O(\sqrt{n})$ rounds with $O(\log n)$ bits at each robot.*

Proof. Each stage of *Grid_Disperse(k)* executes sequentially one after another. Therefore, the correctness of *Grid_Disperse(k)* follows combining the correctness proofs of Lemmas 2–6. The time bound of $O(\sqrt{n})$ rounds also follows summing up the $O(\sqrt{n})$ rounds of each stage. Regarding memory, variables *port_entered*, *port_exited*, *settled*, and *stage* take $O(1)$ bits ($\Delta = 4$ for grids), and *round* takes $O(\log n)$ bits. Moreover, two or more robots at a node can be differentiated using $O(\log k)$ bits, for $k = \Omega(n)$. Therefore, a robot needs in total $O(\log n)$ bits. □

We have the following theorem for $k \leq o(n)$. Details are omitted.

Theorem 4. *Grid_Disperse(k) solves* DISPERSION *correctly for $k \leq o(n)$ robots on a square grid G in $O(\min(k, \sqrt{n}))$ rounds with $O(\log k)$ bits at each robot.*

Proof of Theorem 1: Theorems 3 and 4 together prove Theorem 1 for any $k \leq n$. □

Algorithm 2: *Grid_Disperse_Global(k)* in the global communication model

---

1 **Input:** An n-node square grid G with $k \leq n$ robots positioned arbitrarily on its nodes.

2 **Stage 1:** For each node of G with two or more robots, run a DFS traversal algorithm of Kshemkalyani *et al.* [8] for $W = 48\sqrt{k}\Delta$ rounds;

3 **Stage 2:** If a DFS traversal tree component formed in Stage 1 has $14\sqrt{k}$ or more nodes, then (i) divide the component into sub-components having $6\sqrt{k}$ to $14\sqrt{k} - 1$ nodes, and (ii) collect all the robots on each sub-component into a node in that sub-component;

4 **Stage 3:** The nodes in G with at least $6\sqrt{k}$ robots positioned on them form a square boundary of length $4(\sqrt{k} - 1)$ having $4(\sqrt{k} - 1)$ robots on the boundary (the remaining nodes of G has at most one robot on each);

5 **Stage 4:** If two or more square boundaries overlap, make them non-overlapping by coalescing some square boundaries with others;

6 **Stage 5:** If there are robots, if any, in the interior of each square boundary, collect those robots to a corner of that boundary such that only that corner has multiple robots;

7 **Stage 6:** Disperse the multiple robots on a single corner of each square boundary to the nodes in the interior of that boundary;

---

4   Algorithm in the Global Communication Model (Theorem 2)

We present and analyze an algorithm, $Grid\_Disperse\_Global(k)$, that solves DISPERSION for $k \leq n$ robots on n-node square grid graphs in $O(\sqrt{k})$ time with $O(\log k)$ bits at each robot in the global model. We discuss $Grid\_Disperse\_Global(k)$ for square grids here. The algorithm for rectangular grid graphs is in a full version.

High Level Overview of the Algorithm. $Grid\_Disperse\_Global(k)$ for square grid graphs has six stages, Stage 1–6, which execute sequentially one after another. $Grid\_Disperse\_Global(k)$ works for any $k \leq n$. Each stage runs for $O(\sqrt{k})$ rounds, giving overall $O(\sqrt{k})$ runtime. Stage 1 runs a DFS traversal in parallel for the nodes of G with at least two robots in the initial configuration. Stage 1 forms one or more DFS traversal tree components. Each DFS traversal tree component is a graph. Stage 2 divides each DFS traversal tree component with $14\sqrt{k}$ or more nodes into sub-components containing between $6\sqrt{k}$ to $14\sqrt{k} - 1$ nodes. The robots on a sub-component (say CC) are then collected to a node (say v_{CC}) on that sub-component. Stage 3 is then initiated by the nodes of grid G (for example, node v_{CC} of sub-component CC) having at least $6\sqrt{k}$ robots. They create a square boundary of length $4(\sqrt{k} - 1)$ so that there will be exactly $4(\sqrt{k}-1)$ robots positioned on the square boundary. If there is overlapping between two or more square boundaries constructed in Stage 3, Stage 4 makes them non-overlapping through merging all the square boundaries to one. Stage 5 collects the robots, if any, that are in the interior of each square boundary to a boundary corner node of the square boundary (again the node is v_{CC} for the sub-component CC). Finally, Stage 6 first distributes the robots on the boundary corner node (v_{CC} for sub-component CC) equally to the nodes on a row or a column of the square boundary and then disperses those robots to the nodes inside the square boundary, one on each node, achieving a DISPERSION configuration.

Algorithm for Square Grid Graphs, $k \leq n$. We now describe in detail how Stages 1–6 of $Grid\_Disperse\_Global(k)$ are executed for n-node square grid graphs. The high level pseudocode is given in Algorithm 2. We also abstract some details on what variables are used and how they are used to provide better readability for the overall algorithm. The techniques are highly involved compared to Sect. 3.

Stage 1. Stage 1 is for the nodes of G which have at least two robots positioned on them in the initial configuration. Stage 1 runs a DFS traversal algorithm for arbitrary graphs developed Kshemkalyani *et al.* [8] in the global communication model. Kshemkalyani *et al.* [8] run this DFS traversal until a DISPERSION configuration is achieved, and hence the runtime becomes $\min(2 \cdot 4m, 2 \cdot 2k\Delta)$ rounds for arbitrary graphs. In grid, running this algorithm until reaching a DISPERSION configuration needs $16k$ rounds (in a n-node square grid $m \leq 4n$, $\Delta = 4$, and $k \leq n$), which is clearly sub-optimal. Recall that our goal is to achieve optimal runtime of $O(\sqrt{k})$ rounds in grids. Therefore, we run the DFS traversal of [8] in the global communication model only for $48\sqrt{k}\Delta$ rounds and derive some properties to achieve a $O(\sqrt{k})$-round algorithm for a grid.

Lemma 7. *Let $W = 48\sqrt{k}\Delta$. If the DFS traversal forming a DFS tree component stops before W rounds, there is exactly one robot on each node of that DFS tree component. IF the DFS traversal does not stop until W rounds, there will be $\geq 6\sqrt{k}$ nodes (with at least a robot on each node) on the DFS tree component.*

Stage 2. Consider a connected DFS tree component CC_i with ID CID_i formed in Stage 1. Stage 2 will run for CC_i if it has at least a node with two or more robots on it. For simplicity, we denote such DFS tree components by $CC_i(not\_settled)$. Robots in the component $CC_i(not\_settled)$ have knowledge of the component ID CID_i of $CC_i(not\_settled)$. We have from Lemma 7 that $CC_i(not\_settled)$ must have at least $6\sqrt{k}$ nodes. Suppose $CC_i(not\_settled)$ has $X_i \geq 6\sqrt{k}$ nodes. Then, $CC_i(not\_settled)$ must have $Y_i > X_i$ robots. If $CC_i(not\_settled)$ has $6\sqrt{k} \leq X_i < 14\sqrt{k} - 1$ nodes, we collect all the robots of $CC_i(not\_settled)$ to the node of $CC_i(not\_settled)$ that is CID_i. If there are $X_i \geq 14\sqrt{k}$ nodes on $CC_i(not\_settled)$, we partition $CC_i(not\_settled)$ into sub-components $CC_i^{sub,j}(not\_settled)$ such that each $CC_i^{sub,j}(not\_settled)$ has between $6\sqrt{k}$ and $14\sqrt{k} - 1$ nodes on it. After this, the robots on the nodes of each sub-component $CC_i^{sub,j}(not\_settled)$ are collected to a node in that sub-component. Note that we collect only for sub-components $CC_i^{sub,j}(not\_settled)$ which have $Y_i > X_i$ (or at least a node on that sub-component has two robots on it). It remains to discuss two techniques:

(a) How the partitioning of a connected DFS tree component $CC_i(not\_settled)$ into sub-components $CC_i^{sub,j}(not\_settled)$ is achieved.

(b) How the collection of the robots on the nodes of each sub-component $CC_i^{sub,j}(not\_settled)$ into a node in that sub-component is achieved.

The details are in a full version. We have the following lemma after Stage 2.

Lemma 8. *At the end of Stage 2, there is either no, one, or at least $6\sqrt{k}$ robots on the nodes of grid G. Stage 2 finishes in $O(\sqrt{k})$ rounds.*

Stage 3. Stage 3 is initiated in parallel by the grid nodes which have at least $6\sqrt{k}$ robots positioned on them at the end of Stage 2 (Lemma 8). Consider a node $w \in G$ which has at least $6\sqrt{k}$ robots on it at the end of Stage 2. The goal is to form a square boundary SB_w of length $4(\sqrt{k}-1)$ containing one robot each on the boundary nodes (except node w which will have at least $2\sqrt{k}+1$ robots). This boundary helps to disperse all k robots in the interior. Details are omitted on how the square boundary is formed.

Lemma 9. *At the end of Stage 3, for each node of G with at least $6\sqrt{k}$ robots on it at end of Stage 2, a square boundary of length $4(\sqrt{k}-1)$ is correctly formed with each boundary node having one robot positioned on it, with only exception of a node on the boundary which has at least $2\sqrt{k}+1$ robots on it. Stage 3 finishes in $O(\sqrt{k})$ rounds.*

Stage 4. The goal in Stage 4 is make the square boundaries SB_w non-overlapping. Consider a square boundary SB_i with ID SID_i (the smallest ID robot on node w for the square boundary SB_w). All the robots on the boundary of SB_i know they belong to the boundary SB_i. In the first round of Stage 4, if any robot r_i of SB_i has some other robot r_j with SID_j colocated on the same node, it broadcasts a $Overlap(SID_i, SID_j)$ message. This is to indicate that the square boundary SB_i has met the square boundary $SB_j, j \neq i$. All the robots listen to such broadcasts and build an undirected *square boundary overlapping graph* $SB\_Overlap(B, E')$, where B is the set of square boundary IDs, and edge (SID_i, SID_j) indicates that $Overlap(SID_i, SID_j)$ message has been received. There might be one or more $SB\_Overlap(B, E')$ components and Stage 4 runs in parallel for each of those. A maximal independent set (MIS) is computed for each component and the square boundaries which are not part of the MIS will merge with the neighboring square boundary that is part of the MIS. Details are omitted.

Lemma 10. *At the end of Stage 4, all square boundaries are non-overlapping. The $4(\sqrt{k}-1)$ boundary nodes in each square boundary have a robot each, except a node which has at least $2\sqrt{k}+1$ robots. Stage 4 finishes in $O(\sqrt{k})$ rounds.*

Stage 5. At the end of Stage 4, all the remaining square boundaries are non-overlapping. The only problem might be that there are robots in the interior of the square boundaries. This is particularly because of DFS traversal trees that stopped before W rounds in Stage 1 and the robots of square boundaries from Stage 4 not collected yet. The goal in Stage 5 is to collect those robots, if any, inside each square boundary SB_i to the SID_i node so that at the end of Stage 5, there are robots only the boundary of each SB_i. A challenge is how to collect robots if any in the interior of SB_i in $O(\sqrt{k})$ rounds. For this, $\sqrt{k}-2$ robots first move in a row or column of SB_i and then to the interior of SB_i until reaching the opposite row or column, collecting all the robots that are found on the traversal. They then return to SID_i with all the robots (if any) in the interior of SB_i. Details are in a full version. We have the following lemma.

Lemma 11. *At the end of Stage 5, there will be no robot in the interior of the non-overlapping square boundaries obtained in Stage 4. Only a corner of square boundaries have at least $2\sqrt{k}+1$ robots. Stage 5 finishes in $O(\sqrt{k})$ rounds.*

Stage 6. At the end of Stage 5, we have the following: (i) All square boundaries SB_i are non-overlapping, (ii) There is no robot in the interior of the square boundaries, and (iii) Each boundary node of SB_i has exactly one robot on it except one corner SID_i which has at least $T_i \geq 2\sqrt{k}+1$ robots. The goal in Stage 6 is to disperse $T_i - 1$ robots to the interior nodes in SB_i. Stage 6 uses the ideas as in Stages 4 and 5 of Sect. 3 modified appropriately. Details are in a full version.

Lemma 12. *At the end of Stage 6, all $k \leq n$ robots are distributed such that there is at most one robot on a node of G. Stage 6 finishes in $O(\sqrt{k})$ rounds.*

Theorem 5. *Grid_Disperse_Global(k) solves* DISPERSION *correctly for $k \leq n$ robots on an n-node square grid G in $O(\sqrt{k})$ rounds with $O(\log k)$ bits at each robot.*

Proof. The correctness and runtime of *Grid_Disperse_Global(k)* follows combining Lemmas 7–12. Regarding memory, *Grid_Disperse_Global(k)* uses $O(1)$ number of different variables to be stored by each robot, taking $O(\log k)$ bits for each variable. □

Proof of Theorem 2: Theorem 5 proves Theorem 2 for any $k \leq n$. □

5   Concluding Remarks

We have studied the robot DISPERSION problem on graphs with the object of simultaneously minimizing two fundamental performance metrics: time and memory at each robot. We have presented two algorithms for DISPERSION of $k \leq n$ robots considering for the very first time grid graphs that find applications in many real-life robotic systems and provide simultaneously optimal bounds for both the metrics. The first result is for the local communication model and the second result is for the global communication model. The existing results in the literature were only for trees and arbitrary graphs and they were not able to simultaneously minimize both the metrics.

For future work, it will be interesting to solve DISPERSION on grids with time $O(\sqrt{k})$ in the local model for any $k \leq n$. Another interesting direction will be to extend our algorithms to semi-synchronous and asynchronous settings.

References

1. Augustine, J., Moses Jr., W.K.: Dispersion of mobile robots: a study of memory-time trade-offs. In: ICDCN, pp. 1:1–1:10 (2018)
2. Das, S., Dereniowski, D., Karousatou, C.: Collaborative exploration of trees by energy-constrained mobile robots. Theory Comput. Syst. **62**(5), 1223–1240 (2018)
3. Fraigniaud, P., Gasieniec, L., Kowalski, D.R., Pelc, A.: Collective tree exploration. Networks **48**(3), 166–177 (2006)
4. Hsiang, T., Arkin, E.M., Bender, M.A., Fekete, S.P., Mitchell, J.S.B.: Algorithms for rapidly dispersing robot swarms in unknown environments. In: WAFR, pp. 77–94 (2002)
5. Hsiang, T.-R., Arkin, E.M., Bender, M.A., Fekete, S., Mitchell, J.S.B.: Online dispersion algorithms for swarms of robots. In: SoCG, pp. 382–383 (2003)
6. Kshemkalyani, A.D., Ali, F.: Efficient dispersion of mobile robots on graphs. In: ICDCN, pp. 218–227 (2019)
7. Kshemkalyani, A.D., Molla, A.R., Sharma, G.: Fast dispersion of mobile robots on arbitrary graphs. CoRR, abs/1812.05352 (2018). (Accepted to ALGOSENSORS 2019)

8. Kshemkalyani, A.D., Molla, A.R., Sharma, G.: Dispersion of mobile robots in the global communication model. CoRR, abs/1909.01957 (2019). (Accepted to ICDCN 2020)
9. Martnez, H., Cnovas, J.P., Zamora, M.A., Skarmeta, A.G.: i-Fork: a flexible AGV system using topological and grid maps. In: ICRA, pp. 2147–2152 (2003)
10. Molla, A.R., Moses Jr., W.K.: Dispersion of mobile robots: the power of randomness. In: Gopal, T.V., Watada, J. (eds.) TAMC 2019. LNCS, vol. 11436, pp. 481–500. Springer, Cham (2019). https://doi.org/10.1007/978-3-030-14812-6_30
11. Ortolf, C., Schindelhauer, C.: Online multi-robot exploration of grid graphs with rectangular obstacles. In: SPAA, pp. 27–36 (2012)
12. Sharma, G., Dutta, A., Kim, J.-H.: Optimal online coverage path planning with energy constraints. In: AAMAS, pp. 1189–1197 (2019)

Packing and Covering with Segments

Joseph S. B. Mitchell[1] and Supantha Pandit[2]([✉])

[1] Stony Brook University, Stony Brook, NY, USA
joseph.mitchell@stonybrook.edu
[2] Dhirubhai Ambani Institute of Information and Communication Technology,
Gandhinagar, Gujarat, India
pantha.pandit@gmail.com

Abstract. We study three fundamental geometric optimization problems – independent set, piercing set, and dominating set – on sets of axis-parallel segments in the plane. We consider special cases in which the segments are either unit length or they are anchored on an inclined line (a line with slope -1). When the segments are anchored on both sides, we prove that all three problems are NP-complete (Throughout, we refer to NP-completeness of problems, as the decision versions of the NP-hard optimization problems we consider are all readily seen to be in NP.); NP-completeness was known for the corresponding problems with axis-parallel *rectangles* anchored on an inclined line (Correa et al. [4], Mudgal and Pandit [9], Pandit [10]). Further, we prove that the dominating set problem with unit segments in the plane is NP-complete. When the input segments are anchored on *one* side of the inclined line, there are polynomial-time algorithms for the independent set and piercing set problems.

Keywords: Geometric set cover · Piercing set · Dominating set · Segments · Inclined line · NP-complete

1 Introduction

We study three fundamental network optimization problems – independent set, piercing set and dominating set – in a geometric setting. Specifically, we consider the intersection graph of a set S of axis-parallel segments in the plane, with each segment having one of its endpoints on an inclined line, L (a line with slope -1); we say that such segments are *anchored* on L. We distinguish between the case of S being anchored on *one side* of L (S lies in one of the two closed halfplanes defined by L) or anchored on *both sides* of L (segments of S can be in either of the two closed halfplanes). In the independent set problem, the objective is to find a subset $S' \subseteq S$ such that no two segments in S' intersect. In the piercing set problem, we seek a minimum cardinality set P' of points in the plane such that

Partially supported by the National Science Foundation (CCF-1526406), the US-Israel Binational Science Foundation (project 2016116), and DARPA.

each segment is hit by at least one point from P'. Finally, in the dominating set problem, the objective is to select a minimum-cardinality subset $S' \subseteq S$ such that any segment in $S \setminus S'$ has a nonempty intersection with a segment in S'.

1.1 Related Work

Katz et al. [6] studied various dominating set problems associated with sets of line segments, rays, and lines in the plane. In particular, they considered the dominating set problem on axis-parallel segments in the plane, proving that it is NP-complete. Fekete et al. [5] considered the piercing set problem with different types of line segments (infinite lines, segments, rays, and combinations of them) in the plane. They provide some polynomial-time algorithms and NP-completeness results, as well as efficient approximation algorithms for some variations.

Chepoi and Felsner [3] studied the independent set and piercing set problems on a set of rectangles in the plane, each of which intersects an axis-monotone curve; they provide approximation algorithms for both problems. Later, Correa et al. [4] considered independent set and piercing set problems on rectangles, each intersecting an inclined line; they proved that the independent set problem is NP-complete when the rectangles are anchored on an inclined line from *both* sides (meaning that each rectangle has a vertex on the line and lies in either of the two halfplanes defined by the line), while giving a polynomial-time algorithm when rectangles are all anchored on *one* side (i.e., all rectangles lie in the same halfplane defined by the line, each having a vertex on the line). For the piercing set problem they provide an approximation algorithm. Their results were further extended by Mudgal and Pandit [9], who show that the piercing set problem with rectangles anchored on an inclined line from both sides is NP-complete. In addition, [9] also considered the set cover and hitting set problems with rectangles and squares. Pandit [10] studied the dominating set problem on a set of rectangles each of which intersects an inclined line; he shows that the problem is NP-complete for rectangles, but gives a polynomial-time algorithm in the case of squares (intersecting an inclined line). Recently, Bandyapadhyay et al. [1] considered the dominating set problem with rectangles anchored on an inclined line; actually, their results apply to L-shaped objects, since it can be noted that, in this setting, rectangles and L-shaped objects are indistinguishable.

1.2 Our Contributions

Our contributions include the following:

➤ **Independent Set problem:** We prove that the independent set problem with axis-parallel segments anchored on both sides of an inclined line is NP-complete; in contrast, it is polynomially solvable for axis-parallel segments and rectangles anchored on an inclined line from one side. In fact, for general connected regions (i.e., outerstring graphs, with a given realization) it is also polynomially solvable (see Keil et al. [7]), even for a topological line (there is no need for it to be a straight, inclined line).

➤ **Piercing Set problem:** We prove that the piercing set problem with axis-parallel segments anchored on an inclined line is NP-complete. This problem can be solved in polynomial time if the segments are anchored on an inclined line from one side.

➤ **Dominating Set problem:** We prove that the dominating set problem with axis-parallel segments anchored on an inclined line is NP-complete. We also prove that the dominating set problem with (non-anchored) axis-parallel unit segments in the plane is NP-complete.

2   Preliminaries

We first define the Rectilinear Planar Monotone 3SAT (RPM3SAT) problem. We are given a 3SAT formula ϕ with n variables $x_1, x_2, \ldots, x_n$ and m clauses $C_1, C_2, \ldots, C_m$. Each clause contains exactly 3 literals such that all of them are either positive (positive clause) or negative (negative clause). For each variable or clause we take a horizontal segment. The variable segments are on a horizontal line. The positive clause segments are above the horizontal line, while the negative clause segments are below. The clauses connect to the corresponding three variables by vertical line segments. A formula ϕ is an instance of the RPM3SAT problem if and only if the complete construction can be done while keeping it planar. See Fig. 1(a) for an instance of the RPM3SAT problem. The objective is to determine if there is a truth assignment to the variables that satisfies ϕ. We consider a modified version of the RPM3SAT problem, the Mod-RPM3SAT problem. Instead of placing the variable segments on a horizontal line, we place them on an inclined line L. Let C be a positive clause in the RPM3SAT problem (Fig. 1(a)) that contains variables x_i, x_j, and x_k. We directly connect the clause segment of C with the variable segment of x_i. Further, we extend the vertical segments corresponding to x_j and x_k such that they connect to the variable segments of x_j and x_k respectively. A symmetric construction is done for the negative clauses. See Fig. 1(b) for the modified instance of the RPM3SAT problem instance in Fig. 1(a). de Berg and Khosravi [2] proved that the RPM3SAT problems is NP-complete. This implies that the Mod-RPM3SAT problem is also NP-complete.

We now define some terminology that is used in this paper. Let us consider a variable x_i. Order the positive clauses from left to right that connect to x_i. Now let C_ℓ be a positive clause that contains x_i, x_j, and x_k and assume that the variables are also ordered from left to right. Then by the above ordering of the positive clauses we say that, C_ℓ is a ℓ_1-th, ℓ_2-th, and ℓ_3-th clause for the variables x_i, x_j, and x_k respectively. For example the clause C_2 is a 1-st, 2-nd, and 1-st clause for x_1, x_3, and x_4 respectively. This is true for both RPM3SAT and Mod-RPM3SAT problems. A similar ordering is possible for the negative clauses.

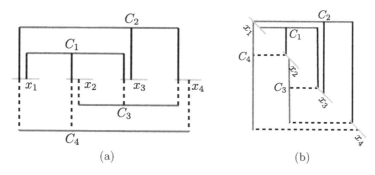

Fig. 1. (a) An instance of the RPM3SAT problem. (b) After modification of Fig. 1(a).

3   Algorithms: Segments Anchored from One Side

3.1   The Independent Set Problem

The independent set problem with segments anchored on an inclined line from one side (ISSIL1 problem) can be solved in polynomial time; in fact, more generally, the maximum independent set problem in outerstring graphs has a polynomial-time solution [7].

We make a simple observation that yields a more direct and efficient solution in the special setting of our problem, by reducing the problem to an instance of independent set in a bipartite graph. Let $S = \{s_1, s_2, \ldots, s_n\}$ be a set of n horizontal/vertical segments that are anchored on an inclined line L from the right side of L. We observe:

Observation 1. *For any two overlapping segments, we can discard the larger of the two segments from the input.*

Assume now that no two segments in S overlap. This implies that no two vertical (resp., horizontal) segments in S intersect. Consider the bipartite graph $G(A, B, E)$, with A (resp., B) being the set of vertical (resp., horizontal) segments of S, and edges E corresponding to intersecting pairs of segments. Now observe that finding a solution to the ISSIL1 problem is equivalent to finding a maximum independent set of vertices in the corresponding bipartite graph instance, which is solved in polynomial time.

3.2   The Piercing Set Problem

The piercing set problem with segments anchored on an inclined line from one side (PSSIL1 problem) can also be solved in polynomial time. We reduce the PSSIL1 problem to an equivalent problem, the edge cover problem in a bipartite graph. Note that Observation 1 holds for the PSSIL1 problem. Then, our problem is exactly that of computing a minimum-size set of edges that covers the vertices in the bipartite graph $G(A, B, E)$. Since the latter problem can be solved in polynomial time (see, e.g., [8]), the PSSIL1 problem also can be solved in polynomial time.

4   Independent Set Problem

In this section we prove that the independent set problem with segments anchored on an inclined line from both sides (ISSIL problem) is NP-complete by giving a reduction from the Mod-RPM3SAT problem.

Variable Gadget: The gadget for variable x_i is shown in Fig. 2(a). Let L be an inclined line. We take $20m + 6$ segments $\{s_1, s_2, \ldots, s_{20m+6}\}$ in a cycle like structure such that one endpoint of each segment touches the line L, and any two consecutive segments along the cycle intersect. From this structure it is easy to observe that there are exactly two optimal independent sets of segments, $S^i = \{s_1, s_3, \ldots, s_{20m+5}\}$ and $T^i = \{s_2, s_4, \ldots, s_{20m+6}\}$, for the gadget of x_i. Note that on the right side of L there are $2m$ gaps $\{g_1, g_2, \ldots, g_{2m}\}$, where the gap g_j, for $1 \leq j \leq 2m$ is the portion of L between the two points p_j and q_j on L, where p_j is the intersection point of the bottom endpoint of s_{5j-3} and L and q_j is the intersection point of the left endpoint of s_{5j+1} and L.

Clause Gadget: For each clause gadget we take 9 segments $\{t_1, t_2, \ldots, t_9\}$ (see Fig. 2(b)). The three segments t_1, t_2, and t_3 are responsible for the interaction with the three literals that the clause contains.

We now explain how the clause gadgets interact with the variable gadgets. As we defined in the preliminaries, for a positive clause C_ℓ that contains the variables x_i, x_j, and x_k, we assume that C_ℓ is the ℓ_1-th, ℓ_2-th, and ℓ_3-th clause for x_i, x_j, and x_k, respectively. For this clause, the 9 segments $\{t_1, t_2, \ldots, t_9\}$ touch the line L (see Fig. 2(b)). The right endpoint of t_4 and the top endpoint of t_5 meet at a point, and the left endpoint of t_4 touches the line L in the gap $g_{2\ell_1-1}$, and the bottom endpoint of t_5 touches the line L in the gap $g_{2\ell_3-1}$. A similar construction is made for t_6 and t_7 (the right and top endpoints of t_6 and t_7, respectively, meet at a point, the left endpoint of t_6 touches L in the gap $g_{2\ell_1-1}$, the bottom endpoint of t_7 touches L in the gap $g_{2\ell_2-1}$) and t_8 and t_9 (the right and top endpoints of t_8 and t_9 respectively meet at a point, the left endpoint of t_8 touches L in the gap $g_{2\ell_2-1}$, the bottom endpoint of t_9 touches L in the gap $g_{2\ell_3-1}$). Also note that the bottom endpoint of t_7 and left endpoint of t_8 meet at a point on L in the gap $g_{2\ell_2-1}$. Now the bottom endpoint of the segment t_1 meet the left endpoint of $s_{10\ell_1-4}$ on L and intersect t_4 and t_6. The bottom endpoint of t_2 touches the segment $s_{10\ell_1-6}$ and the top endpoint meets the intersection point of the bottom endpoint of t_7 and the left endpoint of t_8 on L. Finally, the left endpoint of t_3 touches the line L and intersects the segments $s_{10\ell_3-8}$, t_9, and t_5. See the construction for a positive clause in Fig. 2(b). A similar construction can be made for the negative clauses also.

Clearly, the above construction can be done in polynomial time. We now have the following theorem.

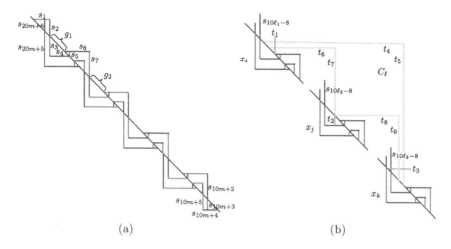

Fig. 2. (a) Structure of a variable gadget. (b) A clause gadget and its interaction with the variable gadgets.

Theorem 1. *The independent set problem with segments anchored on an inclined line from both sides (ISSIL problem) is NP-complete.*

Proof. It is sufficient to prove that ϕ is satisfiable if and only if the ISSIL problem instance S_ϕ has a solution with $n(10m + 3) + 4m$ segments.

Assume that ϕ is satisfiable, i.e., that there is a truth assignment to the variables of ϕ. Now for the gadget of x_i, we select the set S^i if x_i is true. Otherwise, we select the set T^i. Clearly, we select a total of $n(10m+3)$ segments from all of the variables. Now, for any clause gadget, if all three literals are false then none of t_1, t_2, and t_3 can be selected. As a result, there will be at most 3 independent segments in this clause. However, if at least one literal is true, there will be 4 independent segments in that clause. Thus, a total of $n(10m+3) + 4m$ independent segments are selected from a satisfiable assignment of ϕ.

On the other hand, assume that S_ϕ has a solution with $n(10m + 3) + 4m$ segments. Now, for each variable gadget, at most $(10m+3)$ independent segments can be selected. Since the variable gadgets are disjoint we can select at most $n(10m+3)$ independent segments. From each clause gadget, at most 4 segments can be part of an independent set. Also, the clause gadgets are disjoint. Since the solution size is $n(10m+3)+4m$, from each variable gadget exactly $(10m+3)$ and from each clause gadget exactly 4 independent segments can be selected. Clearly there are two solutions of $(10m+3)$ independent segments in the gadget of x_i; either S^i or T^i. We set x_i to be true if S^i is selected in the gadget of x_i; otherwise, set x_i to be false. It is easy to show that this is a satisfying assignment. □

5 Piercing Set Problem

In this section we prove that the piercing set problem with segments anchored on an inclined line from both sides (PSSIL problem) is NP-complete. We give a reduction from the Mod-RPM3SAT problem.

Variable Gadget: The gadget for the variable x_i is shown in Fig. 3(a). We take $8m + 8$ segments in a circular fashion such that all of them touch an inclined line L. Only two consecutive segments along the cycle intersect. In Fig. 3(a) a set C of canonical points $\{p_1, p_2, \ldots, p_{8m+8}\}$ are also shown. Observe that any optimal piercing set can selects points from C. Also observe that, there are exactly two optimal sets of canonical points, $P_i = \{p_1, p_3, \ldots, p_{8m+7}\}$ and $Q_i = \{p_2, p_4, \ldots, p_{8m+8}\}$ that hit all the segments in the gadget of x_i.

Clause Gadgets: Each clause gadget consists of five segments $\{t_1, t_2, t_3, t_4, t_5\}$ (see Fig. 3(b)). The two segments t_4 and t_5 are horizontal and parallel. The three segments t_1, t_2 and t_3 are vertical and responsible for the interaction with the three literals the clause contains. All the segments touch L from right side.

We now describe how the clause gadgets interact with the variable gadgets. Let C_ℓ be a positive clause containing the variables x_i, x_j, and x_k. Also assume that C_ℓ be the ℓ_1-th, ℓ_2-th, and ℓ_3-th clause for x_i, x_j, and x_k respectively. For this clause the two parallel horizontal segments t_4 and t_5 touch the line L from right side between the points $p_{4\ell_1-1}$ and $p_{4\ell_1+1}$ of the gadget of x_i. The lowest end-point of the vertical segment t_1 touches the point $p_{4\ell_1+1}$ of the gadget of x_i and intersect t_4 and t_5. Similarly, the lowest end-point of the vertical segment t_2 (resp. t_3) touches the point $p_{4\ell_2+1}$ (resp $p_{4\ell_3+1}$) of the gadget of x_j (resp x_k) and

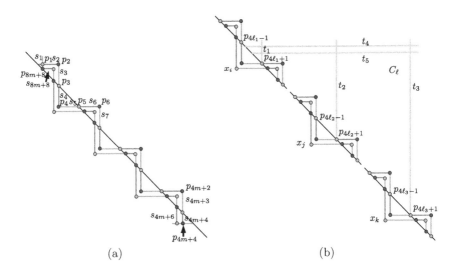

(a)   (b)

Fig. 3. (a) Structure of a variable gadget. (b) A clause gadget and its interaction with the variable gadgets.

intersect t_4 and t_5. See Fig. 3(b) for this construction. Clearly this construction can be done in polynomial time. We now prove the following theorem.

Theorem 2. *The piercing set problem with segments anchored on an inclined line from both sides (PSSIL problem) is NP-complete.*

Proof. We prove that an instance ϕ of the RPM3SAT problem is satisfiable if and only if the instance $\mathcal{S}_\phi$ of the PSSIL problem has a piercing set with $n(4m + 4) + 2m$ points.
($\Leftarrow$) Assume that ϕ is satisfiable and let $\pi : \{x_1, x_2, \ldots, x_n\} \rightarrow \{0, 1\}$ be a satisfying assignment. We select the point set P_i from the gadget of x_i if $\pi(x_i) = 1$. Otherwise select the point set Q_i. Since each clause is satisfiable then at least one of t_1, t_2, and t_3 is hit by these selected points. To hit the remaining segments two points are needed. Hence, overall we need $n(4m + 4) + 2m$ points.
($\Rightarrow$) Assume that $\mathcal{P}_\phi$ has a solution with $n(4m + 4) + 2m$ points. To hit all the segments in a variable gadget requires $4m + 4$ points, either the set P_i or Q_i. Set the variable x_i to be true if P_i is picked from the gadget of x_i. Otherwise picked the set Q_i. Now we show that this is a satisfying assignment. Notice that, in the gadget of C_ℓ, containing x_i, x_j, and x_k, at least 2 points are needed to stab the segments t_4 and t_5. Now if none of t_1, t_2, or t_3 is stabbed by a variable gadget then we need at least 3 points. So we have to choose at least one of the P_i, P_j or P_k from their corresponding gadgets.   □

6   Dominating Set Problem

6.1   Segments Anchored on a Line with Slope -1

In this section we prove that the dominating set problem with segments anchored on an inclined line from both sides (DSSAIL problem) is NP-complete. We give a reduction from the Mod-RPM3SAT problem.

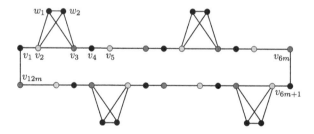

Fig. 4. A graph G.

Variable Gadget: To construct a variable gadget, we first consider an auxiliary result on a graph G that is shown in Fig. 4. The following lemma can be proved easily on G.

Lemma 1. *There exist exactly two minimum dominating sets of vertices in G, either all red colored or all blue colored.*

We encode the graph G as a variable gadget. The gadget for x_i is shown in Fig. 5. Now, using Lemma 1, we say that there are exactly two minimum dominating sets of segments, either all blue ($S^i = \{s_2, s_5, \ldots, s_{12m-1}\}$) or all red ($T^i = \{s_3, s_6, \ldots, s_{12m}\}$), for the gadget of x_i that interpret truth values of x_i.

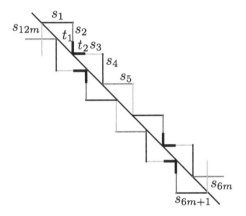

Fig. 5. Variable gadget for the variable x_i.

Clause Gadget: The gadget of C_ℓ, for $1 \le \ell \le m$ consists of a single segment, s_ℓ. This segment coincides with the clause segment of C_ℓ in the RPM3SAT instance. We now describe how the segments corresponding to the positive clauses interact with the segments corresponding to the variables. Since the interaction for negative clauses is similar, we omit its description here.

Let C_ℓ be a positive clause that contains x_i, x_j, and x_k in left to right order. As defined in the preliminaries, let C_ℓ be a ℓ_1-, ℓ_2-, and ℓ_3-th clause for x_i, x_j, and x_k respectively. For the variable x_i, we place the segment s_ℓ such that it intersects with the segment $s_{6\ell_1-4}$ in the gadget of x_i. For x_j, we extend the segment $s_{6\ell_2-4}$ from the gadget of x_j in the upward orientation such that it intersects with s_ℓ. Similarly, for x_k we extend the segment $s_{6\ell_3-4}$ from the gadget of x_k in the upward orientation such that it intersects s_ℓ. See Fig. 6(a) for the construction. A similar/symmetric construction (slightly different and easily visualized) can be done for the negative clauses (see Fig. 6(b)). This completes the construction, and clearly the construction can be done in polynomial time with respect to the number of variables and clauses in ϕ.

Theorem 3. *The dominating set problem with segments anchored on an inclined line from both sides (DSSAIL problem) is NP-complete.*

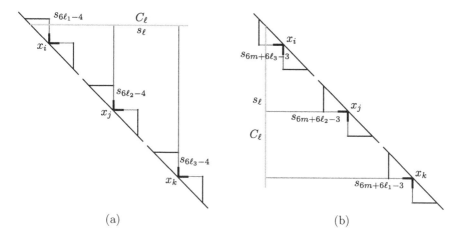

Fig. 6. Variable-clause interaction for (a) a positive clause and (b) a negative clause.

Proof. It is sufficient to prove that the formula ϕ is satisfiable if and only if $\mathcal{S}_\phi$, the DSSAIL instance constructed from ϕ, has a dominating set of $4mn$ segments. ($\Leftarrow$) Assume that ϕ is satisfiable. Let $\pi : \{x_1, x_2, \ldots, x_n\} \to \{0, 1\}$ be a satisfying assignment. We select the set S^i of segments from the gadget of x_i if $\pi(x_i) = 1$. Otherwise, we select the set T^i. Clearly, we select $4mn$ segments. Now since ϕ is a satisfying assignment, by the construction the clause segment is dominated by a selected segment from a variable gadget.
($\Rightarrow$) Assume that $\mathcal{S}_\phi$ has a dominating set of $4mn$ segments. For each variable gadget we need at least $4m$ segments. Since the variable gadgets are disjoint, we need at least $4mn$ segments in total. Since the size of the solution is $4mn$, for each variable gadget exactly $4m$ segments are needed. Note that there are exactly two optimal dominating sets of segments each size exactly $4m$ for each variable gadget. Therefore, we set x_i to be true if S^i is in the solution, otherwise we set x_i to be false. We now show that each clause of ϕ is satisfied. Note that to dominate a clause segment at least one of the three red segments to be selected form the corresponding variable gadgets of that clause, and we select that variable to be true whose red segment is selected. Therefore, the clause is satisfied, and hence the formula is also satisfied. □

6.2 Unit Segments in the Plane

In this section, we prove that the dominating set problem with unit length axis-parallel segments (DSUS problem) is NP-complete. We give a reduction from the RPM3SAT problem (see Fig. 1(a)). We slightly modify the RPM3SAT problem as follows. For each positive clause we take a small square instead of a horizontal segment, and the three legs of shapes "Γ", "⊤", and "⊣" connect the clause square with its corresponding variable horizontal segments (see Fig. 7(a)).

Variable Gadget: We encode the graph in Fig. 4 as a variable gadget for any variable x_i, for $1 \leq i \leq n$. Each gadget has two components, a rectangular chain and at most m leg-chains of segments. Each leg-chain corresponds to a clause leg through which a clause connects to the variable in the RPM3SAT problem. Note that for each variable, there are at most m such clause legs.

Rectangular Chain: In Fig. 7(b), a rectangular chain of segments is depicted. We take $12m$ unit segments $s_1, s_2, \ldots, s_{12m}$ in a rectangular fashion. We also take additional $8m$ segments $t_1, t_2, \ldots, t_{8m}$ such that the two segments t_i and t_{i+1} overlap with the two segments s_{6i-4} and s_{6i-3}, for $1 \leq i \leq 2m$.

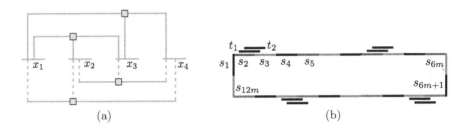

(a)   (b)

Fig. 7. (a) An instance of the RPM3SAT problem. (b) A rectangular chain of segments. The segments t_i are slightly shifted vertically for better visibility.

Leg-Chains: There are three types of leg-chains: left (see Fig. 8(a)), middle (see Fig. 8(b)), and right (vertical flip of left chain). Notice that each leg-chain has three special segments, s', s'', and s'''. The two segments s' and s'' in each chain are responsible to connect with the rectangular chain, and the segment s''' is responsible to interact with the clause gadget.

Connecting a Rectangular and a Leg-Chain: As defined before, for the variable x_i, order the positive clauses left to right that connect to x_i. Let C_ℓ be a ℓ_1-th positive clause for x_i. Then, (i) remove the segment $s_{6\ell-1}$ from the rectangular chain of x_i, (ii) connect the right endpoint of $s_{6\ell-2}$ with the bottom endpoint of s', and (iii) connect the left endpoint of $s_{6\ell}$ with the bottom endpoint of s''.

The variable gadget is the rectangular chain along with at most $2m$ leg-chains attached to it. This makes a big-chain of segments. Observe that this big-chain mimics the graph in Fig. 4. Note that the big-chain of x_i contains $G_i = 8\delta_i$ segments, where $\delta_i > 2m$ is a positive integer. Clearly, there are exactly two optimal solutions, either all red or all blue segments, such that each of their sizes is $2\delta_i$. Let us assume that $G = \sum_{i=1}^{n} G_i$ and $\Delta = \sum_{i=1}^{n} 2\delta_i$.

Clause Gadget: Let $C_\ell = (x_i \vee x_j \vee x_k)$ be a positive clause. For this clause the gadget consists of 7 segments $\{s^i, s^j, s^k, s^1, s^2, s^3, s^4\}$ (see Fig. 8(c)). The 4 segments $\{s^1, s^2, s^3, s^4\}$ forms a square shape corresponding to the square of C_ℓ in the modified RPM3SAT formula. The left endpoint of s^i is attached with s'''

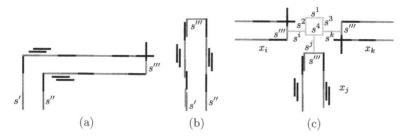

Fig. 8. (a) A left leg-chain. (b) A middle leg-chain. (c) A clause gadget and its interaction with the variable gadgets.

of the gadget of x_i and the right endpoint of s^i is attached with s^2. Similarly, the bottom and right endpoints of s^j and s^k are attached with s''' of the gadgets of x_j and x_k, respectively, and the top and left endpoints of s^j and s^k are attached with s^4 and s^3, respectively. A similar construction is made for negative clauses.

The overall construction takes time polynomial in n and m.

Theorem 4. *The dominating set problem with unit length axis-parallel segments (DSUS problem) is NP-complete.*

Proof. We prove that ϕ, an instance of the RPM3SAT problem, is satisfiable if and only if T_ϕ, the instance of the DSUS problem generated from ϕ, has a solution with $\Delta + 2m$ segments.

Assume that ϕ is satisfiable. For the gadget of x_i, select red segments if x_i is true; otherwise, select blue segments. As a result, in each clause gadget one of s^i, s^j, and s^k is dominated from the variable gadgets. Then, at least 2 segments from each clause gadget dominate the remaining clause segments. Therefore, we select a total of $\Delta + 2m$ segments.

Assume that there is a solution to T_ϕ with $\Delta + 2m$ segments. For the gadget of x_i we need at least $2\delta_i$ segments. For each clause gadget we need at least 2 segments. Since T_ϕ contains $\Delta + 2m$ segments, from the gadget of x_i we pick exactly $2\delta_i$ segments, and from each clause we pick exactly 2 segments. Since there are exactly two optimal solutions for the gadget of x_i, we set variable x_i to be true if all red segments are selected; otherwise, we set x_i to be false. It is easy to see that this assignment satisfies all of the clauses and hence the formula. □

References

1. Bandyapadhyay, S., Maheshwari, A., Mehrabi, S., Suri, S.: Approximating dominating set on intersection graphs of rectangles and L-frames. Comput. Geom. **82**, 32–44 (2019)
2. de Berg, M., Khosravi, A.: Optimal binary space partitions for segments in the plane. Int. J. Comput. Geom. Appl. **22**(03), 187–205 (2012)
3. Chepoi, V., Felsner, S.: Approximating hitting sets of axis-parallel rectangles intersecting a monotone curve. Comput. Geom. **46**(9), 1036–1041 (2013)

4. Correa, J.R., Feuilloley, L., Pérez-Lantero, P., Soto, J.A.: Independent and hitting sets of rectangles intersecting a diagonal line: algorithms and complexity. Discrete Comput. Geom. **53**(2), 344–365 (2015)
5. Fekete, S.P., Huang, K., Mitchell, J.S., Parekh, O., Phillips, C.A.: Geometric hitting set for segments of few orientations. Theory Comput. Syst. **62**(2), 268–303 (2018)
6. Katz, M.J., Mitchell, J.S.B., Nir, Y.: Orthogonal segment stabbing. Comput. Geom. Theory Appl. **30**(2), 197–205 (2005)
7. Keil, J.M., Mitchell, J.S.B., Pradhan, D., Vatshelle, M.: An algorithm for the maximum weight independent set problem on outerstring graphs. Comput. Geom. **60**, 19–25 (2017)
8. Lawler, E.L.: Combinatorial Optimization: Networks and Matroids. Dover, New York (2001)
9. Mudgal, A., Pandit, S.: Covering, hitting, piercing and packing rectangles intersecting an inclined line. In: Lu, Z., Kim, D., Wu, W., Li, W., Du, D.-Z. (eds.) COCOA 2015. LNCS, vol. 9486, pp. 126–137. Springer, Cham (2015). https://doi.org/10.1007/978-3-319-26626-8_10
10. Pandit, S.: Dominating set of rectangles intersecting a straight line. In: Canadian Conference on Computational Geometry, CCCG, pp. 144–149 (2017)

Implicit Enumeration of Topological-Minor-Embeddings and Its Application to Planar Subgraph Enumeration

Yu Nakahata[1]([mark])[ID], Jun Kawahara[1][ID], Takashi Horiyama[2],
and Shin-ichi Minato[1][ID]

[1] Kyoto University, Kyoto, Japan
{nakahata.yu.27e@st,jkawahara@i,minato@i}kyoto-u.ac.jp
[2] Hokkaido University, Sapporo, Japan
horiyama@ist.hokudai.ac.jp

Abstract. Given graphs G and H, we propose a method to implicitly enumerate topological-minor-embeddings of H in G using decision diagrams. We show a useful application of our method to enumerating subgraphs characterized by forbidden topological minors, that is, planar, outerplanar, series-parallel, and cactus subgraphs. Computational experiments show that our method can find all planar subgraphs in a given graph at most five orders of magnitude faster than a naive backtracking-based method.

Keywords: Graph algorithm · Enumeration problem · Decision diagram · Frontier-based search · Topological minor

1 Introduction

Subgraph enumeration is a fundamental task in several areas of computer science. Many researchers have proposed dedicated algorithms for subgraph enumeration to solve various problems. However, the number of subgraphs can be exponentially larger than the size of an input graph. In such cases, it is unrealistic to enumerate all subgraphs explicitly. To overcome this difficulty, we focus on *implicit* enumeration algorithms [7,9,13]. Such an algorithm constructs a *decision diagram* (DD) [2,11] representing the set of subgraphs instead of explicitly enumerating the subgraphs. DDs are tractable representations of families of sets. By regarding a subgraph of a graph as the subset of edges, DDs can represent a set of (edge-induced) subgraphs. The efficiency of implicit enumeration algorithms does not directly depend on the number of subgraphs but rather on the size of the output DD [7]. The size of a DD can be much (exponentially in some cases) smaller than the number of subgraphs, and thus, in such cases, we can expect that the implicit algorithms will work much

This work was supported by JSPS KAKENHI Grant Numbers JP15H05711, JP18-H04091, JP18K04610, JP18K11153, and JP19J21000.

M. S. Rahman et al. (Eds.): WALCOM 2020, LNCS 12049, pp. 211–222, 2020.
https://doi.org/10.1007/978-3-030-39881-1_18

Table 1. Relationship between graph classes and forbidden topological minors. $K_4 - e$ is the graph obtained by removing an arbitrary edge from K_4.

Graph class	Forbidden topological minors
Planar graphs	$K_5, K_{3,3}$
Outerplanar graphs	$K_4, K_{2,3}$
Series-parallel graphs	K_4
Cactus graphs	$K_4 - e$

faster than explicit ones. The framework to construct a DD representing a set of constrained subgraphs is called *frontier-based search* (FBS) [7]. Recently, Kawahara et al. [8] proposed an extension of FBS, *colourful FBS* (CFBS). Especially, they proposed an algorithm to enumerate isomorphic subgraphs, that is, given graphs G and H, it constructs a DD representing the set of subgraphs of G that are isomorphic to H.

To extend types of subgraphs that can be implicitly enumerated, we focus on *topological-minor-embeddings* (TM-embeddings) [3]. A subgraph G' of G is a TM-embedding of H in G if G' is isomorphic to a *subdivision* of H. A subdivision of H is a graph obtained by replacing each edge in H with a path. A TM-embedding is a generalization of an isomorphic subgraph, where each edge must be replaced with a path with length one. An application of TM-embeddings is *forbidden topological minor characterization* (FTM-characterization) [3]. For example, a graph is planar if and only if it contains neither K_5 nor $K_{3,3}$ as a topological minor [3], where K_a is the complete graph with a vertices and $K_{b,c}$ is the complete bipartite graph with the two parts of b and c vertices. Other examples are shown in Table 1 (see [1] for details).

Our Contribution. In this paper, when graphs G and H are given, we propose an algorithm to implicitly enumerate all TM-embeddings of H in G. Using our algorithm and some additional DD operations, we can also implicitly enumerate subgraphs having FTM-characterizations, that is, planar, outerplanar, series-parallel, and cactus subgraphs. Our contributions are:

- Given graphs G and H, we propose an algorithm to implicitly enumerate all TM-embeddings of H in G (Sect. 3.1).
- We propose more efficient algorithms when H is in specific graphs that are used in FTM-characterizations of graph classes in Table 1, that is, complete graphs, complete bipartite graphs, and $K_4 - e$ (Sect. 3.2).
- Combining our algorithm with DD operations, we show how to implicitly enumerate subgraphs having FTM-characterization, that is, planar, outerplanar, series-parallel, and cactus subgraphs (Sect. 3.3).
- We evaluate our method by computational experiments. We apply our method to implicitly enumerating all planar subgraphs in a graph. The results show that our method runs up to five orders of magnitude faster than a naive backtracking-based method (Sect. 4).

Our Techniques. Our key contribution is to extend the algorithm for implicitly enumerating isomorphic subgraphs [8] to TM-embeddings. We first explain the algorithm of Kawahara et al. [8] briefly. They reduced isomorphic subgraph enumeration to enumeration of "colored degree specified subgraphs". Let us consider $K_{2,3}$ in Fig. 1(a). It has the degree multiset $\{2^3, 3^2\}$, where 2^3 means that there are three vertices with degree 2. The graph in Fig. 1(b) has the same degree multiset although it is not isomorphic to $K_{2,3}$. Thus, the degree multiset is not enough to characterize $K_{2,3}$ uniquely. Let us consider the edge-colored graph in Fig. 1(c). For an edge-colored graph with k colors, we consider a *colored degree* of a vertex v, that is, a k-tuple of integers such that its i-th element is the number of color-i edges incident to v. The edge-colored graph in Fig. 1(c) has the colored-degree multiset $M = \{(3,0), (0,3), (1,1)^3\}$, where red and green are respectively colors 1 and 2. Observe that every edge-colored graph with colored-degree multiset M is isomorphic to the graph. Therefore, enumerating subgraphs of G that are isomorphic to $K_{2,3}$ is equivalent to finding all 2-colored subgraphs of G whose degree multisets equal M and then "decolorizing" them.

In TM-embedding enumeration, the size of subgraphs we want to find is not fixed, which makes difficult to identify the subgraphs only by a colored degree multiset. Therefore, we extend the algorithm of Kawahara et al. in two directions: (a) we allow some colored degrees to exist arbitrary many times and (b) we introduce the constraint that each colored subgraph is connected. Using such constraints, we can reduce TM-embeddings enumeration to enumerating subgraphs satisfying the constraints. We propose a framework to enumerate TM-embeddings based on such constraints that can be applied to *every* query graph. Since the complexity of CFBS heavily depends on the number of colors used in the constraint, we discuss how to reduce the number of colors.

2   Preliminaries

2.1   Graphs and Colored Graphs

Let $G = (V(G), E(G))$ be a graph. We define $n = |V(G)|$ and $m = |E(G)|$. Graphs G and H are *isomorphic* if there exists a bijection $f : V(G) \rightarrow V(H)$ such that, for all $u, v \in V(G)$, $\{u, v\} \in E(G) \Leftrightarrow \{f(u), f(v)\} \in E(H)$. For a positive integer k, we define $[k] = \{1, \ldots, k\}$. For a positive integer c and a finite set E, a *c-colored subset* over E is a c-tuple $(D_1, \ldots, D_c)$ such that (a) for each $i \in [c]$, $D_i \subseteq E$ holds and (b) for all $i, j \in [c]$, if $i \neq j$, then $D_i \cap D_j = \emptyset$. A *c-colored graph* H^c is a tuple of a finite set U and a c-colored subset $(D_1, \ldots, D_c)$ over the set $\{\{u, v\} \mid u, v \in U, u \neq v\}$. Here, U is the *vertex set* and $(D_1, \ldots, D_c)$ is the *colored edge set*. The *color-i degree* of $u \in U$ in H^c is the number of color-i edges incident to u. The *colored degree* of u in H^c is a c-tuple $(\delta_1, \ldots, \delta_c)$ of integers, where δ_i is the color-i degree of u. The *colored degree multiset* of H^c, which is denoted by $\mathrm{DS}(H^c)$, is the multiset of the colored degrees of all the vertices in H^c. H^c is a *c-colorized graph* of H if the underlying graph of H^c is isomorphic to H, where the *underlying graph* of H^c is a graph obtained by ignoring the colors of the edges in H^c. A *c-colored subgraph* of G is a *c*-colored graph whose underlying graph is isomorphic to a subgraph of G.

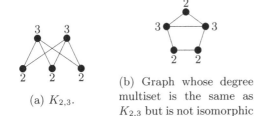

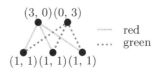

(a) $K_{2,3}$.

(b) Graph whose degree multiset is the same as $K_{2,3}$ but is not isomorphic to $K_{2,3}$.

(c) 2-colorized graph of $K_{2,3}$.

Fig. 1. Graphs and a colored graph.

2.2   Topological Minors and Characterizations of Graphs

Subdividing an edge $\{u, v\}$ of a graph H means removing the edge $\{u, v\}$ from H, introducing a new vertex w, and adding new edges $\{u, w\}$ and $\{v, w\}$. If a graph is obtained by subdividing each edge of H arbitrary times, it is a *subdivision* of H. A graph F is *homeomorphic* to a graph H if F is isomorphic to some subdivision of H.[1] If a graph F is homeomorphic to a graph H, the original vertices of H are the *branch vertices* of F and the other vertices are the *subdividing vertices*. Note that the degree of a branch vertex equals the original degree in H while the degree of a subdividing vertex is 2. (The degree of a branch vertex can be 2 when its original degree in H is 2). For graphs G and H, H is a *topological minor* (TM) of G if G contains a subgraph homeomorphic to H. A subgraph G' of G is a *TM-embedding* of H in G if G' is homeomorphic to H. For families $\mathcal{G}$ and $\mathcal{H}$ of graphs, $\mathcal{G}$ is *forbidden-TM-characterized* (FTM-characterized) by $\mathcal{H}$ when a graph G belongs to $\mathcal{G}$ if and only if, for any subgraph G' of G and $H \in \mathcal{H}$, G' is not homeomorphic to H. For example, the family of planar graphs is FTM-characterized by $\{K_5, K_{3,3}\}$ [10]. The same characterization goes to several graph classes (Table 1).

2.3   Decision Diagram (DD)

A $(c+1)$-*decision diagram* ($(c+1)$-DD) [8] represents a family of *c*-colored subsets over a finite set. Let $E = \{e_1, \ldots, e_m\}$. A $(c+1)$-DD over E is a directed acyclic graph with the unique *root node*[2] and the two *terminal nodes* $\top$ and $\bot$. Nodes

---

[1] In another definition, F is homeomorphic to H if some subdivision of F is isomorphic to some subdivision of H. However, we allow subdividing only for H because H is "contracted enough" when it is a forbidden topological minor, that is, H does not contain redundant vertices with degree 2.

[2] To avoid confusion, we use the terms "node" and "arc" for a $(c+1)$-DD and use "vertex" and "edge" for an input graph.

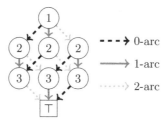

Fig. 2. 3-DD. A square is a terminal node and circles are non-terminal nodes. An integer in a circle is the label of the node. For simplicity, we omit $\bot$ and the arcs pointing at it.

other than the terminal nodes are the *non-terminal nodes*. Each non-terminal node has a *label*, an integer in $[m]$, and $c+1$ arcs. The arcs are the *0-arc*, *1-arc*, ..., and *c-arc*. For each arc, the label of its tail is larger than its head if the tail is non-terminal. In a $(c+1)$-DD, each path from the root node to $\top$ represents a c-colored subset over E. For each path and $j \geq 1$, descending the j-arc of a non-terminal node with label i corresponds to including e_i in a colored subset as a color j element. Descending the 0-arc corresponds to excluding e_i from a colored subset. The set of such paths corresponds to the family of c-colored subsets represented by the $(c+1)$-DD. Figure 2 shows an example of a 3-DD over a set $\{e_1, e_2, e_3\}$. The 3-DD represents the family of the 2-colored subsets each of which contains exactly one element for each color, that is, $\{(\{e_1\}, \{e_2\}),$ $(\{e_1\}, \{e_3\}), (\{e_2\}, \{e_1\}), (\{e_2\}, \{e_3\}), (\{e_3\}, \{e_1\}), (\{e_3\}, \{e_2\})\}$. In the rest of the paper, $\mathbf{Z}^{c+1}$ denotes a $(c+1)$-DD and $[\![\mathbf{Z}^{c+1}]\!]$ denotes the family of c-colored subsets represented by $\mathbf{Z}^{c+1}$. When $c = 1$, we omit the superscript from $\mathbf{Z}^{c+1}$ and write $\mathbf{Z}$, that is, $\mathbf{Z}$ is a 2-DD.

2.4 Colorful Frontier-Based Search (CFBS)

Colorful frontier-based search (CFBS) [8] is a framework of algorithms to construct a DD representing the set of constrained subgraphs. An important application of CFBS is the *isomorphic subgraph enumeration problem*. In the problem, given two graphs, a *host graph* G and a *query graph* H, we output all subgraphs of G that are isomorphic to H. CFBS does not explicitly enumerate subgraphs but constructs a 2-DD representing the set of them. By regarding a subgraph of G as the subset of $E(G)$, 2-DD can represent a set of (edge-induced) subgraphs.

The crux of CFBS is to identify the query graph by a colored degree multiset.

Definition 1 (profile). *A multiset M of c-colored degrees is a profile of H if the following are equivalent:*

- *A graph F is isomorphic to H.*
- *There exists a c-colorized graph F^c of F that satisfies the constraint $\mathcal{C}_M$, where $\mathcal{C}_M$ is a function from c-colored graphs to $\{0, 1\}$ such that $\mathcal{C}_M(F^c) = 1 \Leftrightarrow \mathrm{DS}(F^c) = M$.*

For example, for $K_{2,3}$ in Fig. 1(a), $\{(3,0),(0,3),(1,1)^3\}$ is a profile as shown in Fig. 1(c). In the following, $\mathcal{G}^c(\mathcal{C})$ denotes the family of c-colored subgraphs of G that satisfy the constraint $\mathcal{C}$. $\mathcal{G}(H)$ denotes the family of subgraphs of G that are isomorphic to H. $\mathbf{Z}^{c+1}(\mathcal{C})$ and $\mathbf{Z}(H)$ respectively denote the $(c+1)$-DD representing $\mathcal{G}^c(\mathcal{C})$ and the 2-DD representing $\mathcal{G}(H)$. The framework of algorithms for implicit isomorphic subgraph enumeration based on CFBS is as follows:

1. Find a profile M of H. Let c be the number of the colors used in M.
2. Construct $\mathbf{Z}^{c+1}(\mathcal{C}_M)$.
3. By *decolorizing* $\mathbf{Z}^{c+1}(\mathcal{C}_M)$, we obtain $\mathbf{Z}(H)$.

Here, for a family $\mathcal{F}^c$ of c-colored subsets, its *decolorization* is the family $\mathcal{F} = \left\{ \bigcup_{i \in [c]} D_i \mid (D_1, \ldots, D_c) \in \mathcal{F}^c \right\}$, that is, the family obtained by ignoring the colors from $\mathcal{F}^c$. For DDs, the decolorization of the $(c+1)$-DD representing $\mathcal{F}^c$ is the 2-DD representing $\mathcal{F}$. We can decolorize a DD by a recursive operation utilizing the recursive structure of the DD [8]. To construct a DD efficiently, CFBS uses dynamic programming. The *i-th frontier* F_i is the set of vertices incident to both the edges in $\{e_1, \ldots, e_{i-1}\}$ and $\{e_i, \ldots, e_m\}$. CFBS constructs a DD in a breadth-first manner from the root node and merges two nodes with the same label and states with respect to the frontier. See [8] for details.

CFBS can deal with every finite query graph H in isomorphic subgraph enumeration if we use $|E(H)|$ colors [8]. However, since the complexity of CFBS exponentially depends on the number of colors [8], it is important to find a profile with as few colors as possible. The following theorem states that, for every graph H, it suffices to use $\tau(H)$ colors, where $\tau(H)$ is the minimum size of vertex covers of H. For a graph H, a subset $S \subseteq V(H)$ is a *vertex cover* of H if, for every edge $e \in E(H)$, at least one of its endpoints belongs to S. A star is a graph isomorphic to $K_{1,a}$ for some positive integer a.

Theorem 1 [4]. *Let H be a graph and H^c be a c-colored graph of H such that, for every $i \in [c]$, the subgraph of H^c induced by color-i edges is isomorphic to a star. A graph F is isomorphic to H if and only if there exists a c-colored graph F^c of F such that $\mathrm{DS}(F^c) = \mathrm{DS}(H^c)$. It follows that, for every H, there is a profile using $\tau(H)$-colored degrees.*

3   Algorithms

Due to space limitations, proofs in this section are omitted.

3.1   Implicit Enumeration of TM-Embeddings

Given graphs G and H, we propose an algorithm to construct the 2-DD $\mathbf{Z}(\widehat{H})$ representing the set of all TM-embeddings of H in G. In the following, $\mathcal{S}(H)$ denotes the family of subdivisions of H. In an algorithm to enumerate isomorphic subgraphs based on existing CFBS, finding a profile of H is essential. However, as for TM-embeddings, since $\mathcal{S}(H)$ is an infinite set, it seems difficult to identify $\mathcal{S}(H)$ with a single colored degree multiset. Therefore, we define an *extended profile* for an infinite family $\mathcal{S}(H)$ by extending a profile for a graph H.

Let $\Delta^c = \left\{ \overrightarrow{\delta\langle i \rangle} \mid i \in [c] \right\}$, where $\overrightarrow{\delta\langle i \rangle} = (\delta_1, \ldots, \delta_c), \delta_i = 2, j \neq i \Rightarrow \delta_j = 0$.

Definition 2 (extended profile). *Let c be a positive integer. A multiset M of c-colored degrees is an* extended profile *of $\mathcal{S}(H)$ if the following are equivalent:*

- *A graph F belongs to $\mathcal{S}(H)$.*
- *There exists a c-colorized graph F^c of F that satisfies the constraint $\mathcal{C}_M^*$, where $\mathcal{C}_M^*(F^c) = 1$ if and only if (a) $\mathrm{DS}(F^c)$ is obtained by adding an arbitrary number of elements (allowing duplication) of Δ^c to M and (b) each colored subgraph of F^c is connected.*

Our method of implicit TM-embedding enumeration is written as follows:

1. Find an extended profile M of $\mathcal{S}(H)$. Let c be the number of colors in M.
2. Construct $\mathbf{Z}^{c+1}(\mathcal{C}_M^*)$.
3. By decolorizing $\mathbf{Z}^{c+1}(\mathcal{C}_M^*)$, we obtain $\mathbf{Z}(\widehat{H})$.

Decolorization in Step 3 can be done in the same way as existing CFBS. To construct $\mathbf{Z}^{c+1}(\mathcal{C}_M^*)$ in Step 2, we add the constraints "there are an arbitrary number of vertices whose degrees are in Δ^c" and "each colored subgraph is connected" to existing CFBS. In fact, we propose an algorithm for a more general problem. For a multiset s and a set t of c-colored degrees, let $\mathcal{C}_s^t$ be the corresponding constraint where we replace Δ^c by t in the definition of $\mathcal{C}_s^*$. In other words, for each $\delta \in s$, the number of δ in a subgraph must equal its multiplicity in s. In addition, there can be an arbitrary number of vertices whose colored degrees are in t. Given s and t, we propose an algorithm to construct a DD $\mathbf{Z}^{c+1}(\mathcal{C}_s^t)$.

To assess the efficiency of algorithms based on CFBS, it is usual to analyze the *width* of the output DD [12,13]. The width of a DD is the maximum number of nodes with the same label. It is a measure of both the size of the DD and the time complexity to construct the DD. Let $w = \max_{i \in [m]} |F_i|$. $\mathbb{N}$ denotes the set of non-negative integers. For a c-tuple δ, δ_i denotes the i-th element of δ. For c-tuples δ and γ of integers, we define $\delta \leq \gamma$ if, for all $i \in [c]$, $\delta_i \leq \gamma_i$ holds. When $\delta \leq \gamma$, we say that δ is *dominated* by γ. For a multiset s and a set t of c-tuples, $s \cup t$ denotes the set of tuples that appear in s or t. (When a tuple is contained in s, its multiplicity is ignored in $s \cup t$.) For a set M of c-colored degrees, $\mathcal{D}(M)$ denotes the set of tuples in $\mathbb{N}^c$ that are dominated by a tuple in M, that is, $\mathcal{D}(M) = \{\chi \in \mathbb{N}^c \mid \exists \delta \in M, \chi \leq \delta\}$. For $\delta \in s$, the multiplicity of δ in s is denoted by $s(\delta)$.

Theorem 2. *Given a multiset s and a set t of c-colored degrees, there is an algorithm to construct a DD $\mathbf{Z}^{c+1}(\mathcal{C}_s^t)$ with width*

$$2^{\mathcal{O}(cw \log w)} |\mathcal{D}(s \cup t)|^w \prod_{\delta \in s}(s(\delta) + 1). \tag{1}$$

The theorem shows that the complexity of the algorithm mainly depends on w and c. Although w is determined by the host graph G, we can reduce the value of c by finding an extended profile of the query graph H using as few colors as possible. We discuss how to find such an extended profile for general H in the rest of this subsection and for specific graphs in the right column in Table 1 in the next subsection.

We discuss the complexity for general H. We assume that H has at least two vertices. Similarly to CFBS for isomorphic subgraph enumeration, we show that there is an extended profile for every graph H using $|E(H)|$ and $\tau(H)$ colors. Recall that $\tau(H)$ is the minimum size of vertex covers of H. Although the latter is better in most cases, we show both theorems for comparison.

Theorem 3. *For a graph H, let $H^{|E(H)|}$ be a $|E(H)|$-colorized graph obtained by coloring the edges of H with distinct colors. Then, $M = \mathrm{DS}(H^{|E(H)|})$ is an extended profile of $\mathcal{S}(H)$. There is an algorithm to construct a DD $\mathbf{Z}^{|E(H)|+1}(\mathcal{C}_M^*)$ with width*

$$2^{\mathcal{O}(|E(H)|w \log w)+|V(H)|}\left(2^{|E(H)|}+|E(H)|\right)^w. \tag{2}$$

Theorem 4. *For a graph H, let $H^{\tau(H)}$ be a $\tau(H)$-colorized graph whose each colored subgraph is isomorphic to a star and the set of the centers is a minimum vertex cover of H. Then, $M = \mathrm{DS}(H^{\tau(H)})$ is an extended profile of $\mathcal{S}(H)$. There is an algorithm to construct a DD $\mathbf{Z}^{\tau(H)+1}(\mathcal{C}_M^*)$ with width*

$$2^{\mathcal{O}(\tau(H)w \log w)+|V(H)|}\left(2^{\tau(H)}\tau(H)\right)^w. \tag{3}$$

3.2   Constraints for Forbidden Topological Minors

We derive specific extended profiles for the subdivisions of the graphs in the right column of Table 1: complete graphs, complete bipartite graphs, and $K_4 - e$. While the results for complete bipartite graphs and $K_4 - e$ follow directly from Theorem 4, we can reduce one color for complete graphs. In the following, we discuss complete bipartite graphs first, which is easier than complete graphs.

Theorem 5. *Let a, b $(a \le b)$ be positive integers. A multiset $M_{a,b} = M_{a,b}^1 \cup M_{a,b}^2$ consisting of a-colored degrees is an extended profile of $\mathcal{S}(K_{a,b})$, where*

$$M_{a,b}^1 = \left\{(\delta_1, \ldots, \delta_a) \;\middle|\; \begin{array}{l} \exists i \in [a], \delta_i = b, \\ j \neq i \Rightarrow \delta_j = 0 \end{array}\right\}, \quad M_{a,b}^2 = \left\{\underbrace{(1, \ldots, 1)}_{a}^b\right\}. \tag{4}$$

There is an algorithm to construct a DD $\mathbf{Z}^{a+1}(\mathcal{C}_{M_{a,b}}^*)$ with width

$$2^{\mathcal{O}(aw \log w)}(2^a + ab)^w(b + 1). \tag{5}$$

Figure 3(a) shows a representation of a subdivision of $K_{3,3}$ based on Theorem 5.

Next, we consider the subdivisions of complete graphs. Since the size of a minimum vertex cover of K_a is $a - 1$, there exists an extended profile of $\mathcal{S}(K_a)$

(a) Representation of a subdivision of $K_{3,3}$ based on Theorem 5.

(b) Representation of a subdivision of K_5 based on Theorem 6.

Fig. 3. Representations of subdivisions of $K_{3,3}$ and K_5. In each figure, a filled and non-filled vertex represent a branch and subdividing vertex, respectively. A tuple beside a vertex means the colored degree of the vertex.

with $a-1$ colors by Theorem 4. The extended profile is obtained by decomposing K_a into $K_{1,1}, K_{1,2}, \ldots,$ and $K_{1,a-1}$ and coloring the subgraphs with distinct colors. In this coloring, if we color $K_{1,2}$ with the same color as $K_{1,1}$, the obtained subgraph is K_3. We show that the colored degree multiset obtained from this coloring is also an extended profile of $\mathcal{S}(K_a)$.

Theorem 6. *Let $a \geq 3$ be an integer. A multiset $M_{a-2} = M_{a-2}^1 \cup M_{a-2}^2$ consisting of $(a-2)$-colored degrees is an extended profile of $\mathcal{S}(K_a)$, where*

$$M_{a-2}^1 = \left\{ (2, \underbrace{1, \ldots, 1}_{a-3})^3 \right\}, M_{a-2}^2 = \left\{ (\delta_1, \ldots, \delta_{a-2}) \left| \begin{array}{l} \exists i \in \{2, \ldots, a-2\}, \\ j < i \Rightarrow \delta_j = 0, \\ \delta_i = i + 1, \\ j > i \Rightarrow \delta_j = 1 \end{array} \right. \right\}.$$

$$(6)$$

There is an algorithm to construct a DD $\mathbf{Z}^{a-1}(\mathcal{C}_{M_a}^*)$ with width

$$2^{\mathcal{O}(aw \log w)} \left(3 \cdot 2^{a-2} - a \right)^w.$$

$$(7)$$

Figure 3(b) shows a representation of a subdivision of K_5 based on Theorem 6.

3.3   Enumerating Subgraphs Having FTM-Characterizations

We show how to implicitly enumerate subgraphs having FTM-characterization. We combine DD operations with our algorithm to implicitly enumerate TM-embeddings. UNION [11] is a function whose inputs are two 2-DDs $\mathbf{Z}_1$ and $\mathbf{Z}_2$ and output is the 2-DD representing $[\![\mathbf{Z}_1]\!] \cup [\![\mathbf{Z}_2]\!]$. NONSUPSET [9] is a function whose input is a 2-DD $\mathbf{Z}$ over a finite set E and output is the 2-DD representing the family $\{A \subseteq 2^E \mid \forall B \in [\![\mathbf{Z}]\!], A \not\supseteq B\}$. $\mathcal{G}(\widehat{H})$ denotes the set of subgraphs of G that are homeomorphic to H and $\mathbf{Z}(\widehat{H})$ denotes the 2-DD representing $\mathcal{G}(\widehat{H})$. The following algorithm constructs the 2-DD representing the set of subgraphs of G that is FTM-characterized by $\mathcal{H}$.

1. Initialize a 2-DD $\mathbf{Z}_{\text{subd}}$ by the 2-DD representing the empty set.
2. Choose an arbitrary graph H from $\mathcal{H}$ and remove it from $\mathcal{H}$.
3. Update $\mathbf{Z}_{\text{subd}}$ by $\text{UNION}(\mathbf{Z}_{\text{subd}}, \mathbf{Z}(\widehat{H}))$.
4. If $\mathcal{H}$ is not empty, go back to Step 2. If empty, go on to Step 5.
5. We obtain the final 2-DD $\mathbf{Z}_{\text{ans}}$ by $\text{NONSUPSET}(\mathbf{Z}_{\text{subd}})$.

For example, let us consider the case where we want to implicitly enumerate all planar subgraphs of G. In this case, $\mathcal{H}$ is $\{K_5, K_{3,3}\}$. We construct $\mathbf{Z}(\widehat{K_5})$ and $\mathbf{Z}(\widehat{K_{3,3}})$ and take their union, which is $\mathbf{Z}_{\text{subd}}$. Now $\mathbf{Z}_{\text{subd}}$ represents the set of all subgraphs of G that are homeomorphic to K_5 or $K_{3,3}$. $\mathbf{Z}_{\text{ans}} = \text{NONSUPSET}(\mathbf{Z}_{\text{subd}})$ represents the family of all subgraphs of G that is FTM-characterized by $\mathcal{H} = \{K_5, K_{3,3}\}$. Therefore, $\mathbf{Z}_{\text{ans}}$ represents the family of all planar subgraphs of G. Other types of subgraphs such as outerplanar, series-parallel, and cactus subgraphs can be implicitly enumerated only by changing $\mathcal{H}$ according to Table 1.

4   Computational Experiments

4.1   Settings

We compared several methods to enumerate planar subgraphs. For input graphs, we used complete graphs K_n and $3 \times b$ king graphs $X_{3,b}$ as synthetic data. As real data, we used Rome graph[3], which is often used in studies on graph drawing. We implemented all the code in C++ and compiled them by g++5.4.0 with -O3 option. To handle DDs, we used TdZdd [6] and SAPPORO_BDD inside Graphillion [5]. We used a machine with Intel Xenon E5-2637 v3 CPU and 1 TB RAM. For each case, we set the timeout to one day.

4.2   Comparing Several Methods to Enumerate Planar Subgraphs

We compare the following three methods for planar subgraph enumeration.

- BACKTRACK: It explicitly enumerates subgraphs based on backtracking.
- DDEDGE: It implicitly enumerates subgraphs using DDs. It uses $|E(H)|$ colors based on Theorem 3. In other words, it uses ten colors for $\mathcal{S}(K_5)$ and nine colors for $\mathcal{S}(K_{3,3})$.
- DDVERTEX: It implicitly enumerates subgraphs using DDs. It uses $\tau(H)$ colors based on Theorems 4–6. In other words, it uses three colors both for $\mathcal{S}(K_5)$ and $\mathcal{S}(K_{3,3})$.

As a subroutine of BACKTRACK, we used a planarity test in C++ Boost[4]. For fairness, BACKTRACK does not output solutions but only counts the number of solutions. DDEDGE and DDVERTEX construct DDs representing the set of solutions. Once a DD is constructed, we can count the number of solutions in linear time to the number of nodes in the DD [9].

---

[3] http://www.graphdrawing.org/data.html.
[4] https://www.boost.org/doc/libs/1_71_0/libs/graph/doc/boyer_myrvold.html.

Table 2. Experimental results. Each column shows the name of graphs, the number of vertices and edges, the running time of the three methods (in seconds), and the number of planar subgraphs. "T/O" and "M/O" mean time out and memory out, respectively. "–" means all the methods failed. The number of solutions for K_{10} is from OEIS A066537, which is marked by '*'. We write the fastest time for each input graph in bold.

Graph	$|V|$	$|E|$	BACKTRACK	DDEDGE	DDVERTEX	# Solutions
K_5	5	10	**<0.01**	0.14	**<0.01**	1023
K_6	6	15	**<0.01**	2.12	0.21	32071
K_7	7	21	28.28	35.02	**2.73**	1823707
K_8	8	28	3113.64	620.84	**66.34**	163947848
K_9	9	36	T/O	15623.11	**4694.41**	20402420291
K_{10}	10	45	T/O	T/O	T/O	*3209997749284
$X_{3,4}$	12	29	11029.38	16.83	**0.09**	5.33×10^8
$X_{3,5}$	15	38	T/O	53.93	**1.67**	2.70×10^{11}
$X_{3,10}$	30	83	T/O	665.65	**5.62**	8.93×10^{24}
$X_{3,50}$	150	443	T/O	M/O	**37.28**	1.29×10^{133}
$X_{3,100}$	300	893	T/O	M/O	**76.99**	2.03×10^{268}
$X_{3,500}$	1500	4493	T/O	M/O	**405.04**	7.95×10^{1349}
$X_{3,1000}$	3000	8993	T/O	M/O	M/O	–
G_1 (grafo1764.20)	20	25	792.16	1.09	**0.06**	3.35×10^7
G_2 (grafo1760.28)	28	39	T/O	96.81	**3.76**	5.49×10^{11}
G_3 (grafo10000.38)	38	52	T/O	787.98	**29.43**	4.50×10^{15}
G_4 (grafo10008.42)	42	61	T/O	38647.96	**668.15**	2.30×10^{18}
G_5 (grafo1378.46)	46	62	T/O	M/O	**796.48**	4.61×10^{18}
G_6 (grafo1395.61)	61	78	T/O	M/O	**11992.12**	3.02×10^{23}
G_7 (grafo5287.61)	61	88	T/O	M/O	M/O	–
G_8 (grafo9798.76)	76	91	T/O	M/O	**1709.64**	2.48×10^{27}
G_9 (grafo10006.98)	98	136	T/O	M/O	M/O	–

Table 2 shows the experimental results. In all the cases, all the methods output the same number of solutions. Among the three methods, DDVERTEX ran fastest except for K_6. BACKTRACK finished in a day only when the number of solutions is small (less than 10^9). Although DDEDGE solved more instances than BACKTRACK, it ran out of memory when the size of input or the number of solutions grows. In contrast, DDVERTEX succeeded even for such instances. For example, for $X_{3,4}$, DDVERTEX is 122,544 and 187 times faster than BACKTRACK and DDEDGE. In addition, for $X_{3,500}$, DDVERTEX succeeded in implicitly enumerating 7.95×10^{1349} planar subgraphs only in 405.04 s (less than seven minutes). These results demonstrate the outstanding efficiency of DDVERTEX.

5   Conclusion

Given graphs G and H, we have proposed a method to implicitly enumerate topological-minor-embeddings of H in G using decision diagrams. We also have shown a useful application of our method to enumerating subgraphs characterized by forbidden topological minors, that is, planar, outerplanar, series-parallel, and cactus subgraphs. Computational experiments show that our method can find all planar subgraphs up to 122,544 times faster than a naive backtracking-based method and could solve more problems than the backtracking-based method. Future work is extending our method from topological minors to general minors.

References

1. Brandstädt, A., Le, V.B., Spinrad, J.P.: Graph Classes: A Survey. Society for Industrial and Applied Mathematics (1999)
2. Bryant, R.E.: Graph-based algorithms for Boolean function manipulation. IEEE Trans. Comput. $\mathbf{C-35}$(8), 677–691 (1986)
3. Diestel, R.: Graph Theory. Graduate Texts in Mathematics, vol. 173, 4th edn. Springer, Heidelberg (2012)
4. Horiyama, T., Kawahara, J., ichi Minato, S., Nakahata, Y.: Decomposing a graph into unigraphs (2019). arXiv:1904.09438
5. Inoue, T., Iwashita, H., Kawahara, J., Minato, S.: Graphillion: software library for very large sets of labeled graphs. Int. J. Softw. Tools Technol. Transfer $\mathbf{18}$(1), 57–66 (2016)
6. Iwashita, H., Minato, S.: Efficient top-down ZDD construction techniques using recursive specifications. TCS Technical Reports TCS-TR-A-13-69, pp. 1–28 (2013)
7. Kawahara, J., Inoue, T., Iwashita, H., Minato, S.: Frontier-based search for enumerating all constrained subgraphs with compressed representation. IEICE Trans. Fundam. Electron. Commun. Comput. Sci. $\mathbf{E100-A}$(9), 1773–1784 (2017)
8. Kawahara, J., Saitoh, T., Suzuki, H., Yoshinaka, R.: Colorful frontier-based search: implicit enumeration of chordal and interval subgraphs. In: Kotsireas, I., Pardalos, P., Parsopoulos, K.E., Souravlias, D., Tsokas, A. (eds.) SEA 2019. LNCS, vol. 11544, pp. 125–141. Springer, Cham (2019). https://doi.org/10.1007/978-3-030-34029-2_9
9. Knuth, D.E.: The Art of Computer Programming. Combinatorial Algorithms, Part 1, vol. 4A, 1st edn. Addison-Wesley Professional, Boston (2011)
10. Kuratowski, C.: Sur le problème des courbes gauches en topologie. Fundamenta Mathematicae $\mathbf{15}$(1), 271–283 (1930)
11. Minato, S.: Zero-suppressed BDDs for set manipulation in combinatorial problems. In: Proceedings of the 30th ACM/IEEE Design Automation Conference (DAC 1993), pp. 272–277. ACM, New York (1993)
12. Nakahata, Y., Kawahara, J., Horiyama, T., Kasahara, S.: Enumerating all spanning shortest path forests with distance and capacity constraints. IEICE Trans. Fundam. Electron. Commun. Comput. Sci. $\mathbf{E101-A}$(9), 1363–1374 (2018)
13. Sekine, K., Imai, H., Tani, S.: Computing the Tutte polynomial of a graph of moderate size. In: Staples, J., Eades, P., Katoh, N., Moffat, A. (eds.) ISAAC 1995. LNCS, vol. 1004, pp. 224–233. Springer, Heidelberg (1995). https://doi.org/10.1007/BFb0015427

Partitioning a Graph into Complementary Subgraphs

Julliano Rosa Nascimento[1]($\boxtimes$), Uéverton S. Souza[2], and Jayme L. Szwarcfiter[3,4]

[1] INF, Universidade Federal de Goiás, Goiânia, GO, Brazil
julliano@inf.ufg.br
[2] IC, Universidade Federal Fluminense, Niterói, RJ, Brazil
ueverton@ic.uff.br
[3] IM, COPPE, and NCE, UFRJ, Rio de Janeiro, RJ, Brazil
jayme@nce.ufrj.br
[4] IME, Universidade do Estado do Rio de Janeiro, Rio de Janeiro, RJ, Brazil

Abstract. In the PARTITION INTO COMPLEMENTARY SUBGRAPHS (COMP-SUB) problem we are given a graph $G = (V, E)$, and an edge set property Π, and asked whether G can be decomposed into two graphs, H and its complement $\overline{H}$, for some graph H, in such a way that the edge cut-set (of the cut) $[V(H), V(\overline{H})]$ satisfies property Π. Such a problem is motivated by the fact that several parameterized problems are trivially fixed-parameter tractable when the input graph G is decomposable into two complementary subgraphs. In addition, it generalizes the recognition of complementary prism graphs, and it is related to graph isomorphism when the desired cut-set is empty, COMP-SUB($\emptyset$). In this paper we are particularly interested in the case COMP-SUB($\emptyset$), where the decomposition also partitions the set of edges of G into $E(H)$ and $E(\overline{H})$. We show that COMP-SUB($\emptyset$) is GI-*complete* on chordal graphs, but it becomes more tractable than GRAPH ISOMORPHISM for several subclasses of chordal graphs. We present structural characterizations for split, starlike, block, and unit interval graphs. We also obtain complexity results for permutation graphs, cographs, comparability graphs, co-comparability graphs, interval graphs, co-interval graphs and strongly chordal graphs. Furthermore, we present some remarks when Π is a general edge set property and the case when the cut-set M induces a complete bipartite graph.

Keywords: Partition · Graph partitioning · Complementary subgraphs · Graph isomorphism

1 Introduction

Partition problems in graphs play an indispensable role in Graph Theory. Several algorithms perform steps of decomposing a graph into smaller pieces, in a way that ensures that a large instance becomes smaller and more tractable.

VERTEX COLORING is a classical graph partition problem. To determine whether a graph G has a proper k-vertex coloring is nothing but determining

© Springer Nature Switzerland AG 2020
M. S. Rahman et al. (Eds.): WALCOM 2020, LNCS 12049, pp. 223–235, 2020.
https://doi.org/10.1007/978-3-030-39881-1_19

whether $V(G)$ can be partitioned into k independent sets. Another classical example of a partition problem is CLIQUE COVER, in which we are given a graph $G = (V, E)$ and an integer k, and asked whether there is a set of k subgraphs of G, such that each subgraph is a clique and each vertex of G is contained in exactly one of these subgraphs.

Feder et al. [8] consider the problem of partitioning the vertex set V of a graph G into k parts $V_1, V_2, \ldots, V_k$ with a fixed "pattern" of requirements as to which V_i induces either an independent set or a clique, and which pairs V_i, V_j are completely adjacent or nonadjacent. Churchley and Huang [5] study polarity and monopolarity of graphs. A graph G is called *polar* if its vertex set can be partitioned into A and B in such a way that the subgraphs induced by A and B are respectively a complete multipartite graph and a disjoint union of cliques (i.e. the complement of a complete multipartite graph); when A is an independent set (which is a special complete multipartite graph), G is called *monopolar*.

The notion of partition is present in the definition of several classes of graphs, such as split and bipartite graphs. A *split graph* G is one whose vertex set admits a partition $V(G) = C \cup I$ into a clique C and an independent set I, and a graph G is *bipartite* if its vertex set admits a partition $V(G) = A \cup B$ such that A and B are independent sets. *Complementary prisms*, introduced in 2007 by Haynes et al. [10], also use the underlying concept of partition, in special, complementary partitions. Let G be a graph and $\overline{G}$ its complement. For every vertex $v \in V(G)$ we denote $\overline{v} \in V(\overline{G})$ as its *corresponding vertex*. The *complementary prism* $G\overline{G}$ of a graph G arises from the disjoint union of the graph G and $\overline{G}$ by adding the edges of a perfect matching between the corresponding vertices of G and $\overline{G}$.

In this sense, motivated by the existence of several variations of partition problems, and the definitions of polarity and complementary prisms, we introduce the problem that concerns our study.

Definition 1. *We say that a graph $G = (V, E)$ is decomposed into two graphs G_1 and G_2 if $V(G)$ can be partitioned into V_1 and V_2, where $G[V_1] = G_1$ and $G[V_2] = G_2$. In addition, the cut-set of $[V_1, V_2]$ is said cut-set of this decomposition.*

PARTITION INTO COMPLEMENTARY SUBGRAPHS (COMP-SUB)

Instance: A graph $G = (V, E)$, and an edge set property Π.

Question: Can G be decomposed into two graphs, H and its complement $\overline{H}$, for some graph H, in such a way that the cut-set M of the decomposition satisfies the property Π?

We write COMP-SUB(Π) as a shorthand for PARTITION INTO COMPLEMENTARY SUBGRAPHS with the edge set property Π for the cut-set M. For convenience, we use $G \in$ COMP-SUB(Π) to denote that G is a *yes*-instance of COMP-SUB(Π).

Let $\mathcal{C}$ be the class of graphs decomposable into two graphs, H and its complement $\overline{H}$, for some graph H. Recognizing that a graph G belongs to $\mathcal{C}$ can be of great interest. Suppose we aim to solve k-COLORING, if $G \in \mathcal{C}$ then, by Ramsey's theory, either G is a *no*-instance or G has size bounded by a function of k. More precisely, for every decision problem P where a k-clique, or an independent set of size k is sufficient to certify either *yes* or *no*, it follows that P is trivially solvable in FPT-time when parameterized by k.

By Ramsey's theory we obtain a primary result when Π is an arbitrary edge set property (see Proposition 1). When Π is taken as a perfect matching M between corresponding vertices in H and $\overline{H}$, COMP-SUB(Π) coincides with the recognition of complementary prism graphs, that can be done in polynomial time [2]. In consequence, we are particularly interested in other natural edge set properties Π such as M is empty and M induces a complete bipartite graph.

In the case M is empty, denoted as COMP-SUB($\emptyset$), the decomposition also forms a partition the set of edges of G into two parts, that is, $V(G) = V(H) \cup V(\overline{H})$ and $E(G) = E(H) \cup E(\overline{H})$. We show characterizations for split, starlike, block, and unit interval graphs decomposable into two graphs H and its complement $\overline{H}$, in such a way that the cut-set M of the decomposition is empty. We also obtain complexity results for chordal, permutation and partially for strongly chordal graphs. Furthermore, we present some remarks for COMP-SUB(Π) when Π is M induces a complete bipartite graph.

The rest of the paper is organized as follows. Section 2 consists of general remarks on the problem. Section 3 contains our results on subclasses of chordal graphs. Section 4 contains observations on the problem when M induces a complete bipartite graph.

2   Preliminaries

We consider only finite, simple, and undirected graphs, and we use standard terminology and notation.

Let P and P' be two decision problems. Problem P is *polynomially reducible* to P', denoted $P \propto P'$, if there exists an algorithm R that maps any instance I of P into an equivalent instance I' of P', in which R runs in polynomial time on $|I|$. Problems P and P' are *polynomially equivalent* if $P \propto P'$ and $P' \propto P$. We say that P is *trivial* if every *yes*-instance of P has size bounded by a constant.

Proposition 1. *Let C be a graph class such that either none of whose graphs contain a k-clique or none of whose graphs contain a k-independent set. It holds that* COMP-SUB *restricted to C is trivial.*

Proposition 1 implies that COMP-SUB is trivially solvable in polynomial time on bipartite, planar or cubic graphs, since all of these graph classes are K_5-free. Next we deal with $M = \emptyset$.

2.1 M Is Empty

This subsection contains auxiliary results, intended to map relevant questions on the problem. In Proposition 2, we show necessary conditions related to edges and vertices of the *yes*-instances of COMP-SUB($\emptyset$).

Proposition 2. *Let G be a graph of order $2n$. If $G \in$ COMP-SUB($\emptyset$), then*

1. *G has exactly $\frac{n(n-1)}{2}$ edges.*
2. *G has a connected component with exactly n vertices.*

Given Proposition 2, it is easy to see that COMP-SUB($\emptyset$) on general graphs is equivalent to GRAPH ISOMORPHISM, as stated in Theorem 1. However, we remark a contrast between the complexities of COMP-SUB and GRAPH ISO-MORPHISM considering, for example, bipartite graphs. On this class of graphs, we know that GRAPH ISOMORPHISM is GI-complete [1], while COMP-SUB is trivially solvable in polynomial time (cf. Proposition 1).

Theorem 1. COMP-SUB($\emptyset$) *is GI-complete.*

In Lemmas 1 and 2 we present classes of graphs where the equivalence between COMP-SUB($\emptyset$) and GRAPH ISOMORPHISM holds.

Lemma 1. *Let $\mathcal{C}$ be a hereditary class of graphs closed under complement and closed with respect to taking disjoint union. It holds that* COMP-SUB($\emptyset$) *on $\mathcal{C}$ is polynomially equivalent to* GRAPH ISOMORPHISM *on $\mathcal{C}$.*

Since GRAPH ISOMORPHISM can be solved in linear time for permutation graphs [6], from Lemma 1 the following holds.

Corollary 1. COMP-SUB($\emptyset$) *can be solved in linear time for permutation graphs.*

Lemma 2. *Let $\mathcal{C}$ be a class of graphs closed under complement. The problem of determining whether a graph G can be decomposed into two graphs H and $\overline{H}$ such that $H \in \mathcal{C}$ and $M = \emptyset$ is polynomially equivalent to* GRAPH ISOMORPHISM *on the class $\mathcal{C}$.*

Lemma 2 implies Corollaries 2 and 3.

Corollary 2. COMP-SUB($\emptyset$) *can be solved in linear time for comparability and co-comparability graphs.*

Since the class of permutation graphs contains the cographs [7], and interval (resp. co-interval) graphs is a subclass of co-comparability (resp. comparability) graphs [9], by Corollaries 1 and 2 we remark that COMP-SUB($\emptyset$) can be solved in linear time for cographs, interval graphs, and co-interval graphs.

Corollary 3. COMP-SUB($\emptyset$) *on chordal graphs remains GI-complete.*

Proof. Let G be a chordal graph. Suppose that $G \in$ COMP-SUB($\emptyset$). Then, G can be decomposed into H and its complement $\overline{H}$, for some graph H, such that the cut-set M of the decomposition is empty. Since chordality is a hereditary property, it follows that H and $\overline{H}$ are also chordal. According to Golumbic [9], a graph H and its complement $\overline{H}$ are chordal if and only if H and $\overline{H}$ are split graphs. In addition, H is split if and only if its complement $\overline{H}$ is split [9].

Note that if a graph G can be decomposed into two graphs H and $\overline{H}$ such that H and $\overline{H}$ are both split graphs, then G is chordal. Therefore, by Lemma 2, we have that COMP-SUB($\emptyset$) on chordal graphs is polynomially equivalent to GRAPH ISOMORPHISM on split graphs. Since GRAPH ISOMORPHISM on split graphs is GI-complete [4], then COMP-SUB($\emptyset$) on chordal graphs is GI-complete. $\square$

In view of the hardness result of Corollary 3 we proceed with COMP-SUB($\emptyset$) on subclasses of chordal graphs.

3 COMP-SUB($\emptyset$) on Subclasses of Chordal Graphs

We consider in this section COMP-SUB($\emptyset$) for unit interval, strongly chordal, starlike, block, and split graphs.

3.1 Unit Interval Graphs

A sun_n is a graph on $2n$ vertices, $n \geq 3$, whose vertex set can be partitioned into $W = \{w_1, \ldots, w_n\}$ and $U = \{u_1, \ldots, u_n\}$ such that W is independent, U is a clique and u_i is adjacent to w_j if and only if $i = j$ or $i = (j + 1) \bmod n$. A *unit interval graph* is a $\{C_{n+4}, sun_3, K_{1,3}, net\}$-free graph, for every $n \geq 0$. A *bull graph* is a K_3 with two non-incident pendent edges. The *join* of two graphs G and H is denoted $G + H$.

Proposition 3. *Let G be a unit interval graph with $2n$ vertices. It holds that $G \in$ COMP-SUB($\emptyset$) if and only if $G \in \{(\overline{P}_\ell + K_{n-\ell}) \cup (P_\ell \cup \overline{K}_{n-\ell}), 2bull\}$, for $1 \leq \ell \leq 4$.*

Proof. First, suppose that $G \in \{(\overline{P}_\ell + K_{n-\ell}) \cup (P_\ell \cup \overline{K}_{n-\ell}), 2bull\}$, for $1 \leq \ell \leq 4$. Notice that $\overline{P}_\ell + K_{n-\ell}$ is the complement of $P_\ell \cup \overline{K}_{n-\ell}$, for $1 \leq \ell \leq 4$, and *bull* is self-complementary. Hence $G \in$ COMP-SUB($\emptyset$).

Suppose that $G \in$ COMP-SUB($\emptyset$). By Proposition 2, there exists a connected component of G, say H, with exactly n vertices. Since G is a unit interval graph, then so are H and $\overline{H}$. Then, H (resp. $\overline{H}$) is chordal and $\{sun_3, K_{1,3}, net\}$-free. Consequently, $\overline{H}$ (resp. H) is $\{\overline{K_{1,3}}\} = \{K_3 \cup K_1\}$-free. Since H and $\overline{H}$ are chordal, we have that H and $\overline{H}$ are split. Let $C \cup I$ be a split partition of $V(H)$, where C is a maximal clique and I is an independent set. Since H is a connected split unit interval graph, we have that $|I| \leq 2$ [12]. We consider three cases: $|I| \in \{0, 1, 2\}$.

Case 1. $|I| = 0$.

In this case we have that $H = C = K_n$. Thus, for $\ell = 1$, we have $G = (\overline{P}_\ell + K_{n-\ell}) \cup (P_\ell \cup \overline{K}_{n-\ell})$.

Case 2. $|I| = 1$.

Let $I = \{v\}$. Since H is connected, there exists $x \in C$ such that $vx \in E(H)$. Since C is maximal, we have that there exists $y \in C$ such that $vy \notin E(H)$. We show that $|C \setminus N_H(v)| \leq 2$. Suppose for contradiction that $|C \setminus N_H(v)| \geq 3$. This implies that $(C \setminus N_H(v)) \cup \{v\}$ has an induced subgraph $K_3 \cup K_1$, contradiction.

Suppose that $C \setminus N_H(v) = \{y\}$. We have that $\{v, y\}$ induces a $\overline{P}_2$ and consequently $H = \overline{P}_2 + K_{n-2}$. Hence $(\overline{P}_\ell + K_{n-\ell}) \cup (P_\ell \cup \overline{K}_{n-\ell})$, for $\ell = 2$.

Now suppose that $C \setminus N_H(v) = \{y, y'\}$. We have that $\{v, y, y'\}$ induces a $\overline{P}_3$ and consequently $H = \overline{P}_3 + K_{n-3}$. Hence $G = (\overline{P}_\ell + K_{n-\ell}) \cup (P_\ell \cup \overline{K}_{n-\ell})$, for $\ell = 3$.

Case 3. $|I| = 2$.

Let $I = \{x, y\}$, $X = N_H(x) \setminus N_H(y)$ and $Y = N_H(y) \setminus N_H(x)$. We show first that $|X| \leq 1$ and $|Y| \leq 1$. Suppose, for contradiction, that either $|X| \geq 2$ or $|Y| \geq 2$. We may consider $u, v \in X$. We have that $\{u, v, x, y\}$ induces a $K_3 \cup K_1$, contradiction. The case $|Y| \geq 2$ is analogous. Now, we consider two sub-cases.

Case 3.1. $N_H(x) \cap N_H(y) \neq \emptyset$.

Recall that $|X|, |Y| \in \{0, 1\}$. Since H is $K_{1,3}$-free, we have that $C \setminus N_H(\{x, y\}) = \emptyset$. First, let $|X| = 0$ and $|Y| = 0$. We have that $H = G[\{x, y\}] + K_{n-2} = \overline{P}_2 + K_{n-2}$. Therefore $G = (\overline{P}_\ell + K_{n-\ell}) \cup (P_\ell \cup \overline{K}_{n-\ell})$, for $\ell = 2$.

Let $|X| = 1$ and $|Y| = 0$, with $X = \{v\}$. In this case, we have that $H = G[\{v, x, y\}] + K_{n-3} = \overline{P}_3 + K_{n-3}$. Hence, $G = (\overline{P}_\ell + K_{n-\ell}) \cup (P_\ell \cup \overline{K}_{n-\ell})$, for $\ell = 3$. The case $|X| = 0$ and $|Y| = 1$ is similar.

Now let $|X| = 1$ and $|Y| = 1$, with $X = \{v\}$ and $Y = \{u\}$. We have that $H = G[\{u, v, x, y\}] + K_{n-4} = P_4 + K_{n-4} = \overline{P}_4 + K_{n-4}$. Hence, $G = (\overline{P}_\ell + K_{n-\ell}) \cup (P_\ell \cup \overline{K}_{n-\ell})$, for $\ell = 4$.

Case 3.2. $N_H(x) \cap N_H(y) = \emptyset$.

We know that $|X|, |Y| \leq 1$. Since H is connected, then $|X|, |Y| = 1$.

Let $W = C \setminus N_H(\{x, y\})$. If $|W| = 0$, since $|X|, |Y| = 1$, we have that $H = P_4$. Hence, $G = (\overline{P}_4 + K_0) \cup (P_4 \cup \overline{K}_0)$.

If $|W| = 1$, since $|X|, |Y| = 1$, we have that $H = bull$. Hence $G = 2bull$.

Finally, consider $|W| \geq 2$. Let $w, w' \in W$ and $N_H(y) = \{u\}$. Since $u, w, w' \notin N_H(x)$ and $u, w, w' \in C$, $\{u, w, w', x\}$ induces a $K_3 \cup K_1$, contradiction.   □

3.2   Strongly Chordal Graphs

A graph G is *strongly chordal* if G is chordal and sun_k-free, for every $k \geq 3$.

Proposition 4. COMP-SUB($\emptyset$) *for strongly chordal graphs is polynomially equivalent to* GRAPH ISOMORPHISM *for split* $\{sun_3, sun_4, net\}$*-free graphs.*

Let $G = (C \cup I, E)$ be a split graph. We define the *component graph* of G as the graph G' obtained by removing all edges between the vertices of C from G. We denote by $c_{nt}(G')$ the number of nontrivial components of G'.

To the best of our knowledge, GRAPH ISOMORPHISM on split $\{sun_3, sun_4, net\}$-free graphs remains an open problem. Despite this, we show how to solve GRAPH ISOMORPHISM for split $\{sun_3, sun_4, net\}$-free graphs G_1 and G_2, when the number of nontrivial components of G'_1 and of G'_2 is equal to 2. Such result follows by Theorem 2. We remark that if $c_{nt}(G'_1) \geq 3$, then G'_1 contains $3K_2$ as induced subgraph. Consequently G_1 is not *net*-free, which is not our case.

Theorem 2. *Let $G = H \cup \overline{H}$ be a strongly chordal graph. If $c_{nt}(H') = 2$, then* COMP-SUB($\emptyset$) *can be solved in polynomial time.*

Proof. Let $G = H \cup \overline{H}$ be a strongly chordal graph. Considering that $G \in$ COMP-SUB($\emptyset$), we know from Proposition 4 that H and $\overline{H}$ are also split $\{sun_3, sun_4, net\}$-free graphs. Let $C \cup I$ be a split partition of $V(H)$ such that C is a clique and I is an independent set. Let A and B be the nontrivial components of H'. Since H is *net*-free, it follows that H' is $3K_2$-free. Since $c_{nt}(H') = 2$, we have that A and B are $2K_2$-free graphs. We claim that H is a split permutation graph.

Claim 1. *H is a split permutation graph.*

Let $\mathcal{C}$ be the class of split permutation graphs. Since $\mathcal{C}$ is closed under complement, and GRAPH ISOMORPHISM on $\mathcal{C}$ is solvable in polynomial time [6], Lemma 2 implies that COMP-SUB($\emptyset$) on $\mathcal{C}$ can be solved in polynomial time. □

A graph G is *2-threshold* if it is the edge union of two threshold graphs. We prove in Theorem 4 that split $\{sun_3, sun_4, net\}$-free graphs form a subclass of split 2-threshold graphs. We first present some useful results and define the conflict graph G^* of a graph G. Two edges of a graph G are said to *conflict* if their endpoints induce in G a C_4, a P_4 or a $2K_2$. In a split graph $G = (C \cup I, E)$, if two edges e and f of G are in conflict, then e (and f) has exactly one endpoint in C and one in I.

Definition 2 [11]. *The conflict graph G^* of a graph G is constructed as follows:*

$$V(G^*) = E(G), \quad E(G^*) = \{ef : e \text{ and } f \text{ conflict in } G\}.$$

Theorem 3 [11]. *A graph G is 2-threshold if and only if the conflict graph G^* is bipartite.*

Definition 3. *Let $G = (C \cup I, E)$ be a split graph, G^* the conflict graph of G, and C_k an induced cycle in G^*, for $k \geq 3$. In addition, let $V(C_k) = \{e_1, \ldots, e_k\}$, where $e_i = u_i v_i \in E(G)$, and $u_i \in I$, $v_i \in C$, for every $1 \leq i \leq k$.*

Let $K_{k,k}$ be a complete bipartite graph with vertex set $\{u_i \in I, v_i \in C : 1 \leq i \leq k\}$. We construct a graph $\mathcal{K}_{k,k}$ by removing the edges $\{u_p v_{p+1}, u_{p+1} v_p : 1 \leq p \leq k-1\} \cup \{u_k v_1, u_1 v_k\}$ from $K_{k,k}$. We denote $E(\mathcal{K}_{k,k}) = E_0 \cup E_+ \cup E_-$ in which, for every $i, j \in \{1, \ldots, k\}$, and $\ell \in \{1, \ldots, \frac{k^2-3k}{2}\}$,

$$E_0 = \{e_i = u_i v_j : i = j\}, \; E_+ = \{e_{\ell+} = u_i v_j : i < j\}, \; E_- = \{e_{\ell-} = u_i v_j : i > j\}.$$

Let G' be the component graph of G. The induced subgraph $\mathcal{X}(C_k)$ of G' is defined by:

$$\mathcal{X}(C_k) = G'[E(G') \cap E(\mathcal{K}_{k,k})].$$

Definition 3 of the induced subgraph $\mathcal{X}(C_k)$ of G', for $k \geq 3$, expresses the edges that conflict in an induced cycle C_k in G^*. Figure 1 contains an example of $\mathcal{K}_{5,5}$. In Theorem 4 we analyze mainly edges that can be present or missing in $\mathcal{X}(C_5)$. So, we define a representation of $E(\mathcal{X}(C_5))$ in terms of a vector ν.

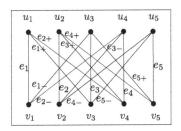

Fig. 1. Induced subgraph $\mathcal{K}_{5,5}$ with respective conflict vector $\nu = [2, 2, 2, 2, 2]$.

Definition 4. *A conflict vector $\nu = [\varepsilon_1, \ldots, \varepsilon_5]$ of the graph $\mathcal{X}(C_5)$ is defined by:*

$$
\varepsilon_\ell_{\ell \in \{1,\ldots,5\}} = \begin{cases} 0, & \text{if } e_{\ell-} \in E(\mathcal{X}(C_5)) \text{ and } e_{\ell+} \notin E(\mathcal{X}(C_5)) \\ 1, & \text{if } e_{\ell+} \in E(\mathcal{X}(C_5)) \text{ and } e_{\ell-} \notin E(\mathcal{X}(C_5)) \\ 2, & \text{if } e_{\ell-}, e_{\ell+} \in E(\mathcal{X}(C_5)). \end{cases}
$$

Now, we define some transformations on a conflict vector ν.

Lemma 3. *Let $\nu = [\varepsilon_1, \ldots, \varepsilon_5]$ be a conflict vector. The following holds:*

(i) Let f be the x-reflection function $f(\nu) = [f'(\varepsilon_1), \ldots, f'(\varepsilon_5)]$ in which

$$
f'(\varepsilon_\ell)_{\ell \in \{1,\ldots,5\}} = \begin{cases} 1, & \text{if } \varepsilon_\ell = 0, \\ 0, & \text{if } \varepsilon_\ell = 1, \\ 2, & \text{otherwise.} \end{cases}
$$

The graphs induced by ν and $f(\nu)$ are isomorphic.
(ii) Let g be the y-reflection function $g(\nu) = [\varepsilon_5, \varepsilon_4, \varepsilon_3, \varepsilon_2, \varepsilon_1]$ where $\nu = [\varepsilon_1, \ldots, \varepsilon_5]$.
The graphs induced by ν and $g(\nu)$ are isomorphic.
(iii) The graphs induced by ν and $f(g(\nu))$ (resp. $g(f(\nu))$) are isomorphic.

Finally, we can proceed with Theorem 4.

Theorem 4. *Let G be a split $\{sun_3, sun_4, net\}$-free graph. Then G is a split 2-threshold graph.*

Proof. Let $G = (C \cup I, E)$ be a split $\{sun_3, sun_4, net\}$-free graph. Notice that G is a split $\{sun_3, sun_4, net\}$-free graph if and only if $G' = G \setminus E(C)$ is a bipartite $\{C_6, C_8, 3K_2\}$-free graph. In view of that, we may argue based on the forbidden subgraphs of G'.

Suppose, for contradiction, that G is not a 2-threshold graph. Then, by Theorem 3, the conflict graph G^* is not bipartite. This implies that G^* contains an induced subgraph C_k, for some odd $k \geq 3$. Let $\mathcal{X}(C_k)$ be an induced subgraph of G' as in Definition 3.

Suppose $k = 3$. It follows that, for every distinct $i, j \in \{1, 2, 3\}$, there is a conflict between $e_i, e_j \in E(\mathcal{X}(C_3))$. Recall that, since G is a split graph, G does not contain $2K_2$ nor C_4 as induced subgraph. Hence, if there is a conflict between two edges of G their endpoints induce a P_4 in G, and a $2K_2$ in G'. Then, we have that the pair of edges e_i, e_j, for every distinct $i, j \in \{1, 2, 3\}$, induces a $2K_2$ in $\mathcal{X}(C_3)$. Consequently, $\{e_1, e_2, e_3\}$ induces a $3K_2$ in $\mathcal{X}(C_3)$. Since G' is $3K_2$-free and $\mathcal{X}(C_3)$ is an induced subgraph of G', we reach a contradiction.

Suppose $k = 5$. This implies that $\{e_i, e_{i+1}\}$ induces a $2K_2$ in $\mathcal{X}(C_5)$, for every $i \in \{1, \ldots, 4\}$, as well as $\{e_1, e_5\}$. We know that there is no conflict between e_i, e_j, for $i \in \{1, \ldots, 4\}$, $j \neq i + 1$, and $i = 5, j \neq 1$. Then, we have that at least one of the edges $e_{\ell+}, e_{\ell-}$ belong to $E(\mathcal{X}(C_5))$, for every $\ell \in \{1, \ldots, 5\}$. For the proof, we consider all the possible cases of edges in $E(\mathcal{X}(C_5))$, expressed in a conflict vector $\nu = [\varepsilon_1 \ldots, \varepsilon_5]$ as in Definition 4.

According to Lemma 3(i), we have that all the cases where $\varepsilon_1 = 0$, that is, $\nu = [0, \varepsilon_2, \ldots, \varepsilon_5]$, have an isomorphic graph induced by $f(\nu) = [1, f'(\varepsilon_2) \ldots, f'(\varepsilon_5)]$. Thus, we disregard the cases when $\varepsilon_1 = 1$. The cases where $\varepsilon_1 = 2$, and $\varepsilon_5 = 0$, which means $\nu = [2, \varepsilon_2, \varepsilon_3, \varepsilon_4, 0]$, reduce to the cases $g(\nu) = [0, g'(\varepsilon_2), g'(\varepsilon_3), g'(\varepsilon_4), 2]$, by Lemma 3(ii). Finally, the cases where $\varepsilon_1 = 2$, and $\varepsilon_5 = 1$, which means $\nu = [2, \varepsilon_2, \varepsilon_3, \varepsilon_4, 1]$, reduce to the cases $f(g(\nu)) = [0, f'(g'(\varepsilon_2)), f'(g'(\varepsilon_3)), f'(g'(\varepsilon_4)), 2]$, by Lemma 3(iii).

Putting everything together, this means evaluating $[0, \varepsilon_2 \ldots, \varepsilon_5]$, with 81 cases, and $[2, \varepsilon_2, \varepsilon_3, \varepsilon_4, 2]$, with 27 cases. We remark that many of these 108 cases could be solved applying functions of Lemma 3, and from the observation that the addition of nonchord edges on the forbidden subgraphs C_6, C_8 in $\mathcal{X}(C_5)$ maintain the same forbidden subgraph in $\mathcal{X}(C_5)$.

For instance, let $\nu_1 = [0, 0, 0, 0, 1]$, see Fig. 2(a). We have that ν_1 induces a C_6 (in bold edges) in $\mathcal{X}(C_5)$. Applying $f(\nu_1)$ (Fig. 2(b)), $g(\nu_1)$ (Fig. 2(c)), $f(g(\nu_1))$ (Fig. 2(d)), and adding edges e_{3-}, e_{5+} to $E(\mathcal{X}(C_5))$ (Fig. 2(e)), we also have an induced C_6 in $\mathcal{X}(C_5)$. Notice that this argumentation is possible since $|C| = |I|$, for any forbidden subgraph $F \in \{C_6, C_8, 3K_2\}$.

We proceed with the 108 cases, written down in Table 1. Notice that, with the abovementioned argumentation, the analysis of the 108 cases could be reduced to the analysis of 10 cases, specified in the first column of Table 1. For short, we omit the brackets and the commas of the notation of $\nu = [\varepsilon_1, \ldots, \varepsilon_5]$. Each case has a corresponding graph $\mathcal{X}(C_5)$, see Fig. 3, illustrating in bold edges the forbidden subgraph $F \in \{C_6, C_8, 3K_2\}$ found. In all cases we reach a contradiction.

Now, let $k \geq 6$. Notice that the set of vertices $\{u_1, u_2, u_3, u_{k-1}, u_k, v_1, v_2, v_3, v_{k-1}, v_k\}$ induces a $\mathcal{X}(C_5)$ in $\mathcal{X}(C_k)$. Then, we find forbidden subgraphs C_6, C_8 or $3K_2$ in $\mathcal{X}(C_k)$ reaching a contradiction as in case $k = 5$. Therefore, we conclude that G is a 2-threshold graph, and additionally G^* is chordal bipartite. $\square$

From Theorems 2 and 4, the following Corollary 4 holds. We remark that, as far as we know, GRAPH ISOMORPHISM on 2-threshold graphs is an open problem.

Corollary 4. *Let G be a strongly chordal graph. There exists a polynomial time algorithm that, either decides whether $G \in$ COMP-SUB$(\emptyset)$, or concludes that G is the disjoint union of two split 2-threshold graphs G_1 and G_2 of the same order, such that $c_{nt}(G_1) = c_{nt}(G_2) = 1$, and G_1^* and G_2^* are chordal bipartite.*

3.3 Starlike, Block, and Split Graphs

Starlike graphs are the intersection graphs of substars of a star [3]. A starlike graph G is a $\{P_5, C_4, C_5, H_1, H_2, 2P_3\}$-free graph (see H_1 and H_2 in Fig. 4) [3].

Proposition 5. *Let G be a starlike graph with $2n$ vertices. It holds that $G \in$ COMP-SUB$(\emptyset)$ if and only if $G = (K_p + \overline{K}_q) \cup (\overline{K}_p \cup K_q)$, in which $p + q = n$.*

Proof. It is clear that $G = (K_p + \overline{K}_q) \cup (\overline{K}_p \cup K_q)$ (for $p + q = n$) is a *yes*-instance of COMP-SUB$(\emptyset)$. Now, consider that G is a *yes*-instance of COMP-SUB$(\emptyset)$. Then, G can be decomposed into two graphs, H and its complement $\overline{H}$, for some graph H, such that the cut-set M of the decomposition is empty. Furthermore, by Proposition 2, there exists a connected component of G, say H, with exactly n vertices.

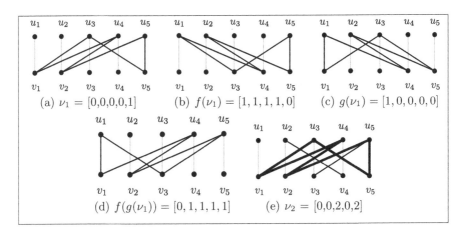

Fig. 2. Cases exemplifying functions of Lemma 3 and nonchord additions.

Table 1. Cases considered for Theorem 4.

Case ν	Forbidden induced subgraph F in $\mathcal{X}(C_5)$	Cases that contain an induced forbidden subgraph isomorphic to F
00000	C_6, Fig. 3(a)	00002, 00020, 00022, 00200, 00202, 00220, 00222, 02000, 02002, 02020, 02022, 02200, 02202, 02220, 02222, 20002, 20022, 20202, 20222, 21112, 21212, 22002, 22022, 22202, 22212, 22222
00001	C_6, Fig. 2(a)	00021, 00201, 01111, 01112, 01121, 01211, 01212, 01221, 02001, 02112, 02121, 02122, 02211, 02212, 02221
00010	C_8, Fig. 3(b)	01000, 01002, 01020, 01200, 01202, 01220, 01222, 02010
00011	C_6, Fig. 3(c)	00012, 00111, 00112, 00121, 00122, 00211, 00212, 00221, 02011, 02012, 02111, 20012, 20112, 20122, 20212, 21002, 21022, 21102, 21122, 21202, 21222, 22012, 22102, 22112, 22122
00100	C_8, Fig. 3(d)	00102, 00120, 02100, 20102
00101	C_8, Fig. 3(e)	01011, 01012, 01021, 01022, 02101, 21012
00110	C_8, Fig. 3(f)	01100, 01102, 01120, 01122, 00210, 02102, 02110, 02120
01001	C_8, Fig. 3(g)	01101, 01201, 02201, 02021
01010	$3K_2$, Fig. 3(h)	$\varnothing$
01110	C_8, Fig. 3(i)	01210, 02210

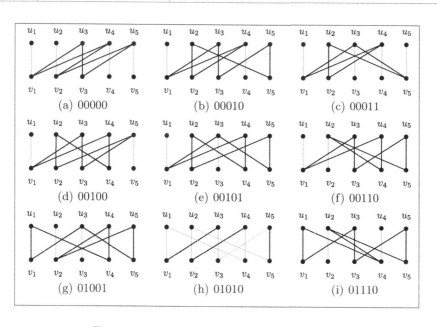

Fig. 3. Forbidden induced subgraphs in $\mathcal{X}(C_5)$.

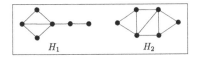

Fig. 4. Two forbidden subgraphs of a starlike graph.

Since G is a starlike graph, then so are H and $\overline{H}$. By definition, H and $\overline{H}$ are $\{P_5, C_4, C_5, H_1, H_2, 2P_3\}$-free. Since $\overline{H}$ is C_4-free, then H is $2K_2$-free. Thus, H is split and so is $\overline{H}$. Consider a split partition $C \cup I$ of $V(H)$, where C is a maximal clique and I is an independent set.

If $I = \emptyset$, then $H = C$ and $\overline{H} = \overline{C}$. Since $|V(H)| = |C| = n$, we have that $H = K_n$ and $\overline{H} = \overline{K}_n$, hence $G = (K_p + \overline{K}_q) \cup (\overline{K}_p \cup K_q)$, for $p = n$ and $q = 0$.

Now, consider that $I \neq \emptyset$. Since C is maximal, $|C| \geq 2$. We show in Claim 2 that $\overline{H}$ is a *cluster graph*, that is, a P_3-free graph, or equivalently, a disjoint union of cliques. Next, we show that $\overline{H}$ has at most one clique with two or more vertices.

Claim 2. $\overline{H}$ *is a cluster graph.*

By Claim 2, we have that $\overline{H}$ is a disjoint union of cliques. Let $\overline{H}_1, \ldots, \overline{H}_\ell$ be the connected components of $\overline{H}$. Since $\overline{H}$ is split, we have that $\overline{H}$ is $2K_2$-free. This implies that there exists at most one $i \in \{1, \ldots, \ell\}$, such that $|V(\overline{H}_i)| \geq 2$. Since $|V(\overline{H})| = n$, for $p + q = n$, we define $\sum_{j \in \{1, \ldots, \ell\} \setminus \{i\}} |V(\overline{H}_j)| = p$ and $|V(\overline{H}_i)| = q$. Hence $\overline{H} = \overline{K}_p \cup K_q$ and $H = \overline{\overline{K}_p \cup K_q} = K_p + \overline{K}_q$. Therefore, $G = (K_p + \overline{K}_q) \cup (\overline{K}_p \cup K_q)$, completing the proof. □

A *diamond graph* is the graph obtained by removing exactly one edge from K_4. A *block graph* is a $\{diamond, C_{n+4}\}$-free graph, for every $n \geq 0$. A *star* S_n is the complete bipartite graph $K_{1,n-1}$. In Proposition 6, we characterize *yes*-instances of COMP-SUB($\emptyset$) when G is a block graph.

Proposition 6. *Let G be a block graph with $2n$ vertices. It holds that $G \in$ COMP-SUB($\emptyset$) if and only if $G \in \{K_n \cup \overline{K}_n, S_n \cup \overline{S}_n, 2P_4, 2bull, (\overline{S_{n-1} \cup K_1}) \cup (S_{n-1} \cup K_1)\}$.*

While GRAPH ISOMORPHISM is GI-complete on split graphs [1], we show in Proposition 7 a characterization of split graphs G in which $G \in$ COMP-SUB($\emptyset$).

Proposition 7. *Let $G \in$ COMP-SUB($\emptyset$) be a split graph. Then $G = K_n \cup \overline{K}_n$.*

4 M Induces a Complete Bipartite Graph

Most of the results of COMP-SUB($\emptyset$) can be applied to COMP-SUB($G[M] = K_{n,n}$). Lemma 4 shows such equivalence.

Lemma 4. *Let $\mathcal{C}$ be a class of graphs. It holds that COMP-SUB($\emptyset$) on $\mathcal{C}$ is polynomially equivalent to COMP-SUB($G[M] = K_{n,n}$) on the complementary class of $\mathcal{C}$.*

References

1. Booth, K.S., Colbourn, C.J.: Problems polynomially equivalent to graph isomorphism. Technical report CS-77-04, University of Waterloo (1979)
2. Cappelle, M.R., Penso, L., Rautenbach, D.: Recognizing some complementary products. Theoret. Comput. Sci. **521**, 1–7 (2014)
3. Cerioli, M.R., Szwarcfiter, J.L.: Characterizing intersection graphs af substars of a star. Ars Combinatoria **79**, 21–31 (2006)
4. Chung, F.R.K.: On the cutwidth and the topological bandwidth of a tree. SIAM J. Algebraic Discrete Methods **6**(2), 268–277 (1985)
5. Churchley, R., Huang, J.: On the polarity and monopolarity of graphs. J. Graph Theory **76**(2), 138–148 (2014)
6. Colbourn, C.J.: On testing isomorphism of permutation graphs. Networks **11**, 13–21 (1981)
7. Corneil, D.G., Perl, Y., Stewart, L.K.: A linear recognition algorithm for cographs. SIAM J. Comput. **14**(4), 926–934 (1985)
8. Feder, T., Hell, P., Klein, S., Motwani, R.: Complexity of graph partition problems. In: Proceedings of the Thirty-First Annual ACM Symposium on Theory of Computing, pp. 464–472. ACM (1999)
9. Golumbic, M.C.: Algorithmic Graph Theory and Perfect Graphs, vol. 57. Elsevier, Amsterdam (2004)
10. Haynes, T.W., Henning, M.A., Slater, P.J., van der Merwe, L.C.: The complementary product of two graphs. Bull. Inst. Comb. Appl. **51**, 21–30 (2007)
11. Mahadev, N.V.R., Peled, U.N.: Threshold Graphs and Related Topics, vol. 56. Elsevier, Amsterdam (1995)
12. Ortiz, Z.C., Maculan, N., Szwarcfiter, J.L.: Characterizing and edge-colouring split-indifference graphs. Discrete Appl. Math. **82**(1–3), 209–217 (1998)

On the Maximum Edge-Pair Embedding Bipartite Matching

Cam Ly Nguyen[1], Vorapong Suppakitpaisarn[2(✉)], Athasit Surarerks[3], and Phanu Vajanopath[3]

[1] Toshiba Corporation, Kawasaki, Japan
camly.nguyen@toshiba.co.jp
[2] The University of Tokyo, Tokyo, Japan
vorapong@is.s.u-tokyo.ac.jp
[3] Chulalongkorn University, Bangkok, Thailand
athasit@cp.eng.chula.ac.th, phanu.vajanopath@gmail.com

Abstract. Given a set of edge pairs in a bipartite graph, we want to find a bipartite matching that includes a maximum number of those edge pairs. While the problem has many applications to wireless localization, to the best of our knowledge, there is no theoretical work for the problem. In this work, unless $P = NP$, we show that there is no constant approximation for the problem. Suppose that k denotes the maximum number of input edge pairs such that a particular node can be in. Inspired by experimental results, we consider the case that k is small. While there is a simple polynomial-time algorithm for the problem when k is one, we show that the problem is NP-hard when k is greater than one. We also devise an efficient $O(k)$-approximation algorithm for the problem.

Keywords: Wireless localization · Computational complexity · Approximation algorithm · Network optimization

1 Introduction

Wireless localization [10] is one of the most important issues in the Internet of Things (IoT). In this problem, one wants to calculate geo-locations of specific wireless devices. Traditional wireless localization systems, for instance, an indoor positioning system, estimate the location of a device through its wireless signals to multiple devices of which locations are known. We can approximate the distance between a wireless access point and a mobile device or between two mobile devices based on wireless signal strength, also called *received signal strength indicator* (*RSSI*), between them. Then, we find locations of the mobiles based on those approximated distances. There are a lot of techniques proposed for the problem including the technique based on semi-definite programming in [2] and the technique based on machine learning in [13].

Recently, the authors of [11,12] have defined a new localization problem found in many IoT applications: the *wireless localization matching problem* (*WLMP*).

© Springer Nature Switzerland AG 2020
M. S. Rahman et al. (Eds.): WALCOM 2020, LNCS 12049, pp. 236–248, 2020.
https://doi.org/10.1007/978-3-030-39881-1_20

Unlike traditional localization problems, the set of locations of wireless devices is known a priori. However, they do not know which location a specific wireless device is at. Because there is at most one device at each location, they want to find a bipartite matching between a set of wireless devices and a set of locations.

WLMP has numerous potential applications. For instance, in wireless lighting systems, a set of bulb locations can be obtained from floor plan blueprints. Workers install bulbs at the specified locations. However, the installation process can be very long with a lot of confirmations if the installed positions for all of the bulbs are specified. Because of that, instead of specifying a position for each bulb, the authors proposed to give a set of bulb positions to the workers, have them install a random bulb at each position, and use the algorithm developed in [12] to automatically match the positions with the devices.

The bipartite matching in this problem is different from the regular maximum bipartite matching problem. For the regular problem, each edge has a score. The problem aims to find a set of edges which forms a bipartite matching and the sum of the edges' scores is maximized [9]. On the other hand, in the setting in [11,12], the authors do not directly know how much likely that a sensor s should be matched to a position p. They know a distance between two positions and, by signal strengths, an approximate distance between two sensors. If the approximate distance between s and s' is close to the distance between p and p', they know that the likelihood of matching s with p and s' with p' should be large. The likelihood score is then not defined on a single edge but on a pair of edges. For each pair of edges e_i, e_j, a score $f(e_i, e_j)$ is given. They want to find a bipartite matching M that maximizes $\sum_{e_i, e_j \in M} f(e_i, e_j)$.

In [7,11,12], the authors proposed several heuristics to solve the problem. However, to the best of our knowledge, there is no theoretical work for it.

1.1 Our Contribution

Informally, our contribution is as follows:

1. We show that there does not exist a constant-factor approximation algorithm for the WLMP.
2. For particular instances that we find in practice, we give an approximation algorithm that almost matches the inapproximability result, and we show that there does not exist a fixed-parameter tractability for those instances.

To be more precise, we consider a simplified version of the problem, where the score $f(e_i, e_j)$ can be only 0 and 1. This simplification is quite intuitive, as we can give $f(\{s(e_i), p(e_i)\}, \{s(e_j), p(e_j)\}) = 1$ if the approximated distance between $p(e_i)$ and $p(e_j)$ is close to the distance between $s(e_i)$ and $s(e_j)$. We then can give $f(\{s(e_i), p(e_i)\}, \{s(e_j), p(e_j)\}) = 0$ otherwise. Our problem is then reduced to be the following:

We want to find a bipartite matching that embeds the maximum number of edge pairs $\{e_i, e_j\}$ such that $f(e_i, e_j) = 1$. In other words, we want to find a bipartite matching M that maximizes $|\{\{e_i, e_j\} \subseteq M : f(e_i, e_j) = 1\}|$.

Our first contribution is, we show that there is no constant-factor approximation algorithm for the problem unless $P = NP$. Also, for all $0 < \epsilon < 1$ there is no $O(2^{\log^{1-\epsilon} n})$-approximation algorithm for this problem unless $NP \subseteq DTIME(n^{\text{polylog}(n)})$. This inapproximability result also applies for the general problem when $f(e_i, e_j)$ can be any real number not just 0 or 1. We have this result by our reduction which preserves approximation ratio from the label cover problem (maximum repetition version) [3] to this problem.

As it is not possible to solve a general problem, we consider particular instances which we found in our previous experimental results in [11]. Let D be a set of edge pairs $\{e_i, e_j\}$ such that $f(e_i, e_j) = 1$, and let k be the maximum number of edge pairs in D that a particular node is incident to, i.e. $k_v := |\{\{e_i, e_j\} \in D : v \in e_i \text{ or } v \in e_j\}|$ for all node v and k is the maximum of k_v. We found from the data that the number k is usually small. Thus, for the second contribution, we consider the problem with a bounded value of k.

We show that the problem is hard for any k greater than one, while it is quite trivial to find a solution for the problem when k is one. Our result is based on the fact that the independent set problem is NP-hard even when input graph is 4-regular [5]. We calculate an Eulerian path from the input graph and, based on the Euler path, construct an input for our problem with $k = 2$. The result illustrates that there is no fixed-parameter tractability for this problem.

One can consider the value of k as a graph degree. Our reduction technique reduces the graph degree from four to two, to show that our problem is hard even when the degree is two. As there are many combinatorial problems of which the hardness on graphs with bounded degree is unknown (such as [15]). We hope that the technique using Eulerian paths introduced in this paper can help to show the hardness of those combinatorial problems.

In addition, we also provide an $O(k)$-approximation algorithm based on an approximation algorithm for the maximum independent set problem.

By the two results for small k, one might think that our problem is similar to the maximum independent set and this problem might not be very interesting for theoretical studies. However, a direct reduction from the maximum independent set, which does not preserve the approximation ratio, only gives us an NP-hardness result for the instance that k is larger than that in practice. Although it is known that there does not exist an $o(n/\log n)$-approximation algorithm for the maximum independent set problem, we cannot easily use that result for our problem. Our reduction works only from instances of the problem with a small degree and, for those instances, there is an approximation algorithm with small approximation ratio. When the reduction starts from a maximum independent set instance with a large degree, the input function f obtained from the reduction does not take only a pair of edges but an arbitrary number of edges. Therefore, for the first contribution, we cannot use the inapproximability of the maximum independent set, to show the inapproximability of our algorithm.

2   Problem Definition

Suppose that there are n sensors and m positions, $S = \{s_1, ..., s_n\}$ is a set of sensors, and $P = \{p_1, ..., p_m\}$ is a set of positions. Consider a complete bipartite graph $G = (S, P, E)$. We are given a set of edge pairs $D \subseteq \{\{e_i, e_j\} : e_i, e_j \in E$ and $e_i \neq e_j\}$. Our problem, called as *the Maximum Edge-pair Embedding Bipartite Matching (MEEBM)*, asks for a bipartite matching $M \subseteq E$ that embeds the maximum number of the edge pairs in D. The edge pair $\{e_i, e_j\}$ is embedded in M if and only if M includes both e_i and e_j. In other words, we want to find

$$\arg \max_{M:\text{matching of } G} |\{d \subseteq M : d \in D\}|.$$

For all node v, define k_v as the number of edge pairs in D that the node v appear at, i.e. $k_v = \{\{e_i, e_j\} \in D : v \in e_i \text{ or } v \in e_j\}|$, and define $k := \max_v k_v$.

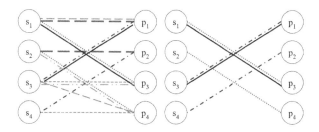

Fig. 1. An example instance of the problem. The instance is shown on the left. We want to find a matching that embeds as many edge pairs (marked by different colors) as possible. An optimal solution is $\{\{s_1, p_3\}, \{s_2, p_4\}, \{s_3, p_1\}, \{s_4, p_2\}\}$ as shown on the right, which embeds the red dotted, the violet dot-dash, and the black solid edge pairs.

Example 1. Consider the graph in Fig. 1 and a set of edge pairs $D = \{d_1, d_2, d_3, d_4, d_5, d_6, d_7\}$, where $d_1 = \{\{s_1, p_1\}, \{s_2, p_2\}\}$, $d_2 = \{\{s_1, p_3\}, \{s_2, p_4\}\}$, $d_3 = \{\{s_3, p_3\}, \{s_2, p_4\}\}$, $d_4 = \{\{s_3, p_3\}, \{s_4, p_4\}\}$, $d_5 = \{\{s_3, p_1\}, \{s_4, p_2\}\}$, $d_6 = \{\{s_1, p_1\}, \{s_3, p_4\}\}$, and $d_7 = \{\{s_1, p_3\}, \{s_3, p_1\}\}$. A bipartite matching M^* that embeds the maximum number of edge pairs in D is $\{\{s_1, p_3\}, \{s_2, p_4\}, \{s_3, p_1\}, \{s_4, p_2\}\}$. It embeds three edge pairs d_2, d_5, and d_7 in the matching. By the definition of k, we have $k = 5$ because the node s_3 appears five times in D.

3   Inapproximability Results

We show a reduction from the label cover problem (maximum repetition version) [3] which preserves an approximation ratio in this section. The label cover problem is defined as in the following definition:

Definition 1. *Given a bipartite graph* $\mathcal{G} = (\mathcal{V}, \mathcal{W}, \mathcal{E})$ *such that* $|\mathcal{V}| = |\mathcal{W}| = \mathcal{N}$, *and an integer* $\mathcal{K}$ *such that* $\mathcal{N}$ *is divisible by* $\mathcal{K}$. *Also, given disjoint partitions of* $\mathcal{V}$ *and* $\mathcal{W}$, *denoted by* $\mathcal{V}_1, \ldots, \mathcal{V}_{\mathcal{K}}$ *and* $\mathcal{W}_1, \ldots, \mathcal{W}_{\mathcal{K}}$, *such that* $|\mathcal{V}_i| = |\mathcal{W}_i| = \mathcal{N}/\mathcal{K}$ *for all* $1 \leq i \leq \mathcal{K}$. *We want to find a set* $\mathcal{V}'$ *and* $\mathcal{W}'$ *such that* $|\mathcal{V}' \cap \mathcal{V}_i| = |\mathcal{W}' \cap \mathcal{W}_i| = 1$ *for all* $1 \leq i \leq \mathcal{K}$ *and the number of edges in the subgraph of* $\mathcal{G}$ *induced by* $\mathcal{V}'$ *and* $\mathcal{W}'$ *is maximized.*

The following results are proved for the label cover problem.

Theorem 1 [1]. *It is not possible to have a constant-factor approximation algorithm for the label cover problem unless* $P = NP$. *Also, for all* $0 < \epsilon < 1$, *it is not possible to have an* $O(2^{\log^{1-\epsilon} \mathcal{N}})$-*approximation algorithm for the problem unless* $NP \subseteq DTIME(n^{polylog(\mathcal{N})})$.

We construct an instance of MEEBM, $(G = (S, P, E), D)$, in the following way. We have $S = \{\mathcal{V}_1, \ldots, \mathcal{V}_{\mathcal{K}}, \mathcal{W}_1, \ldots, \mathcal{W}_{\mathcal{K}}\}$, $P = \mathcal{V} \cup \mathcal{W}$, and $D = \{\{\mathcal{V}_i, v\}, \{\mathcal{W}_j, w\} : v \in \mathcal{V}_i, w \in \mathcal{W}_j$ and $\{v, w\} \in \mathcal{E}\}$. The construction is illustrated in Fig. 2. We show in the following theorem that the reduction in this paragraph preserves approximation ratio.

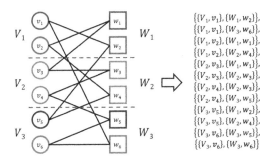

Fig. 2. Reduction from a label cover instance to an MEEBM instance which preserves approximation ratio.

Lemma 1. *If there exists an* α-*approximation algorithm for MEEBM, then there exists an* α-*approximation algorithm for the label cover problem.*

Proof. Suppose that M is a solution of the above MEEBM instance which includes q edge pairs of D. We denote the set of those q edge pairs by D'. Let P' be the set of nodes in P that are incident to edges in M, $\mathcal{V}' = \mathcal{V} \cap P'$, and $\mathcal{W}' = \mathcal{W} \cap P'$. For each $\{\{\mathcal{V}_i, v\}, \{\mathcal{W}_j, w\}\} \in D'$, by the construction of D, we know that $\{v, w\} \in \mathcal{E}$. As $v \in \mathcal{V}'$ and $w \in \mathcal{W}'$, we know that a subgraph of $\mathcal{G}$ induced by $\mathcal{V}' \cup \mathcal{W}'$ contains at least q edges. Because M is a matching on G, the set $\mathcal{V}'$ and $\mathcal{W}'$ contains at most one node in $\mathcal{V}_i$ or $\mathcal{W}_i$ for all $1 \leq i \leq \mathcal{K}$. We can choose to add an arbitrary member from all the set $\mathcal{V}_i$ and $\mathcal{W}_i$ that have no member in $\mathcal{V}'$ and $\mathcal{W}'$. As a result, from the solution of MEEBM M

in polynomial time, we can have a solution of the label cover problem $(\mathcal{V}', \mathcal{W}')$ that contains at least q edges in $\mathcal{E}$.

Conversely, suppose that we have a solution of the label cover problem $(\mathcal{V}', \mathcal{W}')$ containing q edges in $\mathcal{E}$. We denote the set of edges of this solution by $\mathcal{E}'$, for each node $\nu \in \mathcal{V}'$, we set $f(\nu) = i$ if ν is in the partition $\mathcal{V}_i$, and, for each node $w \in \mathcal{W}'$, denote $g(w) = j$ if w is in the partition $\mathcal{W}_j$. Let $M = \{\{\mathcal{V}_{f(\nu)}, \nu\}, \{\mathcal{W}_{g(w)}, w\} : \{\nu, w\} \in \mathcal{E}'\}$. The set M is a matching and a solution of MEEBM because there is exactly one node of $\mathcal{V}_i$ and $\mathcal{W}_j$ in any solution of the label cover $(\mathcal{V}', \mathcal{W}')$. We know that, for each member $\{\nu, w\}$ of $\mathcal{E}'$, the matching M includes an edge pair $\{\mathcal{V}_{f(\nu)}, \nu\}, \{\mathcal{W}_{g(w)}, w\}$ in D. Therefore, from a solution of the label cover that embeds q edges in $\mathcal{E}$, we can construct a solution of MEEBM that embeds q edge pairs in D.

By the discussions in the above paragraph, we can conclude that the optimal value of the label cover and the optimal value of the MEEBM instances are the same. Also, if there is a polynomial-time algorithm which can give an MEEBM solution that can embed q edge pairs in D, we can have a polynomial-time algorithm which can give a label cover solution that include q edges in $\mathcal{E}$. Therefore, our reduction preserves the approximation algorithm. □

By Theorem 1 and Lemma 1, we have the following theorem:

Theorem 2. *It is not possible to have a constant-factor approximation algorithm for MEEBM unless $P = NP$. Also, when n is the number of nodes in G, for all $0 < \epsilon < 1$, it is not possible to have an $O(2^{\log^{1-\epsilon} n})$-approximation algorithm for the problem unless $NP \subseteq DTIME(n^{polylog(n)})$.*

Proof. We know that $\mathcal{N} \leq n = 2\mathcal{N} + 2\mathcal{K} \leq 4\mathcal{N}$. Therefore, $O(2^{\log^{1-\epsilon} \mathcal{N}}) = O(2^{\log^{1-\epsilon} n})$ and $DTIME(\mathcal{N}^{polylog(\mathcal{N})}) = DTIME(n^{polylog(n)})$. □

4 Results for Small k

In this section, we first show that it is trivial to have a polynomial-time algorithm for MEEBM when k is one (Sect. 4.1), then we show that the problem is NP-hard when k is greater than one (Sect. 4.2). Finally, we give an $O(k)$-approximation algorithm for the problem (Sect. 4.3).

4.1 Polynomial Time Algorithm for $k = 1$

If k is one, all nodes appear in D at most once. We know that there exists a bipartite matching $M^* \supseteq \bigcup_{d \in D} d$ which we can find in polynomial time. The matching embeds all edge pairs in D. Hence, it is an optimal solution to this problem.

4.2   NP-Hardness for $k \geq 2$

The problem for any k greater than one is shown to be NP-hard based on the hardness of the maximum independent set problem, which is formally defined as follows:

Definition 2. *Given a graph $\mathcal{G} = (\mathcal{V}, \mathcal{E})$. Find a set $\mathcal{S} \subseteq \mathcal{V}$ such that $|\mathcal{S}|$ is maximized and, for any $\epsilon \in \mathcal{E}$, $\epsilon \not\subseteq \mathcal{S}$.*

In this paper, we use the following hardness result.

Theorem 3 [5]. *The maximum independent set problem is NP-hard even when G is a 4-regular graph.*

Consider an instance of the maximum independent set with a 4-regular graph $\mathcal{G} = (\mathcal{V}, \mathcal{E})$ as an input. We will construct an input of the MEEBM problem, denoted by G and D with $k = 2$, in polynomial time using the following algorithm:

1. Find an Eulerian circuit $\mathcal{Q} = (\nu'_1, \epsilon'_1, \nu'_2, \epsilon'_2, \dots, \nu'_{|\mathcal{E}|}, \epsilon'_{|\mathcal{E}|}, \nu'_1)$ of $\mathcal{G}$ using Fluery's [6] or Hierholzer's [4] algorithm, which are polynomial-time algorithms.
2. Construct a cyclic graph $C_{|\mathcal{E}|}$ with $|\mathcal{E}|$ different nodes $\{\nu_1, \dots, \nu_{|\mathcal{E}|}\}$ and $|\mathcal{E}|$ different edges $\{\epsilon_1 = \{\nu_1, \nu_2\}, \epsilon_2 = \{\nu_2, \nu_3\}, \dots, \epsilon_{|\mathcal{E}|} = \{\nu_{|\mathcal{E}|}, \nu_1\}\}$. Define functions f, g such that $f(\nu_i) = \nu'_i$ and $g(\epsilon_i) = epsilon'_i$ for all i. As $\nu'_i = \nu'_j$ for some $i \neq j$, f is not bijective, while g is bijective as $\epsilon'_1, \dots, \epsilon'_{|\mathcal{E}|}$ are all different. As $\mathcal{G}$ is a 4-regular graph, the number of edges in the graph is $2|\mathcal{V}|$, which is always even. Because any cyclic graph with an even number of nodes is bipartite [14], the graph $C_{|\mathcal{E}|}$ is a bipartite graph.
3. As $\mathcal{G}$ is 4-regular graph, there are exactly 2 edges ϵ'_i and ϵ'_j in Q such that $\nu'_i = \nu'_j$. Let $\mathcal{D} = \{\{\epsilon'_i, \epsilon'_j\} : \nu'_i = \nu'_j\}$. The intersection of any two members of $\mathcal{D}$ is empty set. Because the function g is bijective, the members of the following set D are also disjunctive: $D = \{\{g(\epsilon'_i), g(\epsilon'_j)\} : \nu'_i = \nu'_j\} = \{\{\epsilon_i, \epsilon_j\} : f(\nu_i) = f(\nu_j)\}$. Because all edges appear once in D and all vertices in $C_{|\mathcal{E}|}$ are incident to two edges, all vertices in $C_{|\mathcal{E}|}$ appear twice in D.
4. Our input to the MEEBM problem is a complete bipartite graph $G = (S, P, E)$ such that S and P are two partitions of the graph $C_{|\mathcal{E}|}$, and the set D is previously defined. Because all the nodes in G appear in D twice, k is equal to two for this input.

Example 2. Consider a graph $\mathcal{G}$ as shown in Fig. 3a.

1. A circuit $\mathcal{Q} = (\nu'_1, \{\nu'_1, \nu'_2\}, \nu'_2, \{\nu'_2, \nu'_3\}, \dots, \nu'_{24}, \{\nu'_{24}, \nu'_1\}, \nu'_1)$ where $\nu'_1, \dots, \nu'_{24} = 1, 3, 4, 6, 11, 10, 12, 7, 9, 10, 8, 4, 5, 3, 2, 1, 6, 8, 9, 5, 7, 2, 12, 11$ is one of the Eulerian circuits of the graph $\mathcal{G}$. We show the Eulerian circuit in Fig. 3b.
2. We have the following cyclic graph and function f:
   $C_{24} = (\{\nu_1, \dots, \nu_{24}\}, \{\epsilon_1 = \{\nu_1, \nu_2\}, \dots, \epsilon_{24} = \{\nu_{24}, \nu_1\}\}), f(\nu_1), \dots, f(\nu_{24}) = 1, 3, 4, 6, 11, 10, 12, 7, 9, 10, 8, 4, 5, 3, 2, 1, 6, 8, 9, 5, 7, 2, 12, 11$.

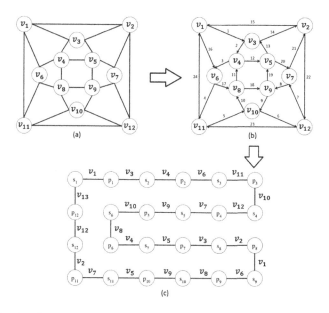

Fig. 3. Reduction from the independent set problem for regular graphs degree four to MEEBM problem when $k = 2$.

3. The set D contains $d_1 = \{\epsilon_1, \epsilon_{16}\}$ as $f(\nu_1) = f(\nu_{16}) = 1$. It also contains $d_2 = \{\epsilon_{15}, \epsilon_{22}\}$, $d_3 = \{\epsilon_2, \epsilon_{14}\}$, $d_4 = \{\epsilon_3, \epsilon_{12}\}$, $d_5 = \{\epsilon_{13}, \epsilon_{20}\}$, $d_6 = \{\epsilon_4, \epsilon_{17}\}$, $d_7 = \{\epsilon_8, \epsilon_{21}\}$, $d_8 = \{\epsilon_{11}, \epsilon_{18}\}$, $d_9 = \{\epsilon_9, \epsilon_{19}\}$, $d_{10} = \{\epsilon_6, \epsilon_{10}\}$, $d_{11} = \{\epsilon_5, \epsilon_{24}\}$, and $d_{12} = \{\epsilon_7, \epsilon_{23}\}$.

4. We have a complete bipartite graph $G = (S, P, E)$ where the vertex set $S = \{s_1 = \nu_1, s_2 = \nu_3, \ldots, s_{12} = \nu_{23}\}$, $P = \{p_1 = \nu_2, p_2 = \nu_4, \ldots, p_{12} = \nu_{24}\}$. We will obtain an instance of MEEBM for $k = 2$, if we replace all ν_i in C_{24} defined in 2 with $s_{\lfloor i/2 \rfloor}$ or $p_{i/2}$. The resulting graph can be seen in Fig. 3c.

In the following lemma, we will prove a property of the previous construction.

Lemma 2. *Let G, D be an input of MEEBM constructed from a 4-regular graph $\mathcal{G} = (\mathcal{V}, \mathcal{E})$ and the optimal solution of MEEBM for the input is M^*. We can construct a maximum independent set of $\mathcal{G}$, denoted by $\mathcal{S}^*$, in polynomial time.*

Proof. Consider an independent set $\mathcal{S}$ of the graph $\mathcal{G}$. We will now show that a bipartite matching of G that contains at least $|\mathcal{S}|$ edge pairs of D can be found in polynomial time. Recall the function f in the construction of G, D. Let $M' = \{\{e_i : f(v_i) = \sigma\} : \sigma \in \mathcal{S}\}$. As all nodes will appear in M' only once, we can use the algorithm in Sect. 4.1 to construct a matching $M \supseteq M'$ in polynomial time if such a matching exists. Also, because for each $\sigma \in \mathcal{V}$, the set $\{e_i : f(v_i) = \sigma\}$ is a different member of D, we know that M and M' contain at least $|\mathcal{S}|$ members of D.

In order to show that M exists, we assume a contradictory statement that there does not exist such an M, and in M', there is more than one edge incident to a node. Let the edges be e_i and e_j and the node be v_p. Because G is a cyclic graph, the edges are either $e_i = e_{|\mathcal{E}|}, e_j = e_0$ or $e_i = e_q, e_j = e_{q+1}$ for some q. We prove for the case that $e_i = e_q$ and $e_j = e_{q+1}$ but we can use the same argument for the case that $e_i = e_{|\mathcal{E}|}$ and $e_j = e_0$. We know that e_q and e_{q+1} are included in M' because $f(v_q), f(v_{q+1}) \in S$. Therefore, there exist $v'_q, v'_{q+1} \in S$. As v'_q, v'_{q+1} are next to each other in the Eulerian path Q, $\{v'_q, v'_{q+1}\} \in \mathcal{E}$. This contradicts the fact that S is an independent set.

Next, consider a bipartite matching M. Suppose that $D' = \{d \in D : d \subseteq M\}$. We can construct an independent set S of $\mathcal{G}$ with size $|D'|$ in polynomial time in the following way: $S = \{f(v_i) : \{e_i, e_j\} \in D'\}$. Note that, as $f(v_i) = f(v_j)$, it does not matter what we choose: $f(v_i)$ or $f(v_j)$. As the value of $f(v_i)$ is different for each edge pair $\{e_i, e_j\}$ in D', $|S|$ is $|D'|$.

We now show that S is an independent set. Assume a contradictory statement that there exists v'_i and v'_j in S such that $\{v'_i, v'_j\} \in \mathcal{E}$. Because all edges appear in the Eulerian circuit Q, we know that one of the following two cases must be satisfied: (1) There exist p, v_p and v_{p+1} such that $f(v_p) = v'_i$ and $f(v_{p+1}) = v'_j$. (2) $f(v_0) = v'_i$ and $f(v_{|\mathcal{E}|}) = v'_j$.

We will work on case 1, but all of the arguments can be also applied to case 2. We know that v'_i and v'_j are both in the set S only if all e_q such that $f(v_q) = v'_i$ or $f(v_q) = v'_j$ are in the set D'. Therefore, we have both e_p and e_{p+1}, which are incident to each other, in M' and also M. This contradicts the assumption that M is a bipartite matching.

As we have an independent set with size K in $\mathcal{G}$ if and only if we can embed K edge pairs in G. We know that the size of the maximum independent set equals the maximum number of edge-pairs embedding. If there exists a bipartite matching that embeds a maximum number of edge pairs K^*, the bipartite matching can be converted to an independent set with size K^*, which is an optimal solution to the independent set problem in polynomial time. □

By the previous lemma, we can prove the following theorem.

Theorem 4. *MEEBM is NP-hard for any k greater than one.*

Proof. As finding a maximum independent set of 4-regular graphs is NP-hard and the input of MEEBM in Lemma 1 has $k = 2$, we know from Lemma 1 that MEEBM is NP-hard when $k = 2$.

We will now show that if MEEBM is NP-hard for $k = k'$ it is also NP-hard for $k = k' + 1$. Thus, MEEBM is also NP-hard for all $k > 2$. Given an input of MEEBM $G_{k'}$ and $D_{k'}$ with $k = k'$. We can construct an input of MEEBM $G_{k'+1}$ and $D_{k'+1}$ with $k = k' + 1$ by applying the following steps.

(1) From $G_{k'} = (S_{k'}, P_{k'}, E_{k'})$ and $D_{k'}$, choose nodes u and v where $u \in S$, $v \in P$ such that one of them occurs k' times in $D_{k'}$. We know that there exists such a node because $k = k'$.

(2) Set $G_{k'+1} = (S_{k'+1} = S_{k'} \cup \{s_x, s_y\}, P_{k'+1} = P_{k'} \cup \{p_x, p_y\}, S_{k'+1} \times P_{k'+1})$.

(3) Let $d_x = \{\{u, p_x\}, \{s_x, v\}\}$ and $d_y = \{\{s_x, p_x\}, \{s_y, p_y\}\}$. Set $D_{k'+1} = D_{k'} \cup \{d_x, d_y\}$.

By the construction, u or v will appear $k' + 1$ times in the input $G_{k'+1}$, $D_{k'+1}$. Thus, we have $k = k' + 1$ here.

Suppose that an optimal solution of MEEBM when its inputs are $G_{k'}$ and $D_{k'}$, denoted by $M^*_{k'}$, embeds $K^*_{k'}$ edge pairs of $D_{k'}$. Then, we know that the bipartite matching $M^*_{k'} \cup \{d_y\}$ embeds $K^*_{k'} + 1$ edge pairs of $D_{k'+1}$. We now argue that an optimal solution of MEEBM when its inputs are $G_{k'+1}$ and $D_{k'+1}$ embeds $K^*_{k'} + 1$ edge pairs of $D_{k'+1}$. Let us assume a contradictory statement that there exists a bipartite matching $M'_{k'+1}$ that embeds at least $K^*_{k'} + 2$ edge pairs in $D_{k'+1}$. As $K^*_{k'}$ is the maximum number of edge pairs in $D_{k'}$ we can embed in a particular matching, we know that $M'_{k'+1}$ must embed both d_x, d_y to have the number of embeddings no smaller than $K^*_{k'} + 2$. However, d_x and d_y cannot be embedded in the same matching. We know it is not possible to have such a matching.

Thus, we know that all optimal solutions of MEEBM can embed $K^*_{k'} + 1$ edge pairs of $D_{k'+1}$. As it cannot embed more than $K^*_{k'}$ edge pairs of D_k, we know that it must include exactly one of the edge pairs d_x or d_y.

If we have an algorithm that can give us an optimal solution when the input has $k = k' + 1$, we can give $G_{k'+1}$, $D_{k'+1}$ to the algorithm and takes d_x or d_y from the solution. By that, we will have a solution that embeds $K^*_{k'}$ edge pairs of $D_{k'}$, an optimal solution when the inputs are $G_{k'}$ and $D_{k'}$. We then can solve the problem for $k = k'$. When solving the problem for $k = k'$ is NP-hard, solving the problem for $k = k' + 1$ is also NP-hard.   □

4.3   Approximation Algorithm

Consider an instance of MEEBM (G, D). Assume that for any member of D, $d = \{\{s_i, p_j\}, \{s_{i'}, p_{j'}\}\}$, there does not exist a case that $s_i = s_{i'}$ and $p_j \neq p_{j'}$. Clearly, it is not possible to have a bipartite matching that embeds both $\{s_i, p_j\}$ and $\{s_i, p_{j'}\}$ for $p_j \neq p_{j'}$. If that happened, the condition for d was not able to be satisfied and it could be excluded from our consideration. Similarly, we can also exclude an edge pair $d = \{\{s_i, p_j\}, \{s_{i'}, p_j\}\}$ such that $s_i \neq s_{i'}$ and $p_j = p_{j'}$ from our consideration. Our reduction is based on the following definition:

Definition 3. *Suppose that* $d = \{\{s_i, p_j\}, \{s_x, p_y\}\}$ *and* $d' = \{\{s_{i'}, p_{j'}\}, \{s_{x'}, p_{y'}\}\}$ *be members of* D. *Let* $\delta = d \cup d'$. *We say that* d *and* d' *conflict with each other if there is* $u \in S \cup P$ *such that* $|\{e \in \delta : u \in e\}| > 1$.

Note that by Definition 3, for edge pairs $d = \{\{u, v\}, \{a, b\}\}$ and $d' = \{\{u, v\}, \{c, d\}\}$, they do not conflict with each other because $d \cup d' = \{\{u, v\}, \{a, b\}, \{c, d\}\}$ which every nodes appear once.

Consider a set $D' \subseteq D$. If there exist two members in D' conflicting with each other, it is straightforward to see that there is no bipartite matching embedding D'. On the other hand, if every two members of D' do not conflict with each other, we can easily find a bipartite matching embedding D'.

Our task is then to find a set $D' \subseteq D$ such that $|D'|$ is maximized and all two members do not conflict with each other. To do that, we construct a graph (D, Σ) where $\Sigma = \{\{d, d'\} \in D : d \text{ and } d' \text{ conflict with each other}\}$. A bipartite matching that embeds all members of $D' \subseteq D$ can be constructed if and only if D' is an independent set. An algorithm for the maximum independent set problem, a problem defined in Definition 2, can be used to find the set D' with the largest size. We show how to construct an instance of the maximum independent set problem from an instance of MEEBM in Fig. 4.

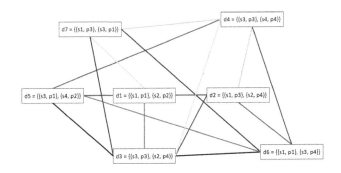

Fig. 4. Instance of the independent set problem constructed from an instance of MEEBM

The following lemma shows that the graph (D, Σ) has small degree.

Lemma 3. *The degree of* (D, Σ) *is not more than* $4k - 4$.

Proof. An edge pair d conflicts with another edge pair d', if there exist $e \in d$ and $e' \in d'$ that have at least one common vertex. Consider an edge pair $d = \{\{s_i, p_j\}, \{s_x, p_y\}\}$. Since s_i can appear at most k times in D, d can violate with at most $k - 1$ other edge pairs due to the vertex s_i. Similarly, d can violate with at most $k - 1$ other edge pairs due to the other vertices associating to d. As there are four nodes in d, degree of d in the instance of the maximum independent set problem is no larger than $4(k - 1) = 4k - 4$. □

It is discussed in [8] that there exists a greedy algorithm with approximation ratio $\frac{\Delta+2}{3}$ when Δ is a maximum degree of the input graph. We thus have the following theorem.

Theorem 5. *There exists a* $\left(\frac{4k-2}{3}\right)$*-approximation algorithm for the MEEBM problem.*

Proof. From Lemma 3, the graph of the maximum independent set problem constructed from MEEBM has the degree at most $4k - 4$. Applying the $\frac{\Delta+2}{3}$ approximation algorithm gives us $\frac{4k-2}{3}$ for the independent set problem. Because the size of the largest independent set equals the maximum number of embedding edge pairs, the application also gives us a $\left(\frac{4k-2}{3}\right)$-approximate solution for the MEEBM problem. □

5   Conclusion

In this work, we consider a problem called the maximum edge-pair embedding bipartite matching (MEEBM) motivated from the wireless localization problems. We show two negative results for this problem. The first result is the inapproximability result where we show that an approximation ratio of MEEBM cannot be better than that of the label cover problem. We currently know that, for all $0 < \epsilon < 1$ it is unlikely to have an $O(2^{\log^{1-\epsilon} n})$-approximation algorithm for both of the problems. As $O(2^{\log^{1-\epsilon} n})$ is $O(n^c)$ when $\epsilon = 1$, it is widely believed that there is a constant c such that $O(n^c)$-approximation algorithm does not exist [3]. The second result shows that there does not exist a fixed-parameter tractability algorithm for the problem for an MEEBM instance with a small degree. The problem is NP-hard even when the degree is only two, but, when the instance degree is k, we propose an $O(k)$-approximation algorithm for the problem.

Acknowledgement. Parts of the work had been done when Phanu Vajanopath was conducting an internship at The University of Tokyo. The internship was supported by the GSI Internship program, Faculty of Science, The University of Tokyo. He was hosted by Prof. Reiji Suda during the program. Also, the authors want to thank reviewers of WALCOM 2020 who kindly gave comments that significantly improve this paper.

References

1. Arora, S., Babai, L., Stern, J., Sweedyk, Z.: The hardness of approximate optima in lattices, codes, and systems of linear equations. J. Comput. Syst. Sci. **54**(2), 317–331 (1997)
2. Biswas, P., Lian, T.C., Wang, T.C., Ye, Y.: Semidefinite programming based algorithms for sensor network localization. ACM Trans. Sens. Netw. (TOSN) **2**(2), 188–220 (2006)
3. Charikar, M., Hajiaghayi, M., Karloff, H.: Improved approximation algorithms for label cover problems. ESA **2009**, 23–34 (2009)
4. Fleischner, H.: X. 1 algorithms for Eulerian trails. In: Eulerian Graphs and Related Topics: Part 1. Annals of Discrete Mathematics, vol. 50, pp. 1–13 (1991)
5. Fleischner, H., Sabidussi, G., Sarvanov, V.I.: Maximum independent sets in 3-and 4-regular Hamiltonian graphs. Discrete Math. **310**(20), 2742–2749 (2010)
6. Fleury, M.: Deux problemes de geometrie de situation. Journal de mathematiques elementaires **2**(2), 257–261 (1883)
7. Ghafourian, A., Georgiou, O., Barter, E., Gross, T.: Wireless localization with diffusion maps. arXiv preprint arXiv:1908.05216 (2019)
8. Halldórsson, M.M., Radhakrishnan, J.: Greed is good: approximating independent sets in sparse and bounded-degree graphs. Algorithmica **18**(1), 145–163 (1997)
9. Hopcroft, J.E., Karp, R.M.: An $n^{5/2}$ algorithm for maximum matchings in bipartite graphs. SIAM J. Comput. **2**(4), 225–231 (1973)
10. Hu, L., Evans, D.: Localization for mobile sensor networks. In: MobiCom 2004, pp. 45–57 (2004)
11. Nguyen, C.L., Georgiou, O., Yonezawa, Y., Doi, Y.: The wireless localisation matching problem and a maximum likelihood based solution. In: ICC 2017, pp. 1–7 (2017)

12. Nguyen, C.L., Georgiou, O., Yonezawa, Y., Doi, Y.: The wireless localization matching problem. IEEE Internet Things J. **4**(5), 1312–1326 (2017)
13. Nguyen, L., Georgiou, O., Suppakitpaisarn, V.: Improved localization accuracy using machine learning and refining RSS measurements. In: GLOBECOM Workshops 2018 (2018)
14. Skiena, S.: Coloring bipartite graphs, chap. 5.5.2. In: Skiena, S. (ed.) Implementing Discrete Mathematics: Combinatorics and Graph Theory with Mathematica, p. 213. Addison-Wesley, Reading (1990)
15. Suppakitpaisarn, V., Dai, W., Baffier, J.F.: Robust network flow against attackers with knowledge of routing method. In: HPSR 2015, pp. 40–47 (2015)

Packing Arc-Disjoint Cycles in Bipartite Tournaments

Ajay Saju Jacob[1] and R. Krithika[2(✉)]

[1] Indian Institute of Technology Madras, Chennai, India
ee16b129@smail.iitm.ac.in
[2] Indian Institute of Technology Palakkad, Palakkad, India
krithika@iitpkd.ac.in

Abstract. An r-partite tournament is a directed graph obtained by assigning a unique orientation to each edge of a complete undirected r-partite simple graph. Given a bipartite tournament T on n vertices, we explore the parameterized complexity of the problem of determining if T has a cycle packing (a set of pairwise arc-disjoint cycles) of size k. Although the maximization version of this problem can be seen as the linear programming dual of the well-studied problem of finding a minimum feedback arc set (a set of arcs whose deletion results in an acyclic graph) in bipartite tournaments, surprisingly no algorithmic results seem to exist. We show that this problem can be solved in $2^{\mathcal{O}(k \log k)} n^{\mathcal{O}(1)}$ time and admits a kernel with $\mathcal{O}(k^2)$ vertices.

1 Introduction

A tournament is a directed graph in which there is exactly one arc between every pair of distinct vertices. Equivalently, a tournament is a directed graph obtained by assigning a unique orientation to each edge of a complete undirected simple graph. More generally, a r-partite tournament is a directed graph obtained by assigning a unique orientation to each edge of a complete undirected r-partite simple graph. These graphs naturally model the results of competitions; tournaments represent round-robin competitions while r-partite tournaments represent competitions between r teams. Due to this modelling ability, tournaments and r-partite torunaments have proven to be tremendously useful in understanding important problems in machine learning, search engine ranking, voting systems, social choice theory and ontologies. In particular, problems on tournaments and bipartite tournaments (r-partite torunaments for $r = 2$) have several applications in these areas. One such problem is FEEDBACK ARC SET. A *feedback arc set* is a set of arcs whose deletion results in an acyclic graph. Given a directed graph and a positive integer k, FEEDBACK ARC SET is the problem of determining if the graph has a set of at most k arcs whose deletion results in an acyclic graph. FEEDBACK ARC SET on tournaments finds applications in rank aggregation and this problem on bipartite tournaments is useful in ontologies mappings [17,18,24].

© Springer Nature Switzerland AG 2020
M. S. Rahman et al. (Eds.): WALCOM 2020, LNCS 12049, pp. 249–260, 2020.
https://doi.org/10.1007/978-3-030-39881-1_21

FEEDBACK ARC SET is known to be NP-hard on tournaments and bipartite tournaments [1,6,7,15]. FEEDBACK ARC SET has a 4-approximation algorithm on bipartite tournaments [27] and a polynomial-time approximation scheme on tournaments [19]. FEEDBACK ARC SET on tournaments and bipartite tournaments is well-studied in the parameterized complexity framework too. In this framework, each problem instance is associated with a non-negative integer k called *parameter*, and a problem is said to be *fixed-parameter tractable* (FPT) if it can be solved in $f(k)n^{\mathcal{O}(1)}$ time for some function f, where n is the input size. For convenience, the running time $f(k)n^{\mathcal{O}(1)}$ where f grows super-polynomially with k is denoted as $\mathcal{O}^\star(f(k))$. A *kernelization algorithm* is a polynomial-time algorithm that transforms an arbitrary instance of the problem to an equivalent instance of the same problem whose size is bounded by some computable function g of the parameter of the original instance. The resulting instance is called a *kernel* and if g is a polynomial function, then it is called a *polynomial kernel* and we say that the problem admits a polynomial kernel. Kernelization typically involves applying a set of rules (called *reduction rules*) to the given instance to produce another instance. A reduction rule is said to be *safe* if it is sound and complete, i.e., applying it to the given instance produces an equivalent instance. In order to classify parameterized problems as being FPT or not, the W-hierarchy is defined: FPT $\subseteq$ W[1] $\subseteq$ W[2] $\subseteq$ ... $\subseteq$ XP. It is believed that the subset relations in this sequence are all strict, and a parameterized problem that is hard for some complexity class above FPT in this hierarchy is said to be fixed-parameter intractable. Further details on parameterized algorithms can be found in [8,9,12].

FEEDBACK ARC SET is FPT when parameterized by k on both tournaments and bipartite tournaments [2,13,16]. Further, FEEDBACK ARC SET has a kernel with $\mathcal{O}(k)$ vertices on tournaments [5] and a kernel with $\mathcal{O}(k^2)$ vertices on bipartite tournaments [23,26]. Though FEEDBACK ARC SET is extensively studied in tournaments and bipartite tournaments, its dual, namely ARC-DISJOINT CYCLES, has surprisingly not been considered in the literature until recently [4]. Given a directed graph G and a positive integer k, the ARC-DISJOINT CYCLES problem is to determine whether G has k arc disjoint cycles. This problem is W[1]-hard in general [20,25]. However, ARC-DISJOINT CYCLES on tournaments is FPT via an $2^{\mathcal{O}(k \log k)}n^{\mathcal{O}(1)}$-time algorithm and admits a kernel with $\mathcal{O}(k)$ vertices [4]. In this paper, we investigate the parameterized complexity of packing arc-disjoint cycles in bipartite tournaments.

ARC-DISJOINT CYCLES IN BIPARTITE TOURNAMENTS (ACBT)
Input: A bipartite tournament B and a positive integer k.
Question: Do there exist k arc-disjoint cycles in B?
Parameter: k

We show that ACBT is FPT and admits a polynomial kernel. En route, we discover an interesting min-max relation analogous to the classical Erdős-Pósa theorem [10]. It is easy to verify that a bipartite tournament B that has k arc-disjoint cycles need not necessarily have k arc-disjoint 4-cycles. However, we can show that if B has k arc-disjoint cycles, then B has k arc-disjoint cycles each of length at most $4k$. This observation is the starting point of our study. Using this observation and the standard color-coding technique [3,8,22], it is easy to show that ACBT is FPT and can be solved in $\mathcal{O}^\star(2^{\mathcal{O}(k^2)})$ time. We improve this result and show the following.

Theorem 1. ACBT *can be solved in* $\mathcal{O}^\star(2^{\mathcal{O}(k \log k)})$ *time.*

This result crucially uses the following Erdös-Pósa type result.

Theorem 2. *Every bipartite tournament B either contains k arc-disjoint 4-cycles or has a feedback arc set of size at most $8(k-1)$.*

Theorem 2 leads to the following kernelization results which is in turn used in the proof of Theorem 1.

Theorem 3. ACBT *admits a kernel with* $\mathcal{O}(k^2)$ *vertices.*

Road Map. The paper is organized as follows. In Sect. 2, we give some definitions related to directed graphs, cycles and bipartite tournaments. In Sect. 3, we show that ACBT is FPT via an $\mathcal{O}^\star(2^{\mathcal{O}(k^2)})$ time algorithm. In Sect. 4, we show the claimed combinatorial result on bipartite tournaments (Theorem 2). In Sect. 5, we show that ACBT admits a quadratic vertex kernel and in Sect. 6, we describe an $\mathcal{O}^\star(2^{\mathcal{O}(k \log k)})$ time algorithm for ACBT using Theorems 2 and 3. Finally, we conclude with some remarks in Sect. 7.

2 Preliminaries

The set $\{1, 2, \ldots, n\}$ is denoted by $[n]$. A *directed graph* (or *digraph*) is a pair consisting of a set V of vertices and a set A of arcs. An arc is specified as an ordered pair of vertices. We will consider only simple unweighted digraphs. For a digraph D, $V(D)$ and $A(D)$ denote the set of its vertices and the set of its arcs, respectively. Two vertices u, v are said to be *adjacent* in D if $(u, v) \in A(D)$ or $(v, u) \in A(D)$. For an arc $e = (u, v)$, $h(e)$ denotes v and $t(e)$ denotes u. For a vertex $v \in V(D)$, its *out-neighborhood*, denoted by $N^+(v)$, is the set $\{u \in V(D) \mid (v, u) \in A(D)\}$ and its *in-neighborhood*, denoted by $N^-(v)$, is the set $\{u \in V(D) \mid (u, v) \in A(D)\}$. For a set of arcs F, $V(F)$ denotes the union of the sets of endpoints of arcs in F. For a set $X \subseteq V(D) \cup A(D)$, $D - X$ denotes the digraph obtained from D by deleting X.

A *path* P in D is a sequence $(v_1, \ldots, v_k)$ of distinct vertices such that for each $i \in [k-1]$, $(v_i, v_{i+1}) \in A(D)$. The set $\{v_1, \ldots, v_k\}$ is denoted by $V(P)$ and the set $\{(v_i, v_{i+1}) \mid i \in [k-1]\}$ is denoted by $A(P)$. A path P is called an *induced* (or *chordless*) path if there is no arc in D that is between two non-consecutive vertices of P. A *cycle* C in D is a sequence $(v_1, \ldots, v_k)$ of distinct vertices such that $(v_1, \ldots, v_k)$ is a path and $(v_k, v_1) \in A(D)$. The set $\{v_1, \ldots, v_k\}$ is denoted by $V(C)$ and the set $\{(v_i, v_{i+1}) \mid i \in [k-1]\} \cup \{(v_k, v_1)\}$ is denoted by $A(C)$.

A cycle $C = (v_1, \ldots, v_k)$ is called an *induced* (or *chordless*) cycle if there is no arc in D that is between two non-consecutive vertices of C with the exception of the arc (v_k, v_1). The length of a path or cycle X is the number of vertices in it and is denoted by $|X|$. For a set $\mathcal{C}$ of paths or cycles, $A(\mathcal{C})$ denotes the set $\{e \in A(D) \mid \exists C \in \mathcal{C}, e \in A(C)\}$. A cycle on three vertices is called a *triangle*. A cycle of length q is called a q-cycle. A digraph is said to be *triangle-free* if it has no triangles. A digraph is called a *directed acyclic graph* if it has no cycles. Any directed acyclic graph D has an ordering σ called *topological ordering* of its vertices such that for each $(u, v) \in A(D)$, $\sigma(u) < \sigma(v)$ holds. A *feedback vertex (arc) set* is a set of vertices (arcs) whose deletion results in an acyclic graph.

A bipartite digraph is a digraph B whose vertex set can be partitioned into two sets X and Y such that every arc in B has one endpoint in X and the other endpoint in Y. Equivalently, a bipartite digraph is a digraph that is obtained by assigning a unique orientation to each edge of an undirected bipartite graph. We denote B as $B[X, Y]$ where X and Y form the bipartition of the underlying bipartite graph. It is easy to see that a bipartite digraph has no triangle and any 4-cycle is an induced 4-cycle. A bipartite tournament is a bipartite digraph $B[X, Y]$ in which for every pair u, v of distinct vertices with $u \in X$, $v \in Y$ either $(u, v) \in A(B)$ or $(v, u) \in A(B)$ but not both.

3   An FPT Algorithm Using Color Coding

Consider an instance $\mathcal{I} = (B, k)$ of ACBT. First, we show that it suffices to determine if B has k arc-disjoint cycles each of length at most $4k$ in order to determine if (B, k) is an yes-instance of ACBT or not.

Observation 4. *Let k and r be positive integers such that $r \leq k$. If a bipartite tournament B contains a set $\mathcal{C}$ of r arc-disjoint cycles, then it also contains a set $\mathcal{C}^*$ of r arc-disjoint cycles each of length at most $4k$.*

Proof. Let $\mathcal{C}$ be a set of r arc-disjoint cycles in B that minimizes $\sum_{C \in \mathcal{C}} |C|$. If every cycle in $\mathcal{C}$ is a 4-cycle, then the claim holds. Otherwise, let C' be the longest cycle in $\mathcal{C}$ and let 2ℓ be its length ($\ell > 2$). Let v_i, v_j be two non-consecutive vertices of the cycle such that $v_i \in X$ and $v_j \in Y$. Then, either $(v_i, v_j) \in \mathcal{A}(B)$ or $(v_j, v_i) \in \mathcal{A}(B)$. In any case, the arc e between v_i and v_j along with $A(C)$ forms a cycle C' of length less than ℓ with $A(C') \setminus \{e\} \subset A(C)$. By our choice of $\mathcal{C}$, this implies that e is an arc in some other cycle $\widehat{C} \in \mathcal{C}$. This property is true for the arc between any pair of non-consecutive vertices in C. Hence, every edge between two non-consecutive vertices of C' must be part of some other cycle in C. That is, $\ell(\ell - 2) \leq 2\ell(k - 1)$ implying that $2\ell \leq 4k$.   □

Let n denote $|V(B)|$ and m denote $|A(B)|$. Suppose $\mathcal{I}$ is a yes-instance and $\mathcal{C}$ is a set of k arc-disjoint cycles in B. From Observation 4, we may assume that the total number of arcs that are in cycles in $\mathcal{C}$ is at most $4k^2$. Using this observation, we proceed as follows. We color the arcs of B uniformly at random from the color set $[\ell]$ where $\ell = 4k^2$. Let $\chi : A(B) \to [\ell]$ denote this coloring.

Proposition 1 ([3]). *If E is a subset of $A(B)$ of size ℓ, then the probability that the arcs in E are colored with pairwise distinct colors is at least $e^{-\ell}$.*

Next, we define two notions of a colorful solution for our problem.

Definition 1 (Colored set of arc-disjoint cycles). *A set $\mathcal{C}$ of arc-disjoint cycles in B that satisfies the property that for any two distinct cycles $C, C' \in \mathcal{C}$ and for any two arcs $e \in A(C), e' \in A(C')$, $\chi(e) \neq \chi(e')$ holds is said to be a colored set of arc-disjoint cycles.*

Definition 2 (Colorful set of arc-disjoint cycles). *A set $\mathcal{C}$ of arc-disjoint cycles in B that satisfies the property that for any two (not necessarily distinct) cycles $C, C' \in \mathcal{C}$ and for any two distinct arcs $e \in A(C), e' \in A(C')$, $\chi(e) \neq \chi(e')$ holds is said to be a colorful set of arc-disjoint cycles.*

Observe that a colorful set of arc-disjoint cycles is also a colored set of arc-disjoint cycles. Rephrasing Proposition 1 in the context of our problem, we have the following observation.

Observation 5. *If $\mathcal{C}$ is a solution of $\mathcal{I}$ with the property that for each $C \in \mathcal{C}$, $|C| \leq 4k$, then $\mathcal{C}$ is a colorful set of arc-disjoint cycles in B with probability at least $e^{-\ell}$.*

Armed with the guarantee that a solution (if one exists) of $\mathcal{I}$ is colorful with sufficiently high probability, we focus on determining if there is a colorful set of cycles in B.

Lemma 1. *If B has a colorful set of k arc-disjoint cycles, then a colored set of k arc-disjoint cycles in B can be obtained in $2^{\mathcal{O}(k^2)}n^{\mathcal{O}(1)}$ time.*

Proof. Let $\ell = 4k^2$. For $i \in [\ell]$, let A_i denote the set of arcs of B that are colored with color i. For a subset S of $[\ell]$, let B_S denote the subgraph of B with $V(B_S) = V(B)$ and $A(B_S) = \bigcup_{i \in S} A_i$. For a subset S of $[\ell]$ and a positive integer $r \leq k$, define $\Gamma(S, r)$ to be 1 if B has a set $\mathcal{C}$ of r cycles such that $A(\mathcal{C}) \subseteq S$ and 0 otherwise. Also, $\Gamma(S, 1)$ is 1 if B_S has a cycle. Further, for $r > 1$, $\Gamma(S, r) = \bigvee_X \Gamma(S \setminus X, r-1)$ where $X \subseteq S$ and B_X has a cycle. Therefore, for each $S \subseteq [\ell]$, $\Gamma(S, k)$ can be computed in $2^{\mathcal{O}(k^2)}n^{\mathcal{O}(1)}$ time. Clearly, B has a colored set of k arc-disjoint cycles if and only if $\Gamma(S, k) = 1$ for some $S \subseteq [\ell]$ and this can be determined in $2^{\mathcal{O}(k^2)}n^{\mathcal{O}(1)}$ time. $\square$

Using the standard technique of derandomization of color coding based algorithms [3,8,22], we have the following result by taking $m = |A(B)|$.

Proposition 2 ([3,8,22]). *Given integers $m, \ell \geq 1$, there is a family $\mathcal{F}_{m,\ell}$ of coloring functions $\chi : A(T) \rightarrow [\ell]$ of size $e^\ell \ell^{\mathcal{O}(\log \ell)} \log m$ that can be constructed in $e^\ell \ell^{\mathcal{O}(\log \ell)} m \log m$ time satisfying the following property: for every set $E \subseteq A(B)$ of size ℓ, there is a function $\chi \in \mathcal{F}_{m,\ell}$ such that $\chi(e) \neq \chi(e')$ for any two distinct arcs $e, e' \in E$.*

Then, we have the following result.

Theorem 6. ACBT *can be solved in* $\mathcal{O}^\star(2^{\mathcal{O}(k^2)})$ *time.*

Proof. Consider an instance $\mathcal{I} = (B, k)$ of ACBT. Let $\ell = 4k^2$. First, we compute the family $\mathcal{F}_{m,\ell}$ of $e^\ell \ell^{\mathcal{O}(\log \ell)} \log m$ coloring functions using Proposition 2 where m is the number of arcs in B. Then, for each coloring function $\chi : A(B) \to [\ell]$ in $\mathcal{F}_{m,\ell}$, we determine if B has a coloredl set of k cycles using Lemma 1. Due to the properties of $\mathcal{F}_{m,\ell}$ guaranteed by Proposition 2, it follows that $\mathcal{I}$ is a yes-instance if and only if B has a set of k cycles that is colorful with respect to at least one of the coloring functions. The overall running time is $\mathcal{O}^\star(2^{\mathcal{O}(k^2)})$. □

4   An Erdös-Pósa Type Theorem

The classical Erdös-Pósa theorem for cycles in undirected graphs states that there exists a function $f(k) = \mathcal{O}(k \log k)$ such that for each non-negative integer k, every undirected graph either contains k vertex-disjoint cycles or has a feedback vertex set consisting of $f(k)$ vertices [10]. An interesting consequence of this theorem is that it leads to an FPT algorithm for VERTEX-DISJOINT CYCLE PACKING (see [21] for more details). Recently, an analogous result for arc-disjoint cycles in tournaments [4] has been used to give an $2^{\mathcal{O}(k \log k)} n^{\mathcal{O}(1)}$ time algorithm and a kernel with $\mathcal{O}(k)$ vertices for finding k arc-disjoint cycles in tournaments.

   In this section, we show that for each non-negative integer k, every bipartite tournament either contains k arc-disjoint cycles or has a feedback arc set consisting of at most $8k$ arcs. We use this result to design an $2^{\mathcal{O}(k \log k)} n^{\mathcal{O}(1)}$ time algorithm and a kernel with $\mathcal{O}(k^2)$ vertices for ACBT.

   For a bipartite digraph $D[X, Y]$, let $\Lambda(D)$ denote the number of pairs u, v of vertices in D with $u \in X$, $v \in Y$ and niether $(u, v) \in A(D)$ nor $(v, u) \in A(D)$.

Lemma 2. *Let* $D[X, Y]$ *be a bipartite digraph in which for every pair* $u \in X$, $v \in Y$ *of distinct vertices, at most one of* (u, v) *or* (v, u) *is in* $A(D)$. *Then, if* D *has no 4-cycle, then we can compute a feedback arc set of* D *of size at most* $\Lambda(D)$ *in polynomial time.*

Proof. We will prove the claim by induction on $|V(D)|$. The claim trivially holds for $|X| < 2$ or $|Y| < 2$ as in these cases, the empty set is a feedback arc set. Hence, assume that $|X| \geq 2$ or $|Y| \geq 2$.

   First we apply a simple preprocessing rule on D. If D has a vertex v that either has no in-neighbours or no out-neighbours then delete v from D. As v is not in any cycle of D, any feedback arc set of D' is an feedback arc set of D.

   Next, for a vertex $v \in V(D)$, define first(v) to be the number of induced paths of length 4 with v as the first vertex. Similarly, define sec(v) to be the number of induced paths of length 4 with v as the second vertex. We claim that $\sum_{v \in V(D)}$ first$(v) = \sum_{v \in V(D)}$ sec(v). Any induced path $P = (u, v, w, z)$ of length 4 in D contributes 1 to first(u) and 1 to sec(v). Further, P does not contribute to first(x) for any $x \neq u$ and P does not contribute to sec(x) for any $x \neq v$. Therefore, P contributes 1 to $\sum_{v \in V(D)}$ first(v) and $\sum_{v \in V(D)}$ sec(v). Hence, $\sum_{v \in V(D)}$ first$(v) = \sum_{v \in V(D)}$ sec(v). It now follows that there is a vertex $u \in V(D)$ such that first$(u) \leq$ sec(u).

Without loss of generality assume that $u \in X$. Consider the following sets of vertices of D: $Y_1 = N^-(u)$, $Y_2 = N^+(u)$, $Y_3 = Y \setminus (Y_1 \cup Y_2)$, $X_2 = N^+(Y_2)$ and $X_1 = X \setminus (X_2 \cup \{u\})$. Following are the properties of these sets.

- Y_1 and Y_2 are non-empty due to the preprocessing.
- There is no arc from a vertex $x \in X_2$ to a vertex $y \in Y_1$. Otherwise, (u, y', x, y) is a 4-cycle where $x \in N^+(y')$ and $y' \in Y_2$.
- By the definition of X_1, there is no arc from a vertex $y \in Y_2$ to a vertex $x \in X_1$.

Let D_1 denote the subgraph $D[X_1, Y_1 \cup Y_3]$ and D_2 denote the subgraph $D[X_2, Y_2]$. As D_1 and D_2 are vertex-disjoint subgraphs of D, we have $\Lambda(D) \geq \Lambda(D_1) + \Lambda(D_2)$. Any induced path $P = (a, u, b, c)$ of length 4 in D with u as the second vertex satisfies the property that $a \in Y_1$ and $c \in X_2$. Further, a and c are non-adjacent as P is an induced path and D has no 4-cycle. Therefore, sec(u) is the number of non-adjacent pairs a, b such that $a \in Y_1$ and $c \in X_2$. As $a \in V(D_1)$ and $c \in V(D_2)$, we have $\Lambda(D) \geq \Lambda(D_1) + \Lambda(D_2) + $ sec(u).

Let E denote the set of arcs (x, y) such that $x \in X_2$ and $y \in Y_3$. Let F_1 and F_2 be feedback arc sets of D_1 and D_2, respectively. We claim that $F = F_1 \cup F_2 \cup E$ is a feedback arc set of D. If there exists a cycle C in the graph obtained from D by deleting the arcs in F, then C has an arc (p, q) with $p \in V(D_1)$ and $q \in V(D_2)$ and an arc (r, s) with $r \in V(D_2)$ and $s \in V(D_1)$. Following are the properties of vertices r and s.

- It is not possible that $r \in Y_2$ and $s \in X_1$ by the definition of X_1.
- It is not possible that $r \in X_2$ and $s \in Y_1$ as D has no 4-cycle.

Therefore, it follows that $(r, s) \in E$ which leads to a contradiction. Therefore, F is a feedback arc set of D of size $|F| = |F_1| + |F_2| + |E|$. As any induced path $P = (u, a, b, c)$ has $b \in X_2$ and $c \in Y_r$, we have $|E| = $ first(u) and by the choice of u, we have first$(u) \leq$ sec(u). Hence, we can conclude that $|F| \leq |F_1| + |F_2| + $ sec(u) and by induction hypothesis, $|F_1| \leq \Lambda(D_1)$ and $|F_2| \leq \Lambda(D_2)$. It now follows that $|F| \leq \Lambda(D)$. □

This leads to the following main result of this section.

Theorem 2 (restated). *For every non-negative integer k, every bipartite tournament B either contains k arc-disjoint 4-cycles or has a feedback arc set of size at most $8(k-1)$ that can be obtained in polynomial time.*

Proof. Suppose C is a maximal set of arc-disjoint 4-cycles in B with $|C| \leq k - 1$. Let D denote the digraph obtained from B by deleting the arcs that are in some 4-cycle in C. Clearly, D has no 4-cycle and $\Lambda(D) \leq 4(k-1)$. From Lemma 2, we know that D has a feedback arc set of size at most $4(k-1)$. Also, if F is a feedback arc set of D, then $F \cup A(C)$ is a feedback arc set of B. Therefore, B has a feedback arc set of size at most $8(k-1)$. □

We will use this result crucially in showing that ACBT can be solved in $\mathcal{O}^\star(2^{\mathcal{O}(k \log k)})$ time and admits a kernel with $\mathcal{O}(k^2)$ vertices.

5   A Polynomial Kernel

In this section, we show that ACBT admits a quadratic vertex kernel. Let $(B[X,Y], k)$ be an instance of ACBT. From Theorem 2, we know that B has either k arc-disjoint 4-cycles or a feedback arc set F of size at most $8(k - 1)$. If the former case holds, then we return a trivial yes-instance of constant size as the kernel. In the latter case, let $S = V(F)$ be a feedback vertex set of B of size at most $16k$. Let D denote the acyclic bipartite tournament $B - S$ and let δ denote a topological ordering of D.

As every cycle in B has at least 2 vertices in S, we have the following observation.

Observation 7. *For every $v \in S$, the subgraph of B induced by $V(D) \cup \{v\}$ is also acyclic.*

As every arc in B is between a vertex in X and a vertex in Y, we have the following observation.

Observation 8. *The subgraphs of B induced by $V(D) \cup (S \cap X)$ and $V(D) \cup (S \cap Y)$ are both acyclic.*

For a vertex $v \in S$, define the following sets.

- $R^+(v)$ is the set of first (with respect to δ) $(2k + 1)$ vertices in $N^+(v)$.
- $R^-(v)$ is the set of last (with respect to δ) $(2k + 1)$ vertices in $N^-(v)$.
- $R(v) = R^+(v) \cup R^-(v)$.

Let B' be the subgraph of B induced by $S \cup \{R(v) \mid v \in S\}$.

Lemma 3. *B has a set of k arc-disjoint cycles if and only if B' has a set of k arc-disjoint cycles.*

Proof. As B' is a subgraph of B, the reverse direction of the claim holds. Suppose B has a set of k arc-disjoint cycles. Among all such sets, let $\mathcal{C}$ be the one that maximizes $\sum_{C \in \mathcal{C}} |V(C) \cap V(B')|$.

Suppose there is a cycle $C \in \mathcal{C}$ that is not in B'. Then, there is a vertex $v_i \in V(C)$ that is not in B'. As the feedback vertex set S is the set of endpoints of the arcs of a feedback arc set of B, any cycle in B has at least two vertices from S. Let a and b be two such vertices in C where $(a, v_1, \ldots, v_i, \ldots, v_q, b)$ is a path in C from a to b such that $v_1, v_2, \ldots, v_q \in V(D)$. Note that $\delta(a) < \delta(v_i) < \delta(b)$ as $B[\{a, v_1, \ldots, v_i, \ldots, v_q\}]$ and $B[\{v_1, \ldots, v_i, \ldots, v_q, b\}]$ are acyclic. Without loss of generality assume that $v_i \in X$.

Case ($a \in X$ and $b \in Y$): In this case, $v_{i-1} \in Y$. Also, $(a, v_{i-1}) \in A(B)$ and $(v_i, b) \in A(B)$. As $v_i \in N^-(b)$ and $v_i \notin V(B')$, we have $|R^-(b)| = 2k + 1$ and there exists a vertex $z \in R^-(b)$ such that arcs (z, b) and (v_{i-1}, z) are not part of any cycle in $\mathcal{C}$. As $z, v_i \in N^-(b)$, $z \in R^-(b)$, $(v_{i-1}, v_i) \in A(B)$ and $v_i \notin R^-(b)$, it follows that $(v_{i-1}, z) \in A(B)$. If $v_{i-1} \in V(B')$ then we can replace the path $(a, v_1, \ldots, v_i, \ldots, v_q, b)$ with (a, v_{i-1}, z, b) to obtain another set $\mathcal{C}'$ of k arc-disjoint cycles such that $\sum_{C \in \mathcal{C}'} |V(C) \cap V(B')| > \sum_{C \in \mathcal{C}} |V(C) \cap V(B')|$ which leads to a contradiction. If $v_{i-1} \notin V(B')$ there exists a vertex $r \in R^+(a)$ such that arcs (a, r) and (r, z) are not part of any cycle in $\mathcal{C}$. As $r \in R^+(a)$ and $v_i \notin R^+(a)$, we have $(r, z) \in A(B)$. Hence, we can replace path $(a, v_1, \ldots, v_i, \ldots, v_q, b)$ with (a, r, z, b) to obtain another set $\mathcal{C}'$ of k arc-disjoint cycles such that $\sum_{C \in \mathcal{C}'} |V(C) \cap V(B')| > \sum_{C \in \mathcal{C}} |V(C) \cap V(B')|$ which leads to a contradiction.

Case ($a \in Y$ and $b \in X$): Then, $v_{i+1} \in Y$, $(v_{i+1}, b) \in A(B)$ and $(a, v_i) \in A(B)$. As $v_i \notin V(B')$ there exists a vertex $z \in R^+(a)$ such that arcs (a, z) and (z, v_{i+1}) are not part of any cycle in $\mathcal{C}$. As $z \in R^+(a)$ and $v_i \notin R^+(a)$, we have $(z, v_{i+1}) \in A(B)$. If $v_{i+1} \in V(B')$ then we can replace the path $(a, v_1, \ldots, v_i, \ldots, v_q, b)$ with (a, z, v_{i+1}, b) to obtain another set $\mathcal{C}'$ of k arc-disjoint cycles such that $\sum_{C \in \mathcal{C}'} |V(C) \cap V(B')| > \sum_{C \in \mathcal{C}} |V(C) \cap V(B')|$ which leads to a contradiction. If $v_{i+1} \notin V(B')$, then there exists a vertex $r \in R^-(b)$ such that arcs (r, b) and (z, r) are not part of any cycle in $\mathcal{C}$. As $r \in R^-(b)$ and $v_i \notin R^-(b)$, we have $(z, r) \in A(B)$. Hence in this case also we can replace path $(a, v_1, \ldots, v_i, \ldots, v_q, b)$ with (a, z, r, b) to obtain another set $\mathcal{C}'$ of k arc-disjoint cycles such that $\sum_{C \in \mathcal{C}'} |V(C) \cap V(B')| > \sum_{C \in \mathcal{C}} |V(C) \cap V(B')|$ which leads to a contradiction.

Case ($a \in Y$ and $b \in Y$): We know that $(v_i, b) \in A(B)$ and $(a, v_i) \in A(B)$. As $v_i \notin V(B')$ there exists a vertex $z \in R^+(a)$ such that arcs (a, z) and (z, b) are not part of any cycle in $\mathcal{C}$. As $z \in R^+(a)$ and $v_i \notin R^+(a)$, we have $(z, b) \in A(B)$. We can replace the path $(a, v_1, \ldots, v_i, \ldots, v_q, b)$ with (a, z, b) to obtain another set $\mathcal{C}'$ of k arc-disjoint cycles such that $\sum_{C \in \mathcal{C}'} |V(C) \cap V(B')| > \sum_{C \in \mathcal{C}} |V(C) \cap V(B')|$ which leads to a contradiction.

Case ($a \in X$ and $b \in X$): In this particular case, we have $v_{i-1}, v_{i+1} \in Y$. Also, $(a, v_{i-1}), (v_{i-1}, b), (a, v_{i+1}), (v_{i+1}, b) \in A(B)$. If $v_{i-1} \in V(B')$ (or $v_{i+1} \in V(B')$) then we can replace $(a, v_1, \ldots, v_i, \ldots, v_q, b)$ with (a, v_{i-1}, b) (or (a, v_{i+1}, b)) to obtain another set $\mathcal{C}'$ of k arc-disjoint cycles such that $\sum_{C \in \mathcal{C}'} |V(C) \cap V(B')| > \sum_{C \in \mathcal{C}} |V(C) \cap V(B')|$ which leads to a contradiction. If $v_{i-1}, v_{i+1} \notin V(B')$, then there exists a vertex $z \in R^+(a)$ such that (a, z) and (z, b) are not part of any cycle in $\mathcal{C}$. As $z \in R^+(a)$, $v_{i-1} \notin R^+(a)$ and $(v_{i-1}, b) \in A(B)$, we have $(z, b) \in A(B)$. Then we can replace $(a, v_1, \ldots, v_i, \ldots, v_q, b)$ with (a, z, b) to obtain another set $\mathcal{C}'$ of k arc-disjoint cycles such that $\sum_{C \in \mathcal{C}'} |V(C) \cap V(B')| > \sum_{C \in \mathcal{C}} |V(C) \cap V(B')|$ which leads to a contradiction. □

Thus, we have the following result.

Theorem 3 (restated). ACBT *admits a kernel with* $\mathcal{O}(k^2)$ *vertices.*

Proof. Given an instance $(B[X, Y], k)$ of ACBT, return (B', k) as the resulting instance where B' is as defined above. From Lemma 3 and from the fact that B' has $\mathcal{O}(k^2)$ vertices, it follows that (B', k) is a kernel of (B, k). □

6   An Improved FPT Algorithm

In this section, we show that ACBT can be solved in $\mathcal{O}^\star(2^{\mathcal{O}(k \log k)})$ time. The idea is to reduce the problem to the following ARC-DISJOINT PATHS problem in directed acyclic graphs.

ARC-DISJOINT PATHS

Input: A digraph D on n vertices and k ordered pairs $(s_1, t_1), \ldots, (s_k, t_k)$ of vertices of D.

Question: Do there exist arc-disjoint paths $P_1, \ldots, P_k$ in D such that P_i is a path from s_i to t_i for each $i \in [k]$?

Parameter: k

On directed acyclic graphs, ARC-DISJOINT PATHS is known to be NP-complete [11], W[1]-hard [25] and solvable in $n^{\mathcal{O}(k)}$ time [14]. Despite its fixed-parameter intractability, we will show that we can use the $n^{\mathcal{O}(k)}$ algorithm to describe another (and faster) FPT algorithm for ACBT.

Theorem 1 (restated). ACBT *can be solved in* $\mathcal{O}^\star(2^{\mathcal{O}(k \log k)})$ *time.*

Proof. Consider an instance (B, k) of ACBT. Using Theorem 3, we obtain a kernel $\mathcal{I} = (\widehat{B}, \widehat{k})$ such that $\widehat{B}$ has $\mathcal{O}(k^2)$ vertices in polynomial time. Further, $\widehat{k} = k$. By definition, (B, k) is a yes-instance if and only if $(\widehat{B}, \widehat{k})$ is a yes-instance. Using Theorem 2, we know that $\widehat{B}$ either contains $\widehat{k}$ arc-disjoint 4-cycles or has a feedback arc set of size at most $8(\widehat{k} - 1)$ that can be obtained in polynomial time. If Theorem 2 returns a set of $\widehat{k}$ arc-disjoint 4-cycles in $\widehat{B}$, then we declare that (B, k) is a yes-instance.

Otherwise, let $\widehat{F}$ be the feedback arc set of size at most $8(\widehat{k} - 1)$ returned by Theorem 2. Let D denote the (acyclic) digraph obtained from $\widehat{B}$ by deleting $\widehat{F}$. Observe that D has $\mathcal{O}(k^2)$ vertices. Suppose $\widehat{B}$ has a set $\mathcal{C} = \{C_1, \ldots, C_{\widehat{k}}\}$ of $\widehat{k}$ arc-disjoint cycles. For each $C \in \mathcal{C}$, we know that $A(C) \cap \widehat{F} \neq \emptyset$ as $\widehat{F}$ is a feedback arc set of $\widehat{B}$. We can guess that subset F of $\widehat{F}$ such that $F = \widehat{F} \cap A(\mathcal{C})$. Then, for each cycle $C_i \in \mathcal{C}$, we can guess the arcs F_i from F that it contains and also the order σ_i in which they appear. This information is captured as a partition $\mathcal{F}$ of F into $\widehat{k}$ sets, F_1 to $F_{\widehat{k}}$ and the set $\{\sigma_1, \ldots, \sigma_{\widehat{k}}\}$ of permutations where σ_i is a permutation of F_i for each $i \in [\widehat{k}]$. Any cycle C_i that has $F_i \subseteq F$ contains a (v, x)-path between every pair (u, v), (x, y) of consecutive arcs of F_i with arcs from $A(D)$. That is, there is a path from $\mathrm{h}(\sigma_i^{-1}(j))$ and $\mathrm{t}(\sigma_i^{-1}((j+1) \bmod |F_i|))$ with arcs from D for each $j \in [|F_i|]$. The total number of such paths in these $\widehat{k}$ cycles is $\mathcal{O}(|F|)$ and the arcs of these paths are contained in D which is a (simple) directed acyclic graph.

The number of choices for F is $2^{|\widehat{F}|}$ and the number of choices for a partition $\mathcal{F} = \{F_1, \ldots, F_{\widehat{k}}\}$ of F and a set $X = \{\sigma_1, \ldots, \sigma_{\widehat{k}}\}$ of permutations is $2^{\mathcal{O}(|\widehat{F}| \log |\widehat{F}|)}$. Once such a choice is made, the problem of finding $\widehat{k}$ arc-disjoint cycles in $\widehat{B}$ reduces to the problem of finding $\widehat{k}$ arc-disjoint cycles

$\mathcal{C} = \{C_1, \ldots, C_{\widehat{k}}\}$ in $\widehat{B}$ such that for each $1 \leq i \leq \widehat{k}$ and for each $1 \leq j \leq |F_i|$, C_i has a path P_{ij} between $\text{h}(\sigma_i^{-1}(j))$ and $\text{t}(\sigma_i^{-1}((j+1) \mod |F_i|))$ with arcs from $D = \widehat{B} - \widehat{F}$. This problem is essentially finding $r = \mathcal{O}(|\widehat{F}|)$ arc-disjoint paths in D and can be solved in $|V(D)|^{\mathcal{O}(r)}$ time using the algorithm in [14]. Therefore, the overall running time of the algorithm is $\mathcal{O}^{\star}(2^{\mathcal{O}(k \log k)})$ as $|V(D)| = \mathcal{O}(k^2)$ and $r = \mathcal{O}(k)$. □

7 Conclusion

We initiated the parameterized complexity study of the problem of packing arc-disjoint cycles in bipartite tournaments. We showed that it is FPT when parameterized by the solution size and admits a quadratic vertex kernel. However, the classical complexity status of the problem is still open, i.e, we do not know if it is NP-hard or not. Resolving the same is a natural future research direction. We conjecture that it is indeed NP-hard.

References

1. Alon, N.: Ranking tournaments. SIAM J. Discrete Math. **20**(1), 137–142 (2006)
2. Alon, N., Lokshtanov, D., Saurabh, S.: Fast FAST. In: 36th International Colloquium on Automata, Languages, and Programming (ICALP), pp. 49–58 (2009)
3. Alon, N., Yuster, R., Zwick, U.: Color-coding. J. ACM **42**(4), 844–856 (1995)
4. Bessy, S., et al.: Packing arc-disjoint cycles in tournaments. In: 44th International Symposium on Mathematical Foundations of Computer Science (MFCS 2019), pp. 27:1–27:14 (2019)
5. Bessy, S., et al.: Kernels for feedback arc set in tournaments. J. Comput. Syst. Sci. **77**(6), 1071–1078 (2011)
6. Charbit, P., Thomassé, S., Yeo, A.: The minimum feedback arc set problem is NP-hard for tournaments. Comb. Probab. Comput. **16**(1), 1–4 (2007)
7. Conitzer, V.: Computing slater rankings using similarities among candidates. In: 21st National Conference on Artificial Intelligence, vol. 1. pp. 613–619 (2006)
8. Cygan, M., et al.: Parameterized Algorithms. Springer, Cham (2015). https://doi.org/10.1007/978-3-319-21275-3
9. Downey, R.G., Fellows, M.R.: Fundamentals of Parameterized Complexity. Springer, London (2013). https://doi.org/10.1007/978-1-4471-5559-1
10. Erdős, P., Pósa, L.: On independent circuits contained in a graph. Can. J. Math. **17**, 347–352 (1965)
11. Even, S., Itai, A., Shamir, A.: On the complexity of timetable and multicommodity flow problems. SIAM J. Comput. **5**(4), 691–703 (1976)
12. Flum, J., Grohe, M.: Parameterized Complexity Theory. Springer, Heidelberg (2006). https://doi.org/10.1007/3-540-29953-X
13. Fomin, F., Pilipczuk, M.: Subexponential parameterized algorithm for computing the cutwidth of a semi-complete digraph. In: 21st Annual European Symposium on Algorithms (ESA 2013), vol. 8125, pp. 505–516 (2013)
14. Fortune, S., Hopcroft, J., Wyllie, J.: The directed subgraph homeomorphism problem. Theor. Comput. Sci. **10**(2), 111–121 (1980)

15. Guo, J., Hüffner, F., Moser, H.: Feedback arc set in bipartite tournaments is NP-complete. Inf. Process. Lett. **102**(2), 62–65 (2007)

16. Karpinski, M., Schudy, W.: Faster algorithms for feedback arc set tournament, kemeny rank aggregation and betweenness tournament. In: Cheong, O., Chwa, K.-Y., Park, K. (eds.) ISAAC 2010. LNCS, vol. 6506, pp. 3–14. Springer, Heidelberg (2010). https://doi.org/10.1007/978-3-642-17517-6_3

17. Kemeny, J.: Mathematics without numbers. Daedalus **88**(4), 577–591 (1959)

18. Kemeny, J., Snell, J.: Mathematical Models in the Social Sciences. Blaisdell, New York (1962)

19. Kenyon-Mathieu, C., Schudy, W.: How to rank with few errors. In: Proceedings of the 39th Annual ACM Symposium on Theory of Computing (STOC), pp. 95–103 (2007)

20. Krivelevich, M., Nutov, Z., Salavatipour, M.R., Yuster, J.V., Yuster, R.: Approximation algorithms and hardness results for cycle packing problems. ACM Trans. Algorithms **3**(4), 48 (2007)

21. Lokshtanov, D., Mouawad, A., Saurabh, S., Zehavi, M.: Packing cycles faster Than Erdős-Pósa. In: 44th International Colloquium on Automata, Languages, and Programming (ICALP), pp. 71:1–71:15 (2017)

22. Naor, M., Schulman, L.J., Srinivasan, A.: Splitters and near-optimal derandomization. In: Proceedings of IEEE 36th Annual Foundations of Computer Science, pp. 182–191 (1995)

23. Paul, C., Perez, A., Thomassé, S.: Conflict Packing yields linear vertex-kernels for Rooted Triplet Inconsistency and other problems. CoRR abs/1101.4491 (2011)

24. Sanghvi, B., Koul, N., Honavar, V.: Identifying and eliminating inconsistencies in mappings across hierarchical ontologies. In: Meersman, R., Dillon, T., Herrero, P. (eds.) OTM 2010. LNCS, vol. 6427, pp. 999–1008. Springer, Heidelberg (2010). https://doi.org/10.1007/978-3-642-16949-6_24

25. Slivkins, A.: Parameterized tractability of edge-disjoint paths on directed acyclic graphs. SIAM J. Discrete Math. **24**(1), 146–157 (2010)

26. Xiao, M., Guo, J.: A quadratic vertex kernel for feedback arc set in bipartite tournaments. Algorithmica **71**(1), 87–97 (2015)

27. van Zuylen, A.: Linear programming based approximation algorithms for feedback set problems in bipartite tournaments. Theoret. Comput. Sci. **412**(23), 2556–2561 (2011)

Matching Random Colored Points
with Rectangles

Josué Corujo[1,2], David Flores-Peñaloza[3], Clemens Huemer[4], Pablo
Pérez-Lantero[5], and Carlos Seara[4(✉)]

[1] Université Paris-Dauphine, Paris, France
`corujo@ceremade.dauphine.fr`
[2] Facultad de Matemática y Computación, Universidad de La Habana,
Havana, Cuba
[3] Departamento de Matemáticas, Facultad de Ciencias, Universidad Nacional
Autónoma de México, Mexico City, Mexico
`dflorespenaloza@gmail.com`
[4] Departament de Matemàtiques, Universitat Politècnica de Catalunya,
Barcelona, Spain
`clemens.huemer@upc.edu, carlos.seara@upc.edu`
[5] Departamento de Matemática y Ciencia de la Computación, USACH,
Santiago, Chile
`pablo.perez.l@usach.cl`

Abstract. Let $S \subset [0,1]^2$ be a set of n points, randomly and uniformly
selected. Let $R \cup B$ be a random partition, or coloring, of S in which each
point of S is included in R uniformly at random with probability $1/2$. We
study the random variable $M(n)$ equal to the number of points of S that
are covered by the rectangles of a maximum strong matching of S with
axis-aligned rectangles. The matching consists of closed rectangles that
cover exactly two points of S of the same color. A matching is strong
if all its rectangles are pairwise disjoint. We prove that almost surely
$M(n) \geq 0.83\,n$ for n large enough. Our approach is based on modeling
a deterministic greedy matching algorithm, that runs over the random
point set, as a Markov chain.

D. Flores-Peñaloza—Research supported by project PAPIIT IN117317 (UNAM, Mexico).

C. Huemer—Research supported by projects MTM2015-63791-R (MINECO/FEDER)
and Gen. Cat. DGR 2017SGR1336.

P. Pérez-Lantero—Partially supported by projects CONICYT FONDECYT/Regular
1160543 (Chile), DICYT 041933PL Vicerrectoría de Investigación, Desarrollo e Innovación USACH (Chile), and Programa Regional STICAMSUD 19-STIC-02.

C. Seara—Research supported by projects MTM2015-63791-R MINECO/FEDER and
Gen. Cat. DGR 2017SGR1640.

 This work has received funding from the European Union's Horizon 2020
research and innovation programme under the Marie Skłodowska-Curie
grant agreement No 734922.

© Springer Nature Switzerland AG 2020
M. S. Rahman et al. (Eds.): WALCOM 2020, LNCS 12049, pp. 261–272, 2020.
https://doi.org/10.1007/978-3-030-39881-1_22

1   Introduction

Given a point set $S \subset \mathbb{R}^2$ of n points, and a class $\mathcal{C}$ of geometric objects, a *geometric matching* of S is a set $M \subseteq \mathcal{C}$ such that each element of M contains exactly two points of S and every point of S lies in at most one element of M. A geometric matching is *strong* if the geometric objects are pairwise disjoint, and *perfect* if every point of S belongs to (or is covered by) some element of M. This type of geometric matching problems was considered by Ábrego et al. [1], who studied the existence and properties of matchings for point sets in the plane when $\mathcal{C}$ is the class of axis-aligned squares, or the class of disks.

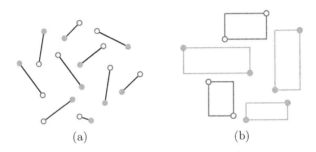

(a)   (b)

Fig. 1. (a) A perfect, strong bichromatic matching of 10 red points and 10 blue points with segments. (b) A perfect, strong monochromatic matching of 6 red points and 4 blue points with axis-aligned rectangles. (Color figure online)

Let $S = R \cup B \subset \mathbb{R}^2$ be a set of n colored points in the plane, each point colored red or blue, where R and B are the sets of red and blue points, respectively. A geometric matching of S is called *monochromatic* if all matching objects cover points of the same color, and *bichromatic* if all matching objects cover points of different colors. For example, monochromatic matchings of two-colored point sets in the plane with straight segments have been studied [4,5]. In the case of bichromatic matchings with straight segments, a classical result in discrete geometry asserts that for any planar point set S consisting of n red points and n blue points in general position (i.e., no three points of S are collinear) there exists a perfect, strong bichromatic matching of S with straight segments [6] (see Fig. 1a).

   In this paper, we consider strong monochromatic matchings with axis-aligned rectangles. Refer to Fig. 1b for an example of a perfect matching of this type. Throughout the paper, every rectangle will be considered axis-aligned and a closed subset of the plane.

   Caraballo et al. [2] studied both monochromatic and bichromatic strong matchings of S with rectangles from the algorithmic point of view. That is, they studied two combinatorial optimization problems for given $S = R \cup B$: find a monochromatic strong matching of S with the maximum number of rectangles, and find a bichromatic strong matching of S with the maximum number of rectangles; proving that both problems are NP-hard and giving a polynomial-time 4-approximation algorithm in each case. As noted by Caraballo et al., these two problems are special cases of the Maximum Independent Set of Rectangles problem (MISR): Given a finite set $\mathcal{R}$ of rectangles in the plane, find a subset $\mathcal{R}' \subseteq \mathcal{R}$ of maximum cardinality, denoted $\alpha(\mathcal{R})$, such that every pair of rectangles in $\mathcal{R}'$ are disjoint.

Indeed, suppose that we want to find a monochromatic matching of S with the maximum number of rectangles. For every distinct $p, q \in \mathbb{R}^2$, let $D(p, q)$ denote the minimum axis-aligned rectangle (i.e., the rectangle such that both dimensions are minimum) that encloses p and q. Let $\mathcal{R}(S)$ be the set of all rectangles $D(p, q)$ such that $p, q \in S$, p and q have the same color, and $D(p, q)$ contains no points of S different from p and q. Finding a monochromatic strong matching of S with the maximum number of rectangles is equivalent to finding in $\mathcal{R}(S)$ a maximum subset of pairwise disjoint rectangles, whose size is $\alpha(\mathcal{R}(S))$, that is, solving the MISR problem in $\mathcal{R}(S)$.

In this paper, we study monochromatic strong matchings of S with rectangles from the combinatorial point of view. From this point forward, every rectangle will cover precisely two points of S. Point sets $S = R \cup B$ exist in which no matching rectangle is possible (e.g., S is a color-alternating sequence of points on the line $y = x$), and point sets in which a perfect strong matching with rectangles exists (e.g., an even number of red points in the negative part of the line $y = x$, and an even number of blue points in the positive part). These two extreme cases show that it is not worth studying the number $\alpha(\mathcal{R}(S))$ for fixed, or given, colored point sets S. This does not happen, for example, for monochromatic strong matchings with segments for fixed sets of red and blue points: Dumitrescu and Kaye [4] proved that every two-colored point set $S = R \cup B$ of n points in general position admits a strong matching with segments that covers at least $\frac{6}{7}n - O(1)$ of the points; furthermore, there exist n-point sets such that every strong matching with segments covers at most $\frac{94}{95}n + O(1)$ points. Instead, we want to study $\alpha(\mathcal{R}(S))$ when S is a random point set in the square $[0, 1]^2$, in which the positions of the n points of S are random and the color of each point of S is also random. Formally:

Let $n > 0$ be an integer, and let $S \subset [0, 1]^2$ be a set of n points, randomly and uniformly selected. Let $R \cup B$ be a random partition (i.e., coloring) of S in which each point of S is included in R uniformly at random with probability $1/2$. We study the random variable $M(n) = 2 \cdot \alpha(\mathcal{R}(S))$ equal to the number of points of S that are covered by the rectangles of a maximum monochromatic strong matching of S with rectangles.

Given a set S of n points, randomly and uniformly selected in the square $[0, 1]^2$, Chen et al. [3] studied a similar variable: the random variable $\alpha(D(S))$, where $D(S)$ is the random graph with vertex set S and two points $p, q \in S$ define an edge if and only if $D(p, q) \cap S = \{p, q\}$. Here, $\alpha(D(S))$ denotes the size of a maximum independent set of $D(S)$.

One result of Chen et al. [3, Theorem 1] states that if n tends to infinity, then $\alpha(D(S)) = O(n(\log^2 \log n)/\log n)$ with probability tending to 1. This result implies that if $C(n)$ denotes the number of points of S that are covered by a maximum monochromatic matching of S with rectangles, where the rectangles may overlap (i.e., the matching is not necessarily strong), then $C(n) = n - o(n)$ with probability tending to 1. In fact, let M' be a maximum monochromatic matching of S with rectangles, where M' is not necessarily strong, and let $S' \subset S$ be the points not covered by M'. Note that at least $|S'|/2$ points of S' have the same color, and they form an independent set in the graph $D(S)$. Then, with probability tending to 1, we have that M' covers at least $n - |S'| = n - O(n(\log^2 \log n)/\log n) = n - o(n)$ points.

2   Preliminaries

Since for matching S with rectangles, only the left-to-right and bottom-to-top orders of S are relevant, and since the probability that two points of S are in the same vertical or horizontal line is zero, we consider S equal to the point set $S_\pi = \{(i, \pi(i)) \mid i = 1, 2, \ldots, n\}$, where $\pi : \{1, 2, \ldots, n\} \to \{1, 2, \ldots, n\}$ is a randomly and uniformly selected permutation. This assumption was also done by Chen et al. [3].

We have implemented a **Python** program that, given n, generates a uniform random permutation π, and selects the color of each $p \in S_\pi$ (red or blue) randomly and uniformly with probability $1/2$. The program then runs a deterministic algorithm on $S_\pi = R \cup B$ that greedily finds a maximum independent subset of rectangles in $\mathcal{R}(S_\pi)$. The greedy algorithm iterates the points of S_π from left to right, and for each point p in the iteration, it performs the following action: If p is not matched with any point prior to p in the iteration, it finds (if it exists) the first point q to the right of p such that $D(p, q) \in \mathcal{R}(S_\pi)$ and $D(p, q)$ has empty intersection with all matching rectangles already reported. If q exists, then the algorithm reports $D(p, q)$ as a matching rectangle. In any case, regardless of whether q exists, the algorithm continues the iteration to the next unmatched point p.

For large n, say $n = 10000$, the implemented algorithm reports a matching covering approximately $\frac{97}{100}n$ of the points. In fact, we run the algorithm $N = 100$ times for $n = 10000$, and average the outputs (i.e., the percentage of matched points) and computed the standard deviation. The average output is 0.978 and the standard deviation 0.0022. See in Table 1 the row $k = \infty$, where k is a parameter that will be explained later.

Then, it seems that $M(n) \geq \frac{97}{100}n$ for n large enough and probability close to 1. More formally, using the Central Limit theorem, we have that

$$\left(0.9780 - \frac{0.0022}{\sqrt{N}} z_{0.99}, 1\right] \subset (0.97, 1]$$

is a 99% confidence interval for the expected value of $M(n)/n$. We denote by $z_{0.99} \approx 2.33$ the real value which satisfies $\mathrm{Prob}(Z \leq z_{0.99}) = 0.99$, for Z a normal random variable with mean 0 and variance 1.

Table 1. The table shows the experimental results obtained when running the greedy matching algorithm for $n \in \{1000, 10000\}$ points, parameterized with $k \in \{1, 2, 3, \ldots, 8\}$, or not parameterized ($k = \infty$). For each combination of n, k, we run the algorithm 100 times, and measured the mean and standard deviation of the ratio between the total number of matched points and n.

k	$n = 1000$		$n = 10000$	
	mean	sdev	mean	sdev
1	0.6653	0.0175	0.6673	0.0052
2	0.7948	0.0104	0.7934	0.0036
3	0.8301	0.0097	0.8304	0.0034
4	0.8555	0.0094	0.8562	0.0028
5	0.8727	0.0090	0.8736	0.0026
6	0.8860	0.0087	0.8864	0.0026
7	0.8953	0.0084	0.8962	0.0026
8	0.9031	0.0079	0.9041	0.0025
∞	0.9724	0.0062	0.9780	0.0022

Analyzing the algorithm, when run over the random S_π, seems to be a good approach for obtaining a high lower bound for $M(n)$. One way to analyze the algorithm is to consider a parameterized version of it, with a parameter k, such that each unmatched point p finds its match point q among only the next k points of S_π to the right of p. Let $\mathcal{A}_k$ denote this parameterized algorithm. For further experimental results, see Table 1.

In the next two sections, we show how to model (an adaptation of) $\mathcal{A}_k$ as a Markov chain, for any fixed $k \in \{1, 2, 3, \ldots\}$. Then, we show that the algorithm $\mathcal{A}_3$ almost surely guarantees $M(n) \geq \frac{83}{100}n$, for n large enough, by computing the stationary distribution of the Markov chain and applying the Ergodic theorem. For the theory on Markov chains, refer to Norris [7].

3   The Markov Chains

From this point forward, we also consider $S = S_\pi$, and whenever we say point i, for $i \in \{1, 2, \ldots, n\}$, or just i when it is clear from the context, we are referring to the point $p_i := (i, \pi(i)) \in S$. Let $color(i) \in \{R, B\}$ be the color of point i.

Let $k \in \{1, 2, 3, \ldots\}$ be a constant, and let $\tilde{\mathcal{A}}_k$ be the following adaptation of algorithm $\mathcal{A}_k$, consisting in the next idea:

Suppose that $\mathcal{A}_k$ matches points i and j, with $i < j \leq i + k$, when the iteration of S_π is on point i. Algorithm $\tilde{\mathcal{A}}_k$ iterates S_π from left to right, and will also match i and j but, in contrast with $\mathcal{A}_k$, when the iteration is on j, or on a point to the right of j. Using $\tilde{\mathcal{A}}_k$ instead of $\mathcal{A}_k$, allows us to describe in a more compact way the states of the memory of the algorithm during the iteration of the elements of S_π.

Let $E(j)$ be the data structure associated with point $j \in \{1, 2, \ldots, n\}$, that is maintained by $\tilde{\mathcal{A}}_k$ during the iteration of S_π. For any j, let $i = i(j)$ be the smallest element in the set $\{\max(1, j - (k - 1)), \ldots, j\}$ such that the point i is not matched, and each point in $\{i + 1, \ldots, j\}$ is matched with a point to the left of i or is not yet matched. If i exists, then $E(j)$ consists of the following elements:

- The set $U(j) \subseteq \{i, i + 1, \ldots, j\}$ of the points that are not matched, with $i \in U(j)$.
- The set $\mathrm{Rect}(j)$ of the (pairwise disjoint) rectangles that match the points in $\{i + 1, \ldots, j\} \backslash U(j)$ with points to the left of i.
- The number $f(j)$ of points of S_π that are matched while the iteration is at point j.

Otherwise, if i does not exist, then $E(j)$ consists of the same three above elements with $U(j) = \emptyset$ and $\mathrm{Rect}(j) = \emptyset$.

For $j = 1$, we have $U(1) = \{1\}$, $\mathrm{Rect}(1) = \emptyset$, and $f(1) = 0$. We show now how to obtain $E(j + 1)$ from $E(j)$, for any $j \in \{1, \ldots, n - 1\}$.

First, we match points i and $j + 1$ if and only if $j + 1 \leq i + k$, $\mathrm{color}(i) = \mathrm{color}(j + 1)$, and the rectangle $D(p_i, p_{j+1})$ does not overlap any rectangle in $\mathrm{Rect}(j)$.

After that, we match other points in $(U(j) \backslash \{i\}) \cup \{j + 1\}$ if and only if i was matched in the previous step, or we have finished with point i. We say that we have *finished* with point i if there do not exist more chances for point i to be matched, which is equivalent to $i + k \leq j + 1$. This final matching procedure consists in running the original algorithm $\mathcal{A}_k$ with input the points $\{i + 1, \ldots, j, j + 1\}$, but with the extra condition that the algorithm terminates if the current point t on the iteration of $\{i + 1, \ldots, j, j + 1\}$ from left to right, cannot be matched with any other one to its right (i.e., the matching on points in $(U(j) \backslash \{i\}) \cup \{j + 1\}$ is performed by running the original algorithm $\mathcal{A}_k$). This is because t must find its match among the points in $\{j + 2, \ldots, t + k\}$, before any matching between points in $\{t + 1, \ldots, j + 1\}$ occurs.

We set $f(j + 1)$ equal to the total number of points matched at iteration j. Obtaining $U(j + 1)$ and $\mathrm{Rect}(j + 1)$ is straightforward.

Let $j \in \{1, 2, \ldots, n\}$. The *state* of $E(j)$ is a 2-tuple formed by:

As first component, (a certificate of) the relative positions between the points of $U(j)$ and the rectangles of $\mathrm{Rect}(j)$, together with the color of each point of $U(j)$. If the leftmost point is colored blue, then we switch the color of every point such that the leftmost one is always red.

As second component, the number $f(j)$ of matched points. We say that two states e and e' are *equal* (i.e., $e = e'$) if: (i) the first components are equal, or one first component is symmetric to the other in the vertical direction, and (ii) the second components are equal.

Let $\mathcal{E} = \{e_1, e_2, \ldots, e_N\}$ denote the set of all possible states of $E(j)$, which is a finite set, and let $X_j \in \mathcal{E}$ be the random variable equal to the state of $E(j)$. Let $e \in \mathcal{E}$ be a state, and assume that e is the state of $E(j)$ for some j.

Let $f(e) = f(j)$ (with abuse of notation), and let $N(e)$ be the *neighborhood* of e, which is the multiset consisting of the state of $E(j+1)$ for every color and every different relative position, with respect to the elements of both $U(j)$ and $\text{Rect}(j)$, of point $j+1$. See for example Fig. 2.

Lemma 1. *Let $e, e' \in \mathcal{E}$ be two states. For every $j \geq 2$, we have:*

$$\text{Prob}(X_{j+1} = e' \mid X_j = e) = \frac{m}{2\left(|U(j)| + 2|\text{Rect}(j)| + 1\right)},$$

where m is the multiplicity of e' in $N(e)$.

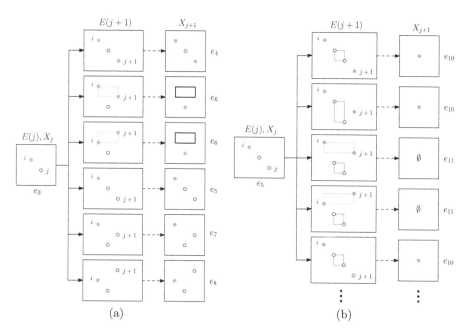

(a)   (b)

Fig. 2. (a) Example of the data structure $E(j)$, its state $X_j = e_3$, and the states in the neighborhood $N(e_3) = \{e_4, e_5, e_6, e_6, e_7, e_8\}$ corresponding to $E(j+1)$, for each position and color of point $j+1$. Note that $f(e_6) = 2$, and $f(e_i) = 0$ for all $e_i \in \{e_4, e_5, e_7, e_8\}$. (b) Example of $E(j)$ and its state $X_j = e_5$, with $N(e_5) = \{e_{10}, e_{10}, e_{10}, e_{10}, e_{10}, e_{10}, e_{11}, e_{11}\}$. Note that $f(e_{10}) = 2$ and $f(e_{11}) = 4$. For every of the four positions of the blue point $j+1$, the resulting state is e_{10}.

Proof. Through each point of $U(j)$ draw a horizontal line, and for each rectangle of $\text{Rect}(j)$ draw a horizontal line through the top side and a horizontal line through the bottom side. Each of these $K = |U(j)| + 2|\text{Rect}(j)|$ lines goes through a different element of S_π, and they subdivide the plane into $K + 1$ strips. Since the point $j+1$ is to the right of both every point of $U(j)$ and every rectangle of

Rect(j), the relative position of point $j+1$ with respect to the elements of $U(j)$ and Rect(j) is to be in one of these strips, and this happens with probability $1/(K+1)$. Furthermore, the color of point $j+1$ is given with probability $\frac{1}{2}$. The lemma follows.   □

Note that the value of X_{j+1} depends on the value of X_j, and does not depend in any of the values of $X_1, X_2, \ldots, X_{j-1}$. Formally,

$$\text{Prob}(X_{j+1} = x_{j+1} \mid X_j = x_j, \ldots, X_1 = x_1) = \text{Prob}(X_{j+1} = x_{j+1} \mid X_j = x_j)$$

for all $x_1, \ldots, x_{j+1} \in \mathcal{E}$ such that $\text{Prob}(X_j = x_j, \ldots, X_1 = x_1) > 0$.

Thus, the sequence $(X_n)_{n \geq 1}$ is a Markov chain, denoted $\mathcal{C}_k$, over the set $\mathcal{E} = \{e_1, e_2, \ldots, e_N\}$ of states. Let P denote the transition matrix, of dimensions $N \times N$, such that $P_{i,j} = \text{Prob}(X_{\ell+1} = e_j \mid X_\ell = e_i)$. The key observation is that the total number of points matched by the algorithm $\tilde{A}_k$, denoted $M_k(n)$, is precisely

$$M_k(n) = \sum_{j=1}^{n} f(X_j).$$

A Markov chain is *irreducible* if with positive probability any state can be reached from any other state [7]. We need that this property holds in $\mathcal{C}_k$, as stated in the next lemma.

Lemma 2. *The Markov chain $\mathcal{C}_k$ is irreducible.*

Proof. Assume without loss of generality that e_1 is the state of $E(1)$, which consists of a single red point and ensures $f(e_1) = 0$. Let $e \in \mathcal{E} \setminus \{e_1\}$ be any other state, which by definition is the state of $E(j)$ for some j. Then, in $\mathcal{C}_k$ the state e can be reached from e_1 with positive probability (Lemma 1).

We prove now that also with positive probability, the state e_1 can be reached from e, which implies that $\mathcal{C}_k$ is irreducible. Note that with positive probability, the point $j+1$ may be matched with point $i(j)$ in $E(j+1)$. Then, for some point $t \geq j+1$ we have with positive probability that for $\ell = j+1, \ldots, t$ the points $i(\ell-1)$ and ℓ are matched in $E(\ell)$, and $U(t) = \emptyset$ and Rect(t) = $\emptyset$.

Let e' denote the state of $E(t)$, and we have $\text{Prob}(X_{t+1} = e_1 \mid X_t = e') = 1$. Hence, the state e_1 can be reached from e with positive probability, and $\mathcal{C}_k$ is thus irreducible.   □

Since $\mathcal{C}_k$ is irreducible (Lemma 2) and has a finite set of states, it has a unique stationary distribution $s = (s_1, s_2, \ldots, s_N)$, which is the solution of the system

$$s = s \cdot P, \quad s_1 + s_2 + \cdots + s_N = 1$$

of linear equations [7]. Furthermore, since $f(e) \in \{0, 2, 4, \ldots, 2\lceil \frac{k+1}{2} \rceil\}$ for all $e \in \mathcal{E}$, the function f is bounded and then the Ergodic theorem ensures

$$\lim_{n \to \infty} \frac{M_k(n)}{n} = \lim_{n \to \infty} \frac{1}{n} \sum_{j=1}^{n} f(X_j) = \sum_{i=1}^{N} s_i f(e_i),$$

almost surely [7]. Let $\alpha_k = \sum_{i=1}^{N} s_i f(e_i)$. We then arrive to the main result of this paper:

Theorem 1. *Let $\pi : \{1, 2, \ldots, n\} \to \{1, 2, \ldots, n\}$ be a uniform random permutation. Let*

$$S_\pi = \{(i, \pi(i)) \mid i = 1, 2, \ldots, n\}$$

be a random point set, where the color (red or blue) of each point of S_π is selected randomly and uniformly with probability $1/2$. Let $k \in \{1, 2, 3, \ldots\}$ be a constant. For all constant $\varepsilon > 0$ and n large enough, almost surely the number $M_k(n)$ of points of S_π that are matched by the algorithm $\tilde{A}_k$ satisfies $M_k(n) \geq (\alpha_k - \varepsilon)n$.

4   The Markov Chain for $k = 3$

In this section, we consider the algorithm $\tilde{A}_3$ and give a precise value for α_3. In Table 2, we describe the states, and the transitions between the states, of the Markov chain $\mathcal{C}_3$. The transition matrix P is in Fig. 3.

Since $f(e) = 2$ for all $e \in \{e_2, e_6, e_9, e_{10}, e_{16}, e_{17}, e_{18}\}$, $f(e_{11}) = 4$, $f(e) = 0$ for all other state e, and the stationary distribution $s = (s_1, \ldots, s_{18})$ satisfies

$$s_2 = \frac{167959}{816233}, \quad s_6 = \frac{69640}{816233}, \quad s_9 = \frac{6800}{816233}, \quad s_{10} = \frac{58650}{816233},$$

$$s_{11} = \frac{13600}{816233}, \quad s_{16} = \frac{5950}{816233}, \quad s_{17} = \frac{1360}{816233}, \quad s_{18} = \frac{1190}{816233},$$

we obtain

$$\alpha_3 = 2(s_2 + s_6 + s_9 + s_{10} + s_{16} + s_{17} + s_{18}) + 4s_{11} = \frac{677498}{816233} \approx 0.830030151.$$

By Theorem 1, taking $\varepsilon = \alpha_3 - 0.83 > 0$, for n large enough we have almost surely that

$$M(n) \geq M_3(n) \geq 0.83\, n.$$

It can be noted in Table 1 that in practice this lower bound is satisfied. Then, we obtain our second result:

Theorem 2. *Let $n > 0$ be an integer, and let $S \subset [0, 1]^2$ be a set of n points, randomly and uniformly selected. Let $R \cup B$ be a random partition (i.e., coloring) of S in which each point of S is included in R uniformly at random with probability $1/2$. For n large enough, almost surely we have that the maximum number $M(n)$ of points of S that are covered by a monochromatic strong matching of S with rectangles satisfies $M(n) \geq 0.83\, n$.*

Table 2. The table shows the 18 states of the Markov chain for $k = 3$. In the second column we show the first component of e_i, and in the third column we show the second component $f(e_i)$. In the last column we show the neighbor states of e_i as a list of tuples of the form $(e_j, P_{i,j})$, where $P_{i,j} = \mathrm{Prob}(X_{\ell+1} = e_j \mid X_\ell = e_i) > 0$ is the transition probability from e_i to e_j.

e_i	elem. of e_i	$f(e_i)$	neighbors of e_i
e_1		0	$(e_2, 1/2), (e_3, 1/2)$
e_2	$\emptyset$	2	$(e_1, 1)$
e_3		0	$(e_4, 1/6), (e_5, 1/6), (e_6, 1/3), (e_7, 1/6), (e_8, 1/6)$
e_4		0	$(e_4, 1/8), (e_5, 1/8), (e_6, 3/8), (e_7, 1/8), (e_9, 1/4)$
e_5		0	$(e_{10}, 3/4), (e_{11}, 1/4)$
e_6		2	$(e_2, 1/4), (e_{12}, 1/8), (e_{13}, 1/8), (e_{14}, 1/4), (e_{15}, 1/4)$
e_7		0	$(e_{10}, 3/4), (e_{11}, 1/4)$
e_8		0	$(e_{10}, 3/4), (e_{16}, 1/4)$
e_9		2	$(e_2, 3/10), (e_{12}, 1/10), (e_{14}, 1/5), (e_{15}, 1/5), (e_{17}, 1/5)$
e_{10}		2	$(e_2, 1/2), (e_3, 1/2)$
e_{11}	$\emptyset$	4	$(e_1, 1)$
e_{12}		0	$(e_2, 1/2), (e_3, 3/10), (e_6, 1/5)$
e_{13}		0	$(e_2, 1/2), (e_3, 3/10), (e_6, 1/5)$
e_{14}		0	$(e_2, 3/10), (e_3, 1/2), (e_6, 1/5)$
e_{15}		0	$(e_2, 1/2), (e_3, 3/10), (e_6, 1/5)$
e_{16}		2	$(e_2, 1/5), (e_{12}, 1/10), (e_{13}, 1/10),$ $(e_{14}, 1/10), (e_{15}, 3/10), (e_{18}, 1/5)$
e_{17}		2	$(e_1, 5/6), (e_2, 1/6)$
e_{18}		2	$(e_1, 5/6), (e_2, 1/6)$

	1	2	3	4	5	6	7	8	9	10	11	12	13	14	15	16	17	18
1	0	1/2	1/2	0	0	0	0	0	0	0	0	0	0	0	0	0	0	0
2	1/1	0	0	0	0	0	0	0	0	0	0	0	0	0	0	0	0	0
3	0	0	0	1/6	1/6	1/3	1/6	1/6	0	0	0	0	0	0	0	0	0	0
4	0	0	0	1/8	1/8	3/8	1/8	0	1/4	0	0	0	0	0	0	0	0	0
5	0	0	0	0	0	0	0	0	0	3/4	1/4	0	0	0	0	0	0	0
6	0	1/4	0	0	0	0	0	0	0	0	0	1/8	1/8	1/4	1/4	0	0	0
7	0	0	0	0	0	0	0	0	0	3/4	1/4	0	0	0	0	0	0	0
8	0	0	0	0	0	0	0	0	0	3/4	0	0	0	0	0	1/4	0	0
9	0	3/10	0	0	0	0	0	0	0	0	0	1/10	0	1/5	1/5	0	1/5	0
10	0	1/2	1/2	0	0	0	0	0	0	0	0	0	0	0	0	0	0	0
11	1/1	0	0	0	0	0	0	0	0	0	0	0	0	0	0	0	0	0
12	0	1/2	3/10	0	0	1/5	0	0	0	0	0	0	0	0	0	0	0	0
13	0	1/2	3/10	0	0	1/5	0	0	0	0	0	0	0	0	0	0	0	0
14	0	3/10	1/2	0	0	1/5	0	0	0	0	0	0	0	0	0	0	0	0
15	0	1/2	3/10	0	0	1/5	0	0	0	0	0	0	0	0	0	0	0	0
16	0	1/5	0	0	0	0	0	0	0	0	0	1/10	1/10	1/10	3/10	0	0	1/5
17	5/6	1/6	0	0	0	0	0	0	0	0	0	0	0	0	0	0	0	0
18	5/6	1/6	0	0	0	0	0	0	0	0	0	0	0	0	0	0	0	0

Fig. 3. The transition matrix P of the Markov chain for $k = 3$.

5   Discussion and Open Problems

Using the theory of Markov chains, by modeling a deterministic greedy matching algorithm that runs over the random colored n-point set, we have proved that almost surely at least $0.83\,n$ points can be matched with pairwise disjoint axis-aligned rectangles, that is, $M(n) \geq 0.83\,n$. This lower bound was obtained from the stationary distribution of a Markov chain for the parameter $k = 3$. Building the Markov chain for any $k \geq 4$ and computing its stationary distribution will give, by Theorem 1, higher lower bounds, as the results of Table 1 suggest. Experimental results suggest that we may be able to prove the bound $M(n) \geq 0.97\,n$.

The trivial upper bound for $M(n)$ is $M(n) \leq n$. Obtaining tighter lower and upper bounds for $M(n)$ seems to be more challenging. There are cases in which we must consider matching strategies more general than that of the greedy algorithm. Hence, the main open question here is whether $\lim_{n\to\infty} M(n)/n = 1$ or $\lim_{n\to\infty} M(n)/n < 1$.

References

1. Ábrego, B.M., et al.: Matching points with squares. Discrete Comput. Geom. **41**(1), 77–95 (2009)
2. Caraballo, L.E., Ochoa, C., Pérez-Lantero, P., Rojas-Ledesma, J.: Matching colored points with rectangles. J. Comb. Optim. **33**(2), 403–421 (2017)
3. Chen, X., Pach, J., Szegedy, M., Tardos, G.: Delaunay graphs of point sets in the plane with respect to axis-parallel rectangles. Random Struct. Algorithms **34**(1), 11–23 (2009)

4. Dumitrescu, A., Kaye, R.: Matching colored points in the plane: some new results. Comput. Geom. **19**(1), 69–85 (2001)
5. Dumitrescu, A., Steiger, W.L.: On a matching problem in the plane. Discrete Math. **211**, 183–195 (2000)
6. Larson, L.C.: Problem-Solving Through Problems. Springer, New York (1983). https://doi.org/10.1007/978-1-4612-5498-0
7. Norris, J.R.: Markov Chains, 2nd edn. Cambridge University Press, Cambridge (1998)

Designing Survivable Networks with Zero-Suppressed Binary Decision Diagrams

Hirofumi Suzuki[1,4P](✉), Masakazu Ishihata[2], and Shin-ichi Minato[3]

[1] Graduate School of Information Science and Technology, Hokkaido University, Sapporo, Japan
h-suzuki@ist.hokudai.ac.jp,
[2] NTT Communication Science Laboratories, Kyoto, Japan
ishihata.masakazu@lab.ntt.co.jp
[3] Graduate School of Informatics, Kyoto University, Kyoto, Japan
minato@i.kyoto-u.ac.jp
[4] Fujitsu Laboratories Ltd., Kanagawa, Japan
suzuki-hirofumi@fujitsu.com

Abstract. Various network systems such as communication networks require *survivability* that is tolerance of attacks, failures, and accidents. Designing a network with high survivability is formulated as a *survivable network design problem* (SNDP). The input of the SNDP is a pair of an edge-weighted graph and a *requirement* of topology and survivability. For an edge subset of the graph, if it satisfies the requirement, we call it a *desired edge subset* (DES). The output of the SNDP is the minimum weight DES. Although the SNDP is an optimization problem, to simply solve it is not always desired in terms of the practical use: Designers sometimes want to test multiple DESs including non-optimal DESs, because the theoretical optimal DES is not always the practical best.

In this paper, instead of the optimization, we propose a method to enumerate all DESs with a compact data structure, called the *zero-suppressed binary decision diagram* (ZDD). Obtained ZDDs support practical network design by performing optimization, sampling, and filtering of DESs. The proposed method combines two typical techniques constructing ZDDs, called the *frontier-based search* (FBS) and the *family algebra*, and includes a novel operation on ZDDs. We demonstrate that our method works on various real-world instances of practical scales.

Keywords: Network design · Survivable network · Decision diagram

1 Introduction

Network systems are necessary in various real-life scenarios such as communication, power distribution, and transportation. Generally, they are susceptible to attacks, failures, and accidents. For example, communication networks are demanded to work even if some equipment failure. In the area of network design,

© Springer Nature Switzerland AG 2020
M. S. Rahman et al. (Eds.): WALCOM 2020, LNCS 12049, pp. 273–285, 2020.
https://doi.org/10.1007/978-3-030-39881-1_23

the tolerance to such incidents is defined as the *survivability* [1]. Namely, real-world network systems require high survivability.

The *survivable network design problem* (SNDP) is to design a network system with high survivablity [2]. The input of the SNDP is a pair of an edge-weighted graph and a *requirement*: the edge-weighted graph represents available resources and their costs, and the requirement is a graph property representing required topology and survivability. For an edge subset of the input graph, if it satisfies the requirement, we call it a *desired edge subset* (DES). The output of the SNDP is the minimum weight DES. The literature [2] introduces various requirements of the SNDP for example, *hop-constrained paths*, *edge-disjoint paths*, and *bounded rings*. Figure 1 shows an example SNDP with bounded rings.

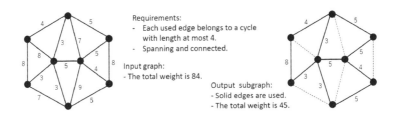

Fig. 1. Example SNDP with bounded rings.

The integer linear programming (ILP) formulation and the branch-and-cut algorithm have been commonly used to solve the SNDP. Such algorithms have been proposed for various settings independently [3–7]. However, they are not always desired in terms of the practical use: Actual network design is an iterative process including topological design, network-synthesis, and network-realization [8], and the SNDP applies to topological design. This implies the following issues.

– In many cases, because mathematical models include some inaccuracy and approximations, the theoretical optimal DES is not always the practical best.
– In many cases, because some requirements are too vague and complex to formalize, it is hard to incorporate such requirements in the SNDP.

Therefore, designers often want to test multiple DESs which may not be optimal.

In this study, instead of finding an optimal DES, we aim to enumerate all the DESs. Since the number of DESs is exponentially huge in general, we should avoid the explicit enumeration and management of DESs. Therefore, we implicitly enumerate DESs by using the *zero-suppressed binary decision diagram* (ZDD) [9], which is a compressed and reusable data structure representing set families including DESs. Once a ZDD of all DESs is obtained, we can use various practical operations over the ZDD such as obtaining an optimal DES, uniform sampling of DESs, and filtering DESs by specified conditions. They are useful to solve not only the SNDP but also the issues above as follows.

– By sampling of DESs, we obtain various DESs which may not be optimal.

- By filtering DESs iteratively, we have a chance to obtain much better DESs fitting non-formalized requirements.

Therefore, designer can easily test multiple DESs with various possibilities.

Our main contribution is to propose a practical method to construct ZDDs for DESs. For the ZDD construction, there are two representative approaches which are based on the *family algebra* [10] and the *frontier-based search* (FBS) [11], respectively. The family algebra is the general term of set family operations which are efficiently executed on ZDDs. The FBS is a direct construction method for ZDDs representing specific constrained subgraphs. Historically, they have often been used for separate purposes, for example, the maximal clique enumeration by the family algebra [12] and the graph partition enumeration by the FBS [13]. In contrast, the proposed method uses both of the family algebra and the FBS. The main idea is to describe DESs by the set family operations in the family algebra where the FBS can construct a ZDD for each set family. Moreover, to partially accelerate the method, we also propose a novel set family operation executed on ZDDs named the *multinomial disjoint join*, which is the multinomial version of a binary operation called the *disjoint join* [14].

We conduct computational experiments to evaluate the proposed method. In our purpose, the multinomial disjoint join is empirically faster than iterative version. The other results show that, for various real-world instances of practical scales, our method succeeds to construct ZDDs of all DESs in reasonable time.

2   Preliminaries

Here, we introduce the notation for the graph structure used in this paper. In addition, we formulate the SNDP and introduce the ZDD that is a data structure used in our approach. Here, for a positive integer $k \in \mathbb{N}$, let $[k] := \{1, \ldots, k\}$.

2.1   Notation

Let $G = (V, E, w)$ be a weighted graph with a vertex set V, an edge set E, and an edge weight function $w \colon E \to \mathbb{R}_+$[1]. For any edge subset $X \subseteq E$, $V[X]$ denotes the induced vertices that is the set of the end points of each edge in X, i.e., $V[X] := \bigcup_{\{u,v\} \in X} \{u, v\}$. Similarly, $G[X]$ denotes the edge induced subgraph such that $G[X] := (V[X], X)$. Let $w(X)$ be the total weight of all the elements in X, i.e., $w(X) := \sum_{e \in X} w(e)$.

An edge subset $P \subseteq E$ denotes a path if it restores a vertex sequence $(v_1, \ldots, v_{|P|+1})$ where $v_i \neq v_j$ $(i \neq j)$ and $\{v_i, v_{i+1}\} \in P$ for all $i \in [|P|]$. Suppose that $v_1 = s$ and $v_{p+1} = t$, P is called an s-t path. An edge subset $R \subseteq E$ denotes a cycle if it restores a vertex sequence $(v_1, \ldots, v_{|R|})$ where $v_i \neq v_j$ $(i \neq j)$, $\{v_i, v_{i+1}\} \in R$ for all $i \in [|R| - 1]$, and $\{v_r, v_1\} \in R$.

---

[1] Although our approach can handle both directed and undirected graphs, we describe only the undirected version (i.e., $E \subseteq \{\{u, v\} \mid u, v \in V\}$) in this paper.

2.2 Survivable Network Design Problem

Given a weighted graph G and a requirement $C: 2^E \rightarrow \{True, False\}$, if $C(X) = True$ for an edge subset $X \subseteq E$, we say X satisfies C. If X satisfies C, we call X *desired edge subset* (DES). Let $\mathcal{E}_C$ be the set family of all the DESs, i.e., $\mathcal{E}_C := \{X \subseteq E \mid C(X) = True\}$.

Let us suppose that C can be decomposed into two distinct requirements C_{top} and C_{sur} where $\mathcal{E}_C = \mathcal{E}_{C_{\text{top}}} \cap \mathcal{E}_{C_{\text{sur}}}$. We call C_{top} and C_{sur} the *topology require-ment* and the *survivability requirement*, respectively. A topology requirement describes a base structure of subgraphs. In contrast, a survivability requirement describes connectivity or path length among specified vertices. Then the SNDP is a problem to find a DES defined as follows:

$$X^* \in \underset{X \in \mathcal{E}_{C_{\text{top}}} \cap \mathcal{E}_{C_{\text{sur}}}}{\arg \min} \; w(X). \tag{1}$$

Although the SNDP is an optimization problem, we aim to enumerate all the DESs $\mathcal{E}_C$. Generally, the number of DESs (i.e., $|\mathcal{E}_C|$, $|\mathcal{E}_{C_{\text{top}}}|$, and $|\mathcal{E}_{C_{\text{sur}}}|$) is expo-nentially huge in $|E|$. Therefore, naive algorithms such as an exhaustive search often do not work. In the following, we introduce some examples of topology requirements and survivability requirements.

Examples of The Topology Requirement

- Steiner Subgraph: If G has an *s-t* path for any vertex pair $s, t \in V$ ($s \neq t$), G is said to be *connected*. Given a vertex subset $T \subseteq V$, if a subgraph contains T, it is said to be *Steiner*. Let C_T^{ss} be a requirement defined as:

$$C_T^{\text{ss}}(X) := T \subseteq V[X], \; (G[X] \text{ is connected}). \tag{2}$$

- Steiner Tree: If G is connected and has no cycles, G is said to be a *tree*. For a vertex subset $T \subseteq V$, if a subgraph contains T and is a tree, it is said to be a *Steiner tree*. Let C_T^{st} be a requirement defined as:

$$C_T^{\text{st}}(X) := T \subseteq V[X], \; (G[X] \text{ is a tree}). \tag{3}$$

Examples of The Survivability Requirement

- Hop-Constrained Paths: For a positive integer $h \in \mathbb{N}$, a path $P \subseteq E$ is said to be *h-hop-constrained* iff $|P| \leq h$. For two distinct vertices $s, t \in V$ and an edge subset $X \subseteq E$, let $\mathcal{P}_{s,t}[X] \subseteq 2^X$ be the set of all the *s-t* paths on $G[X]$. Given a set of tuples $Q \subseteq V \times V \times \mathbb{N}$, let C_Q^{hcp} be a requirement defined as:

$$C_Q^{\text{hcp}}(X) := \forall (s, t, h) \in Q \; (\exists P \in \mathcal{P}_{s,t}[X] \; (|P| \leq h)). \tag{4}$$

- Edge-Disjoint Paths: Given a positive integer $k \in \mathbb{N}$, k paths $P_1, \ldots, P_k \subseteq E$ are said to be *k-edge-disjoint* iff they are pairwise disjoint, i.e., $P_i \cap P_j = \emptyset$

for all $i, j \in [k]$ $(i \neq j)$. Here, we introduce the following notation for a tuple of k sets $F = (F_1, \ldots, F_k)$:

$$\text{disjoint}(F) := \forall i, j \in [k] \ (i \neq j \iff F_i \cap F_j = \emptyset). \tag{5}$$

Given a set of tuples $Q \subseteq V \times V \times \mathbb{N}$, let C_Q^{edp} be a requirement defined as:

$$C_Q^{\text{edp}}(X) := \forall (s, t, k) \in Q \ (\exists P \in \mathcal{P}_{s,t}[X]^k \ (\text{disjoint}(P))). \tag{6}$$

– Bounded Rings: Given a positive integer $r \in \mathbb{N}$, a cycle $R \subseteq E$ is said to be r-bounded iff $|R| \leq r$. For an edge subset $X \subseteq E$, let $\mathcal{R}[X] \subseteq 2^X$ be the set of all the cycles on $G[X]$. Let C_r^{br} be a requirement defined as:

$$C_r^{\text{br}}(X) := \forall e \in X \ (\exists R \in \mathcal{R}[X] \ (e \in R)). \tag{7}$$

2.3 Zero-Suppressed Binary Decision Diagrams

As mentioned above, our aim is to enumerate all the DESs. Therefore, we handle exponentially huge set families of DESs by using the ZDD, which is a compressed data structure representing set families. If the universal set is U which is totally ordered, ZDDs can represent any set families $\mathcal{F} \subseteq 2^U$. Here we assume that $U = E = \{e_1, \ldots, e_m\}$ $(i < j \iff e_i < e_j)$.

A ZDD is a rooted directed acyclic graph denoted by $Z = (N, A)$ with a node set N, and an arc set A. It has exactly one root node ρ and exactly two terminal nodes $\bot$ and $\top$. Each non-terminal node $\alpha \in N$ has a label $\ell(\alpha) \in [m]$, which indicates that the node α is associated with an edge $e_{\ell(\alpha)}$. Each non-terminal node also has exactly two descending arcs called the 0-arc and the 1-arc. The node pointed by the x-arc of α is called the x-child, and denoted by α_x where $\ell(\alpha) < \ell(\alpha_x)$ if α_x is not a terminal.

A ZDD represents a set family as follows: For any node $\alpha \in N$, a directed path from a node α to $\top$ represents a subset. If the path descends the 1-arc of α, the edge $e_{\ell(\alpha)}$ is included, otherwise excluded.

Any ZDD has the unique *reduced* form. A ZDD is reduced if the following two rules are applied as long as possible: (i) Delete α if $\alpha_1 = \top$. If α has been the head of an arc, its new head becomes $\bot$. (ii) Share any two nodes β, β' where $\ell(\beta) = \ell(\beta')$, $\beta_0 = \beta_0'$ and $\beta_1 = \beta_1'$. These rules eliminate the redundant nodes in the ZDD. Hereinafter, we use the terms "ZDD" for reduced ZDDs unless otherwise noted. Figures 2 and 3 shows examples of ZDDs.

As discussed in Sect. 3, we need to manage multiple ZDDs. We can share any nodes, whose descendants are equivalent, among managed ZDDs [9,15].

3 Proposed Method

In this section, we present a method to construct ZDDs of all the DESs with various requirements.

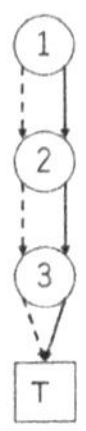

Fig. 2. ZDD of a power set

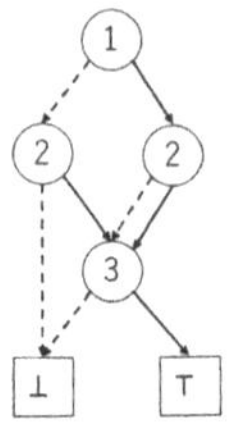

Fig. 3. ZDD of $\{\{e_1, e_2, e_3\}, \{e_1, e_3\}, \{e_2, e_3\}\}$

3.1 ZDD Construction

We propose a three-step method constructing a ZDD of $\mathcal{E}_C$. The method uses two important techniques over the ZDD, which are called the *frontier-based search* (FBS) and the *family algebra*, respectively. Each step of the method is as follows:

1. Construct ZDDs of DESs with the topology requirement and some supporting ZDDs for the survivability requirement by using the FBS.
2. Construct a ZDD of DESs with the survivability requirement by using the family algebra.
3. Construct a ZDD of DESs with both of the topology requirement and the surivivability requirement by using the family algebra.

Frontier-Based Search. The FBS is a technique for constructing ZDDs of various constrained edge subsets in a given graph, such as edge subsets with less than or equal to h edges $\mathcal{E}_h := \{X \subseteq E \mid |X| \leq h\}$, s-t paths $\mathcal{P}_{s,t}[E]$ $(s, t \in V)$, cycles $\mathcal{R}[E]$, steiner trees $\mathcal{E}_{C_T^{st}}$, and steiner subgraphs $\mathcal{E}_{C_T^{ss}}$. Therefore, we can obtain the ZDDs of DESs with the topology requirements C_T^{st} and C_T^{ss}. The ZDDs for the subgraphs above are used in the next step.

As for the time complexity of the FBS, the results in the literature [16] includes the following observation: If the constrained subgraphs are described by the class of *monadic-second order* formula (MSO) and the input graph has *bounded path-width*, the FBS is performed with linear time in m. Here, s-t paths, cycles, steiner trees, and steiner subgraphs are described by MSO. On the other hand, although edge subsets with h edges cannot be described by MSO, the ZDD construction for $\mathcal{E}_h$ is performed with pseudo polynomial time $O(hm)$.

Family Algebra. Various set family operations, which are collectively called the family algebra, can be efficiently executed on ZDDs. Let two ZDDs Z_1 and Z_2 represent $\mathcal{F}_1$ and $\mathcal{F}_2$, respectively, and $\circ$ is an operator for a set family operation. Then, we can efficiently obtain the new ZDD of $\mathcal{F}_1 \circ \mathcal{F}_2$ by the family algebra on Z_1 and Z_2. The family algebra includes the union $\cup$, the intersection $\cap$, the difference $\setminus$, and the rest. The following two operations are characteristic operations included in the family algebra, and used in the proposed method.

– Restriction $\lhd$: $\mathcal{F}_1 \lhd \mathcal{F}_2 := \{F_1 \in \mathcal{F}_1 \mid \exists F_2 \in \mathcal{F}_2 \, (F_2 \subseteq F_1)\}$
– Disjoint Join $\bowtie$: $\mathcal{F}_1 \bowtie \mathcal{F}_2 := \{F_1 \cup F_2 \mid F_1 \in \mathcal{F}_1, \, F_2 \in \mathcal{F}_2, \, F_1 \cap F_2 = \emptyset\}$

As for the time complexity of the family algebra, the results in [17] say that the union, the intersection, and the difference can be computed with polynomial time in the number of nodes in the input ZDDs. Although restriction and disjoint join have no known results for the time complexity, the family algebra tends to work in fast if the ZDDs are small enough.

In the following lemmas, we show some formulas using the family algebra to obtain the ZDDs of $\mathcal{E}_{C_{\text{top}}}$ with the survivability requirement C_Q^{hcp}, C_Q^{edp}, and C_r^{br}. We omit the proofs of the lemmas, because they are straightforward.

Lemma 1. *For any* $Q \subseteq V \times V \times \mathbb{N}$, *we have*

$$\mathcal{E}_{C_Q^{\text{hcp}}} = \bigcap_{(s,t,h) \in Q} (2^E \lhd (\mathcal{P}_{s,t}[E] \cap \mathcal{E}_h)). \tag{8}$$

Lemma 2. *We define the disjoint join among* k *same set families* $\delta(\mathcal{F}, k)$ *as:*

$$\delta(\mathcal{F}, k) = \begin{cases} \mathcal{F} & (k = 1), \\ \mathcal{F} \bowtie \delta(\mathcal{F}, k-1) & (\text{otherwise}). \end{cases} \tag{9}$$

Then, for any $Q \subseteq V \times V \times \mathbb{N}$, *we have*

$$\mathcal{E}_{C_Q^{\text{edp}}} = \bigcap_{(s,t,k) \in Q} (2^E \lhd \delta(\mathcal{P}_{s,t}[E], k)). \tag{10}$$

Lemma 3. *Let us simply denote the restriction with a singleton* $\mathcal{F} \lhd \{\{x\}\}$ *by* $\mathcal{F} \lhd x$. *For any* $r \in \mathbb{N}$, *we have*

$$\mathcal{E}_{C_r^{\text{br}}} = \bigcap_{e \in E} (2^{E \setminus \{e\}} \cup (2^E \lhd ((\mathcal{R}[E] \cap \mathcal{E}_r) \lhd e))). \tag{11}$$

Therefore, since we can obtain all ZDDs of set families in the lemmas, we also obtain ZDDs of $\mathcal{E}_{C_{\text{sur}}}$ with the survivability requirements C_Q^{hcp}, C_Q^{edp}, and C_r^{br}. Finally, we obtain a ZDD of $\mathcal{E}_C = \mathcal{E}_{C_{\text{top}}} \cap \mathcal{E}_{C_{\text{sur}}}$ by the intersection on ZDDs.

3.2   Multinomial Disjoint Join

Here we propose a new operation over the ZDD, named the *multinomial disjoint join*. Its concept is to compute the disjoint join among k set families at a time, and it empirically improves the performance of the computation of $\mathcal{E}_{C_Q^{\text{edp}}}$.

Let S be a totally ordered set and $\mathcal{F} = (\mathcal{F}_1, \ldots, \mathcal{F}_k)$ be a tuple of set families where $k \geq 2$, $\mathcal{F}_i \subseteq 2^S$ for all $i \in [k]$. Then we define the multinomial disjoint join $\Delta(\mathcal{F})$ as follows:

$$\Delta(\mathcal{F}) := \{\cup_{i \in [k]} F_i \mid \mathrm{F} \in \prod_{i \in [k]} \mathcal{F}_i, \, \text{disjoint}(\mathrm{F})\} \tag{12}$$

Then we compute $\Delta(\mathcal{F})$ recursively. Let $e = \max\{s \in S \mid \exists i \in [k], \exists F \in \mathcal{F}_i, s \in F\}$. We define $\mathcal{F}_i^- := \{F \mid F \in \mathcal{F}_i, e \notin F\}$, $\mathcal{F}_i^+ := \{F \setminus \{e\} \mid F \in \mathcal{F}_i, e \in F\}$, $\mathcal{F}^- := (\mathcal{F}_1^-, \ldots, \mathcal{F}_k^-)$, and $\mathcal{F}_i^+ = (\mathcal{F}_1^-, \ldots, \mathcal{F}_{i-1}^-, \mathcal{F}_i^+, \mathcal{F}_{i+1}^-, \ldots, \mathcal{F}_k^-)$ for all $i \in [k]$. Here we have the following lemma.

Lemma 4. $\Delta(\mathcal{F})$ *is described by the recursive formula as follows:*

$$\Delta(\mathcal{F}) = \begin{cases} \emptyset & (\exists i \in [k] \ (\mathcal{F}_i = \emptyset)) \\ \{\emptyset\} & (\forall i \in [k] \ (\mathcal{F}_i = \{\emptyset\})) \\ \Delta(\mathcal{F}^-) \cup (e \cdot \bigcup_{i \in [k]} \Delta(\mathcal{F}_i^+)) & (otherwise). \end{cases} \quad (13)$$

The proof of Lemma 4 is easily done by an induction.

Lemma 4 indicates that the multinomial disjoint join can be computed recursively on ZDDs. Suppose that the ZDD Z_i represents $\mathcal{F}_i$ for all $i \in [k]$. Let Z_i^- and Z_i^+ be a ZDD of $\mathcal{F}_i^-$ and $\mathcal{F}_i^+$. We obtain Z_i^- and Z_i^+ in a constant time because they are equal to Z_i or the descendants of the root node of Z_i. Let $\mathbf{Z} = (Z_1, \ldots, Z_k)$ be a tuple of the ZDDs and $\Delta'(\mathbf{Z})$ be the multinomial disjoint join on $\mathbf{Z}$. We define $\mathbf{Z}^- := (Z_1^-, \ldots, Z_k^-)$ and $\mathbf{Z}_i^+ = (Z_1^-, \ldots, Z_{i-1}^-, Z_i^+, Z_{i+1}^-, \ldots, Z_k^-)$ for all $i \in [k]$. Then $\Delta'(\mathbf{Z})$ can be computed by the following recursive formula:

$$\Delta'(\mathbf{Z}) = \begin{cases} \text{A ZDD having only } \bot & (\exists i \in [k] \ (Z_i \text{ is equal to } \bot)) \\ \text{A ZDD having only } \top & (\forall i \in [k] \ (Z_i \text{ is equal to } \top)) \\ \Delta'(\mathbf{Z}^-) \cup (e \cdot \bigcup_{i \in [k]} \Delta'(\mathbf{Z}_i^+)) & (otherwise). \end{cases} \quad (14)$$

For two ZDDs Z_1 and Z_2, The operation $Z_1 \cup Z_2$ computes the ZDD of their union. For an element e and a ZDD Z, the operation $e \cdot Z$ computes the ZDD with a new root node whose label, 0-child, and 1-child are e, $\bot$, and the root node of Z, respectively.

Here we can use a memo cache technique to reduce redundant computations. Let $\lambda(\mathcal{F}) := \{\mathcal{F}_i \mid i \in [k]\}$ and $\lambda'(\mathbf{Z}) := \{\text{The root node of } Z_i \mid i \in [k]\}$. For two tuples of set families $\mathcal{F}_1$ and $\mathcal{F}_2$, $\Delta(\mathcal{F}_1) = \Delta(\mathcal{F}_2)$ if $\lambda(\mathcal{F}_1) = \lambda(\mathcal{F}_2)$. This implies that, for two tuples of ZDDs $\mathbf{Z}_1$ and $\mathbf{Z}_2$, $\Delta'(\mathbf{Z}_1) = \Delta'(\mathbf{Z}_2)$ if $\lambda'(\mathbf{Z}_1) = \lambda'(\mathbf{Z}_2)$. Therefore, the result of $\Delta'(\mathbf{Z})$ can be memorized with the key $\lambda'(\mathbf{Z})$.

4   Experiments

We conducted computational experiments to evaluate the proposed method. All the code was implemented in C++ (g++7.4.0 with the -O3 option). All the experiments were conducted on a 64-bit Ubuntu 18.04 LTS with an Intel(R) Xeon(R) E-2174G 3.80 GHz CPU and 64 GB RAM. All the graphs used in the experiments were obtained from SNDlib (http://sndlib. zib.de/home.action) that provides various medium scale real-world graphs for the SNDP. We select the graphs where the exhaustive search does rather not work on it realistically ($|E| \geq 40$) and it has no bridges. Table 1 shows the data of all the graphs.

Table 1. The data of all the graphs used in the experiments

Name	$|V|$	$|E|$	Name	$|V|$	$|E|$	Name	$|V|$	$|E|$	Name	$|V|$	$|E|$
cost266	37	57	france	25	45	janos-us	26	42	norway	27	51
dfn-bwin	10	45	germany50	50	88	janos-us-ca	39	61	pioro40	40	89
dfn-gwin	11	47	giul39	39	86	newyork	16	49	sun	27	51
di-yuan	11	42	india35	35	80	nobel-eu	28	41	ta1	24	51

4.1 Performance of Multinomial Disjoint Join

First we evaluated the performance of the multinomial disjoint join for computing ZDDs of k-edge-disjoint paths. We used eight graphs dfn-bwin, dfn-gwin, di-yuan, germany50, giul39, india35, newyork, and pioro40 each of which has minimum cuts with enough size. For each $k \in \{3, 4, 5\}$, we randomly selected distinct 50 vertex pairs. Then we tried to compute the k-edge disjoint paths of each pair by the multinomial disjoint join and the iterative disjoint join, respectively. The time limit is 300 s on all the cases.

The results are shown in Table 2. For three graphs dfn-bwin, dfn-gwin, and di-yuan, we failed the computation on all the cases. This suggests that the proposed method does not work on dense graphs. For the graph giul39, whereas the iterative disjoint join failed on almost all the cases, the multinomial disjoint join succeeded on almost all the cases. All the failures were caused because of time limitation. On other many cases, the multinomial disjoint join works equal to or faster than the iterative disjoint join. This result shows an advantage of the multinomial disjoint join.

Table 2. Results of computing the multinomial disjoint join. #S and "Time" denotes the number of success cases and the median of computation time (s) on them. The format is "iterative : multinomial".

Name	$k = 3$		$k = 4$		$k = 5$	
	#S	Time	#S	Time	#S	Time
germany50	50 : 50	0.46 : 0.23	50 : 50	0.70 : 0.34	50 : 50	0.48 : 0.11
giul39	0 : 50	– : 17.92	1 : 50	28.63 : 33.34	1 : 48	145.75 : 43.50
india35	48 : 50	3.76 : 1.63	50 : 50	7.08 : 4.62	50 : 50	6.48 : 9.26
newyork	50 : 50	1.48 : 0.83	50 : 50	3.12 : 3.73	50 : 50	4.24 : 8.90
pioro40	50 : 50	0.13 : 0.10	50 : 50	0.18 : 0.17	50 : 50	0.16 : 0.07

4.2 Performance of Proposed Method

Second we evaluated the performance of the proposed method for constructing ZDDs of DESs. The target DESs are as follows: (a) $\mathcal{E}_{C_T^{\mathrm{st}}} \cap \mathcal{E}_{C_Q^{\mathrm{hcp}}}$ where $|Q| \in$

Table 3. Results for $\mathcal{E}_{C_T^{st}} \cap \mathcal{E}_{C_Q^{hcp}}$. #S, Time, and #Sol denotes the number of success cases, the median of computation time (s) in success cases, the median of the number of solutions in success cases, respectively.

Name	$|Q| = 5$			$|Q| = 10$			$|Q| = 20$		
	#S	Time	#Sol	#S	Time	#Sol	#S	Time	#Sol
cost266	50	0.006	2.6e+12	50	0.037	1.4e+10	50	0.098	4.4e+04
dfn-bwin	50	0.062	3.1e+05	50	0.202	0	50	0.841	0
dfn-gwin	48	0.070	1.2e+06	48	0.606	4.7e+02	47	5.591	0
di-yuan	50	0.060	1.4e+06	50	0.235	7.5e+02	50	1.969	0
france	50	0.003	1.4e+09	50	0.008	3.0e+07	50	0.021	4.7e+04
germany50	47	0.236	4.7e+19	43	5.752	3.3e+17	40	12.609	1.3e+12
giul39	42	1.178	3.1e+18	32	6.551	1.7e+16	13	29.533	0
india35	46	0.175	1.1e+17	39	14.509	2.3e+14	29	38.921	2.2e+09
janos-us	50	0.003	2.7e+08	50	0.008	3.4e+06	50	0.021	3.8e+03
janos-us-ca	50	0.013	7.8e+12	50	0.063	1.1e+11	50	0.345	2.1e+07
newyork	50	0.064	2.3e+08	50	0.286	7.5e+05	50	2.628	0
nobel-eu	50	0.002	2.0e+08	50	0.005	5.4e+06	50	0.011	4.8e+03
norway	50	0.012	3.5e+10	50	0.080	9.3e+08	50	0.145	8.6e+04
pioro40	47	0.146	2.2e+19	34	12.732	4.6e+16	27	66.810	1.2e+10
sun	50	0.010	5.3e+10	50	0.065	3.2e+08	50	0.332	1.2e+05
ta1	50	0.004	1.7e+10	50	0.012	1.7e+08	50	0.036	2.6e+05

$\{5, 10, 20\}$ and $T = \bigcup_{(s,t,h) \in Q} \{s, t\}$. Let $d(s, t)$ be the distance between s and t. We randomly generated 50 cases where $d(s, t) \le h \le 2d(s, t)$ for all $(s, t, h) \in Q$. (b) $\mathcal{E}_{C_T^{ss}} \cap \mathcal{E}_{C_Q^{edp}}$ where $|Q| \in \{5, 10, 20\}$ and $T = \bigcup_{(s,t,k) \in Q} \{s, t\}$. We randomly generated 50 cases where $2 \le k \le 5$ and the size of the minimum cut between s and t is not less than k for all $(s, t, k) \in Q$. (c) $\mathcal{E}_{C_V^{ss}} \cap \mathcal{E}_{C_r^{br}}$ where $r \in \{5, 10, \infty\}$. The time limit is 300 s on all the cases.

The results are shown in Tables 3, 4, and 5. On almost all cases of four graphs dfn-bwin, dfn-gwin, di-yuan, and giul39, we failed the computation of $\mathcal{E}_{C_T^{ss}} \cap \mathcal{E}_{C_Q^{edp}}$ and $\mathcal{E}_{C_V^{ss}} \cap \mathcal{E}_{C_r^{br}}$. In contrast, on other many cases, we succeeded the computation in a few minutes. Especially, the number of their solutions is exponentially huge. Thus, except for dense graphs, the proposed method tends to work enough on medium scale real-world graphs.

Table 4. Results for $\mathcal{E}_{C_T^{ss}} \cap \mathcal{E}_{C_Q^{edp}}$. #S, Time, and #Sol denotes the number of success cases, the median of computation time (s) in success cases, the median of the number of solutions in success cases, respectively.

Name	$|Q| = 5$			$|Q| = 10$			$|Q| = 20$		
	#S	Time	#Sol	#S	Time	#Sol	#S	Time	#Sol
cost266	50	0.012	1.9e+08	50	0.021	4.4e+04	50	0.039	4.5e+02
france	50	0.005	6.1e+06	50	0.009	6.0e+03	50	0.018	4.6e+01
germany50	50	2.818	6.5e+15	50	4.234	1.0e+11	50	6.364	1.4e+06
giul39	2	245.390	3.9e+18	0	–	–	0	–	–
india35	50	54.712	9.0e+17	50	91.423	1.6e+13	50	178.498	4.3e+09
janos-us	50	0.003	1.6e+05	50	0.005	1.2e+03	50	0.008	3.2e+01
janos-us-ca	50	0.009	5.0e+08	50	0.016	4.3e+05	50	0.027	1.6e+03
newyork	50	41.622	3.4e+11	50	65.939	8.8e+09	47	146.826	3.1e+07
nobel-eu	50	0.002	1.5e+04	50	0.003	1.2e+02	50	0.004	2.0e+00
norway	50	0.033	5.8e+07	50	0.058	3.4e+04	50	0.107	4.2e+02
pioro40	50	1.382	2.6e+18	50	2.063	3.0e+13	50	3.555	1.6e+08
sun	50	0.033	5.4e+07	50	0.059	4.4e+04	50	0.109	4.4e+02
ta1	50	0.008	3.1e+09	50	0.013	5.5e+06	50	0.023	3.2e+04

Table 5. Results for computing $\mathcal{E}_{C_V^{ss} \wedge C_r^{br}}$. Time and #Sol denotes the computation time (s) and the number of solutions, respectively.

Name	$r = 5$		$r = 10$		$r = \infty$	
	Time	#Sol	Time	#Sol	Time	#Sol
cost266	0.012	1.6e−03	0.043	1.1e−07	0.051	4.2e−07
france	0.009	4.8e−05	0.035	3.2e−06	0.015	3.3e−06
germany50	0.098	0	6.681	1.7e−15	5.918	2.0e−16
giul39	1.088	3.3e−06	–	–	48.473	5.8e−20
india35	1.175	5.3e−16	–	–	7.677	6.3e−18
janos-us	0.005	3.2e−03	0.016	9.8e−05	0.015	1.9e−06
janos-us-ca	0.013	0	0.047	2.0e−08	0.065	1.0e−09
newyork	2.830	1.8e−13	146.643	2.7e−13	1.322	2.7e−13
nobel-eu	0.004	1.2e−02	0.010	9.2e−04	0.011	1.6e−05
norway	0.016	0	0.196	2.9e−10	0.069	5.4e−10
pioro40	0.121	7.2e−16	55.087	1.5e−22	4.187	2.4e−22
sun	0.016	0	0.221	2.9e−10	0.069	5.4e−10
ta1	0.014	1.0e−10	0.071	3.8e−10	0.022	3.8e−10

5   Conclusion

In this paper, we proposed a practical method to enumerate all DESs of the SNDPs by using ZDDs. The proposed method uses both of the FBS and the family algebra, which are independently used for constructing ZDDs historically. We also proposed a novel operation over ZDDs named the multinomial disjoint join. Moreover, obtained ZDDs can be used for various applications such as optimization, sampling, and filtering. They support much better network design. The results of computational experiments showed that our method efficiently constructed ZDDs of all DESs on various medium scale real-world instances.

Acknowledgement. This work was supported by JSPS KAKENHI Grant Number 15H05711.

References

1. Ellison, R., Linger, R., Longstaff, T., Mead, N.: Survivable network system analysis: a case study. IEEE Software **16**, 70–77 (1999)
2. Kerivin, H., Mahjoub, A.R.: Design of survivable networks: a survey. Networks **46**(1), 1–21 (2005)
3. Akgün, I.: New formulations for the hop-constrained minimum spanning tree problem via sherali and driscoll's tightened miller-tucker-zemlin constraints. Comput. Oper. Res. **38**(1), 277–286 (2011)
4. Diarrassouba, I., Gabrel, V., Ridha Mahjoub, A., Gouveia, L., Pesneau, P.: Integer programming formulations for the k-edge-connected 3-hop-constrained network design problem. Networks **67**, 148–169 (2015)
5. Fortz, B.: Design of survivable networks with bounded rings. In: Network Theory and Applications (2000)
6. Gouveia, L., Simonetti, L., Uchoa, E.: Modeling hop-constrained and diameter-constrained minimum spanning tree problems as steiner tree problems over layered graphs. Math. Program. **128**(1), 123–148 (2011)
7. Rodriguez-Martin, I., Salazar Gonzlez, J.J., Yaman, H.: The ring/κ-rings network design problem: model and branch-and-cut algorithm. Networks **68**, 130–140 (2016)
8. Penttinen, A.: Chapter 10 - Network planning and dimensioning. Lecture Notes: S-38.145 - Introduction to Teletraffic Theory. Helsinki University of Technology (1999)
9. Minato, S.: Zero-suppressed BDDS for set manipulation in combinatorial problems. In: DAC, pp. 272–277 (1993)
10. Knuth, D.E.: The Art of Computer Programming: Bitwise tricks & Techniques. Binary Decision Diagrams, vol. 4, fascicle 1. Addison-Wesley Professional, Boston (2009)
11. Kawahara, J., Inoue, T., Iwashita, H., Minato, S.: Frontier-based search for enumerating all constrained subgraphs with compressed representation. IEICE Trans. Fundam. **E100–A**(9), 1773–1784 (2017)
12. Coudert, O.: Solving graph optimization problems with ZBDDS. In: Proceedings of the 1997 European Conference on Design and Test, EDTC 1997, p. 224 (1997)

13. Kawahara, J., Horiyama, T., Hotta, K., Minato, S.I.: Generating all patterns of graph partitions within a disparity bound. In: WALCOM: Algorithms and Computation, pp. 119–131 (2017)

14. Kawahara, J., Saitoh, T., Suzuki, H., Yoshinaka, R.: Solving the longest oneway-ticket problem and enumerating letter graphs by augmenting the two representative approaches with ZDDs. In: Phon-Amnuaisuk, S., Au, T.-W., Omar, S. (eds.) CIIS 2016. AISC, vol. 532, pp. 294–305. Springer, Cham (2017). https://doi.org/10.1007/978-3-319-48517-1_26

15. Minato, S., Ishiura, N., Yajima, S.: Shared binary decision diagram with attributed edges for efficient boolean function manipulation. In: 27th ACM/IEEE Design Automation Conference, pp. 52–57 (1990)

16. Amarilli, A., Bourhis, P., Jachiet, L., Mengel, S.: A circuit-based approach to efficient enumeration. In: 44th International Colloquium on Automata, Languages, and Programming (ICALP 2017), vol. 80, pp. 111:1–111:15 (2017)

17. Yoshinaka, R., Kawahara, J., Denzumi, S., Arimura, H., Ichi-Minato, S.: Counterexamples to the long-standing conjecture on the complexity of BDD binary operations. Inf. Process. Lett. **112**, 636–640 (2012)

Approximability of the Independent Feedback Vertex Set Problem for Bipartite Graphs

Yuma Tamura$^{(\boxtimes)}$, Takehiro Ito, and Xiao Zhou

Graduate School of Information Sciences, Tohoku University, Aoba-yama 6-6-05, Sendai 980-8579, Japan
yuma.tamura.t5@dc.tohoku.ac.jp, {takehiro,zhou}@ecei.tohoku.ac.jp

Abstract. Given a graph G with n vertices, the independent feedback vertex set problem is to find a vertex subset F of G with the minimum number of vertices such that F is both an independent set and a feedback vertex set of G, if it exists. This problem is known to be NP-hard for bipartite planar graphs. In this paper, we study the approximability of the problem. We first show that, for any fixed $\varepsilon > 0$, unless P = NP, there exists no polynomial-time $n^{1-\varepsilon}$-approximation algorithm even for bipartite planar graphs. This gives a contrast to the existence of a polynomial-time 2-approximation algorithm for the original feedback vertex set problem on general graphs. We then give an $\alpha(\Delta - 1)/2$-approximation algorithm for bipartite graphs G of maximum degree Δ, which runs in $O(t(G) + \Delta n)$ time, under the assumption that there is an α-approximation algorithm for the original feedback vertex set problem on bipartite graphs which runs in $O(t(G))$ time.

1 Introduction

A *feedback vertex set* F of an undirected graph $G = (V, E)$ is a vertex subset of G such that the subgraph of G induced by $V \setminus F$ is a forest. (See Fig. 1(b) as an example.) For a given graph G, the *feedback vertex set problem* is to find a feedback vertex set of G with the minimum number of vertices. The feedback vertex set problem is one of the most classical NP-hard problems, and many algorithms have been developed from various viewpoints over the years.

Misra et al. [13] introduced an independence variant of the feedback vertex set problem. An *independent set* I of a graph $G = (V, E)$ is a vertex subset of G such that the subgraph of G induced by I contains no edge. A vertex set F' of G is said to be an *independent feedback vertex set* of G if it is both an independent set and a feedback vertex set of G. (See Fig. 1(c)) Note that an independent feedback vertex set of a graph does not always exist; for example, consider a

T. Ito—Partially supported by JST CREST Grant Number JPMJCR1402, and JSPS KAKENHI Grant Numbers JP18H04091 and JP19K11814, Japan.
X. Zhou—Partially supported by JSPS KAKENHI Grant Number JP19K11813, Japan.

M. S. Rahman et al. (Eds.): WALCOM 2020, LNCS 12049, pp. 286–295, 2020.
https://doi.org/10.1007/978-3-030-39881-1_24

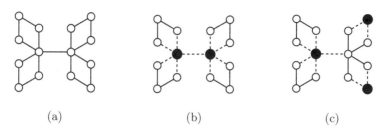

Fig. 1. (a) A graph G, (b) a minimum feedback vertex set of G, and (c) a minimum independent feedback vertex set of G, where each vertex in the feedback vertex sets is depicted by a black circle.

complete graph with four or more vertices. For a given graph G, the *independent feedback vertex set problem* is to find an independent feedback vertex set of G with the minimum number of vertices, if it exists. For convenience, we sometimes call the feedback vertex set problem the *original problem*, and the independent feedback vertex set problem the *independence variant*.

1.1   Related Results and Known Results

The original problem is APX-complete for general graphs [2]. This means that the problem is unlikely to have a polynomial-time approximation scheme (PTAS). Moreover, it remains NP-hard even for bipartite planar graphs of maximum degree four [14]. The original problem has been intensively studied from various viewpoints, such as of approximation [1,2,4], fixed-parameter tractability (FPT) [10], and tractability on special graph classes [8,9,15].

As Misra et al. pointed out in [13], by inserting a new vertex in every edge of a graph, the original problem can be reduced to the independence variant without changing the size of optimal solutions. This implies that the independence variant is APX-hard for bipartite graphs, and remains NP-hard even for bipartite planar graphs of maximum degree four. In the same paper, Misra et al. also developed a fixed-parameter algorithm whose running time is $O(5^k n^{O(1)})$, where n is the number of vertices in a graph and k is the solution size as the parameter. Recently, Li and Pilipczuk improved this running time to $O(3.619^k n^{O(1)})$ [11]. The independence variant has been studied also from the viewpoint of graph classes; for example, it is solvable in polynomial time for bounded treewidth graphs [16], chordal graphs [16], P_5-free graphs [7], and graphs of diameter two [6]. Interestingly, for the latter two graph classes, their polynomial-time solvabilities of the original problem remain open.

The independence variant is strongly related to the *near-bipartiteness problem* [5,7,17]. In the problem, for a given graph G, our task is to decide whether G has *at least one* independent feedback vertex set. Therefore, the intractability of the independence variant is inherited from the near-bipartiteness problem. It is known that the near-bipartiteness problem is NP-complete even for graphs of maximum degree four [17], graphs of diameter at most three [5], and for line graphs of bipartite planar subcubic graphs [7]. Note that, since any bipartite graph has an independent feedback vertex set, the near-bipartiteness problem is trivially solvable for bipartite graphs.

1.2 Our Contribution

In this paper, we study the approximability of the independent feedback vertex set problem for bipartite graphs.

We first show that, unless P = NP, the independence variant admits no polynomial-time approximation algorithm within a factor $n^{1-\varepsilon}$ for any fixed $\varepsilon > 0$, even on bipartite planar graphs, where n is the number of vertices in a graph. This gives a contrast to the existence of polynomial-time 2-approximation algorithms for the original problem on general graphs [2,4]. One might think that this hardness result is straightforward, because the independence variant has the constraint of independent sets. In fact, the independent set problem, which finds a maximum-size independent set of a given graph, is also hard to approximate within a factor $n^{1-\varepsilon}$ in polynomial time, for any fixed $\varepsilon > 0$ [18]. However, since the independent set problem is a maximization problem whereas the independent feedback vertex set problem is a minimization problem, it is not straightforward to give our hardness result. We also point out that the independent set problem admits a PTAS for planar graphs [3], and is solvable in polynomial time for bipartite graphs (from König's theorem).

We then give an $\alpha(\Delta - 1)/2$-approximation algorithm for bipartite graphs G of n vertices and maximum degree Δ, which runs in $O(t(G) + \Delta n)$ time, under the assumption that there is an α-approximation algorithm for the original problem on bipartite graphs which runs in $O(t(G))$ time. As we will show later, this result yields a polynomial-time $O(\Delta)$-approximation algorithm for the independence variant on bipartite graphs. Notice that, from our inapproximability result, unless P = NP, there is no polynomial-time $\Delta^{1-\varepsilon}$-approximation algorithm for any fixed $\varepsilon > 0$. In this sense, our approximation factor is best possible with respect to the exponent of Δ.

The proofs for the claims marked with (∗) are postponed to the full version.

2   Preliminaries

In this paper, we assume that graphs are undirected, unweighted, simple and connected. Let $G = (V, E)$ be a graph; we sometimes denote by $V(G)$ and $E(G)$ the vertex set and edge set of G, respectively. For a vertex v in G, we denote by $N(v)$ the set of all neighbors of v in G, that is, $N(v) = \{w \in V(G) : vw \in E(G)\}$. For a vertex subset V' of a graph $G = (V, E)$, let $G[V']$ be the subgraph of G induced by V'. For a subset $W \subseteq V$, we denote simply by $G - W$ the induced subgraph $G[V \setminus W]$.

For a graph G, a vertex subset I of G is called an *independent set of G* if $G[I]$ contains no edge, and a vertex subset F of G is called a *feedback vertex set* *of* G if $G - F$ is a forest. We sometimes say that a feedback vertex set F of G is *independent* if $G[F]$ forms an independent set of G. Let

$$\mathsf{OPT}(G) = \min\{|F| : F \text{ is an independent feedback vertex set of } G\};$$

we let $\mathsf{OPT}(G) = +\infty$ if G has no independent feedback vertex set. Given a graph G, the *independent feedback vertex set problem* is to find an independent feedback vertex set F of G such that $|F| = \mathsf{OPT}(G)$. Analogously, we define $\mathsf{OPT}_{\mathrm{FVS}}(G)$ for feedback vertex sets; notice that $\mathsf{OPT}_{\mathrm{FVS}}(G) < |V(G)| - 1$ always holds.

3 Inapproximability

As mentioned before, the independent feedback vertex set problem is APX-hard even for bipartite graphs. In this section, we give the following stronger result.

Theorem 1. *Let* $\varepsilon > 0$ *be any fixed constant. The independent feedback vertex set problem admits no polynomial-time approximation algorithm within a factor* $n^{1-\varepsilon}$ *for bipartite planar graphs of* n *vertices, unless* $\mathrm{P} = \mathrm{NP}$.

We prove the theorem in the remainder of this section, by giving a gap-producing reduction from the planar 3-satisfiability problem.

Recall that the *3-satisfiability problem* (3-SAT, for short) is the problem of asking if there exists a satisfying assignment for a given 3-CNF formula ϕ. The *associated graph* $G_\phi = (X \cup C, E)$ of ϕ is a bipartite graph such that

(i) each vertex in X corresponds to a variable in ϕ, and each vertex in C corresponds to a clause of ϕ; and

(ii) two vertices $v \in X$ and $w \in C$ are joined by an edge in E if and only if the variable corresponding to v appears in the clause corresponding to w.

The formula ϕ is said to be *planar* if its associated graph is planar. Given a planar 3-CNF formula ϕ, the *planar 3-satisfiability problem* (PLANAR 3-SAT, for short) is to determine whether there exists a satisfying assignment of ϕ. PLANAR 3-SAT is known to be NP-complete [12].

3.1 Gadgets

In this subsection, we construct five kinds of gadgets for our reduction.

We first define a *forbidding gadget*, which consists of three parts: a *root vertex*, a *core vertex*, and *petals*. (See Fig. 2(a)) The root vertex will be identified with a vertex of another gadget defined later. Each petal is a cycle of four vertices, and the forbidding gadget have p petals that share the core vertex. We use the simplified illustration shown in Fig. 2(b) for the forbidding gadget. The forbidding gadget forbids adding the root vertex to a minimum independent feedback vertex set, and forces to choose the center vertex. To see this, notice that we cannot choose both root and center vertices at the same time, because they are adjacent. Therefore, if we choose the root vertex, we must choose at

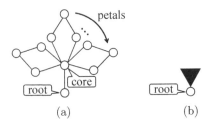

(a)   (b)

Fig. 2. (a) A forbidding gadget, where some petals are omitted. (b) The simplified illustration of a forbidding gadget, where the core vertex and all petals are simply illustrated as a black triangle.

least p vertices additionally, each of which comes from a petal in the forbidding gadget. An independent feedback vertex set is said to be *proper* if it contains the core vertex for every forbidding gadget.

Using the forbidding gadget, we will define the other gadgets. The *variable gadget* X_i is illustrated in Fig. 3(a), and corresponds to a variable x_i of a given 3-CNF formula ϕ. Every proper independent feedback vertex set must contain v_i or v_i' of X_i. As we will explain later, we regard $v_i \in F$ as setting $x_i =$ true, and say that X_i is the true-*state*. Conversely, we regard $v_i \notin F$ as setting $x_i =$ false, and say that X_i is the false-*state*.

The *clause gadget* C_j is illustrated in Fig. 3(b). This gadget corresponds to a clause $(c_{j,1} \vee c_{j,2} \vee c_{j,3})$ in ϕ, and the three vertices $u_{j,1}, u_{j,2}$ and $u_{j,3}$ of C_j correspond to the three literals in the clause. Every proper independent feedback vertex set must contain at least one of $u_{j,1}, u_{j,2}$, and $u_{j,3}$.

The *positive edge gadget* and the *negative edge gadget* are illustrated in Fig. 3(c) and (d), respectively. The purpose of these edge gadgets is to propagate the state of a variable gadget to a clause gadget; note that the negative edge gadget propagates the opposite state of the variable gadget. The positive edge gadget has only two proper independent feedback vertex sets, as shown in Fig. 4. Similarly, the negative edge gadget has only two proper independent feedback vertex sets $\{v_\ell\}$ and $\{u_\ell\}$ (together with all core vertices of forbidding gadgets). Observe that for every proper independent feedback vertex set F, $v_\ell \in F$ if and only if $u_\ell \in F$ for the positive edge gadget, and $v_\ell \in F$ if and only if $u_\ell \notin F$ for the negative edge gadget.

3.2   Reduction

In this subsection, we construct the corresponding graph $G_{\phi,p}$ for the independent feedback vertex set problem, from a given instance ϕ of PLANAR 3-SAT.

Let $G_\phi = (X \cup C, E)$ be the associated graph of a given 3-CNF formula ϕ for PLANAR 3-SAT. We fix a plane embedding of G_ϕ arbitrarily. Then, we replace each vertex $x_i \in X$ with a variable gadget X_i, and each vertex $c_j \in C$ with a clause gadget C_j. Next, consider an edge $x_i c_j \in E$, and assume that $x_i c_j$ appears as the k-th edge, where $k \in \{1, 2, 3\}$, if we see the three edges incident to

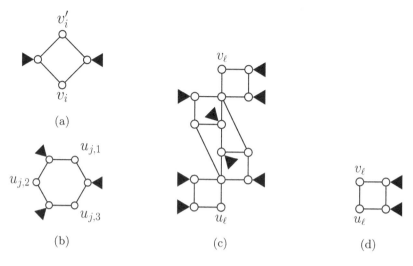

Fig. 3. (a) A variable gadget, (b) a clause gadget, (c) a positive edge gadget, and (d) a negative edge gadget.

c_j in the clockwise direction. We replace the edge $x_i c_j \in E$ with a positive edge gadget if the corresponding variable of x_i appears as a positive literal; otherwise we replace $x_i c_j$ with a negative edge gadget. In either case, we identify the vertex v_ℓ in the edge gadget with v_i of X_i, and also identify the vertex u_ℓ with $u_{j,k}$ of C_j. Let $G_{\phi,p}$ be the resulting graph, where p is the number of petals in each forbidding gadget. Since G_ϕ and all gadgets are bipartite and planar, $G_{\phi,p}$ is also a bipartite planar graph. In addition, we can construct $G_{\phi,p}$ in polynomial time.

Let q_ϕ be an arbitrary integer such that $q_\phi \geq 81(s+t)$, where s and t are the numbers of variables and clauses in ϕ, respectively. We denote by $n_{\phi,p}$ the number of vertices in $G_{\phi,p}$. Then, we have the following lemma.

Lemma 1 (∗). *It holds that* $n_{\phi,p} < (p+1)q_\phi$.

The following lemma is the key for the proof of Theorem 1.

Lemma 2 (∗). *For any integer* $p \geq q_\phi$, *the following* (I) *and* (II) *hold:*

(I) *if* $\mathsf{OPT}(G_{\phi,p}) < p+1$, *then* ϕ *has a satisfying assignment; and*
(II) *if* $\mathsf{OPT}(G_{\phi,p}) > q_\phi$, *then* ϕ *has no satisfying assignment.*

We now prove Theorem 1. Assume for a contradiction that there exists a polynomial-time approximation algorithm within a factor $n^{1-\varepsilon}$ for some fixed $\varepsilon > 0$, where n is the number of vertices in a given graph. Observe that $\varepsilon \leq 1$ must hold. We denote by $\mathsf{APX}(G_{\phi,p})$ the size of a solution for $G_{\phi,p}$ produced by the approximation algorithm. Then, we have

$$\mathsf{OPT}(G_{\phi,p}) \leq \mathsf{APX}(G_{\phi,p}) \leq n_{\phi,p}^{1-\varepsilon} \cdot \mathsf{OPT}(G_{\phi,p}). \tag{1}$$

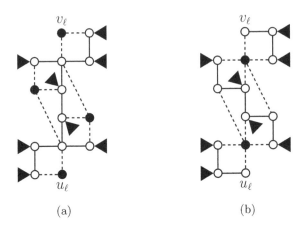

(a)   (b)

Fig. 4. The two proper independent feedback vertex sets of the positive edge gadget, formed by the black vertices together with all core vertices of forbidding gadgets.

We set

$$p = \left\lceil q_\phi^{(2-\varepsilon)/\varepsilon} \right\rceil - 1.$$

Consider the case where $\mathsf{APX}(G_{\phi,p}) < p+1$ holds. By (1) it holds in this case that $\mathsf{OPT}(G_{\phi,p}) < p+1$. Then, Lemma 2(I) says that ϕ has a satisfying assignment. Consider the other case, where $\mathsf{APX}(G_{\phi,p}) \geq p+1$ holds. By (1) it holds in this case that $n_{\phi,p}^{1-\varepsilon} \cdot \mathsf{OPT}(G_{\phi,p}) \geq p+1$. Then, by Lemma 1 we have

$$\mathsf{OPT}(G_{\phi,p}) \geq \frac{p+1}{n_{\phi,p}^{1-\varepsilon}} > \frac{p+1}{((p+1)q_\phi)^{1-\varepsilon}} = \frac{(p+1)^\varepsilon}{q_\phi^{1-\varepsilon}} \geq \frac{\left(q_\phi^{(2-\varepsilon)/\varepsilon}\right)^\varepsilon}{q_\phi^{1-\varepsilon}} = q_\phi.$$

Then, Lemma 2(II) says that ϕ has no satisfying assignment.

In this way, $\mathsf{APX}(G_{\phi,p}) < p+1$ if and only if ϕ has a satisfying assignment. Since we have assumed that $\mathsf{APX}(G_{\phi,p})$ can be computed in polynomial time, this means that we can solve PLANAR 3-SAT in polynomial time. This is a contradiction unless P = NP. This completes the proof of Theorem 1.   □

4   Approximation Algorithm

We note that the independent feedback vertex set problem is solvable in linear time for graphs of maximum degree at most two, because each connected component of such a graph is either a path or a cycle. In this section, we give the following theorem.

Theorem 2. *Let G be a bipartite graph of n vertices and maximum degree $\Delta \geq 3$. Suppose that, for the feedback vertex set problem, there is an α-approximation algorithm which outputs a solution of G in $O(t(G))$ time.*

Algorithm 1

Input: A bipartite graph $G = (A \cup B, E)$
Output: An independent feedback vertex set F of G

1: Apply the α-approximation algorithm for the original problem to G, and obtain a feedback vertex set F' of G
2: Let $F_A = F' \cap A$ and $F_B = F' \cap B$; assume that $|F_A| \leq |F_B|$ without loss of generality
3: **while** there is an edge $v_A v_B$ such that $v_A \in F_A$ and $v_B \in F_B$ **do**
4:      $F_A \leftarrow F_A \setminus \{v_A\}$
5:      **if** $N(v_A) \setminus F_B \neq \emptyset$ **then**
6:          Choose an arbitrary vertex $u \in N(v_A) \setminus F_B$
7:          $F_B \leftarrow (F_B \cup N(v_A)) \setminus \{u\}$
8:      **end if**
9: **end while**
10: Return $F = F_A \cup F_B$

Then, for the independent feedback vertex set problem, there exists an approximation algorithm which outputs a solution of G within a factor $\alpha(\Delta - 1)/2$ in $O(t(G) + \Delta n)$ time.

As a proof of Theorem 2, we give Algorithm 1 below. To avoid confusion, we sometimes denote by F_A^i and F_B^i the sets F_A and F_B after Algorithm 1 executes the while loop of the lines 3–9 exactly i times, respectively. For notational convenience, let $F_A^0 = F' \cap A$ and $F_B^0 = F' \cap B$. Furthermore, let F_A^f and F_B^f be the sets F_A and F_B when Algorithm 1 terminates, respectively. Let $F^i = F_A^i \cup F_B^i$, and hence Algorithm 1 outputs $F^f = F_A^f \cup F_B^f$.

Lemma 3 ($*$). Algorithm 1 *outputs an independent feedback vertex set F of the input bipartite graph G.*

We now estimate the approximation factor of Algorithm 1. Specifically, we will prove that the output F of Algorithm 1 satisfies

$$|F| \leq \alpha \cdot \frac{\Delta - 1}{2} \cdot \mathsf{OPT}(G). \tag{2}$$

To verify (2), we first note that Algorithm 1 adds at most $|N(v_A)| - 2 \leq \Delta - 2$ vertices to F_B in line 7, for each execution of the whole loop; recall that two vertices $v_B, u \in N(v_A)$ were not added to F_B. Since Algorithm 1 executes the whole loop exactly $|F_A^0 \setminus F_A^f|$ times, we have $|F_B^f| \leq |F_B^0| + (\Delta - 2)|F_A^0 \setminus F_A^f|$ and hence

$$
\begin{aligned}
|F| &= |F_A^f| + |F_B^f| \\
&\leq (|F_A^0| - |F_A^0 \setminus F_A^f|) + (|F_B^0| + (\Delta - 2)|F_A^0 \setminus F_A^f|) \\
&= |F_A^0| + |F_B^0| + (\Delta - 3)|F_A^0 \setminus F_A^f| \\
&\leq |F'| + (\Delta - 3)|F_A^0|.
\end{aligned}
\tag{3}
$$

Since we have assumed that $|F_A^0| \leq |F_B^0|$ without loss of generality, it holds that $|F_A^0| \leq |F'|/2$. Therefore, since $\Delta - 3 \geq 0$, by (3) we have

$$|F| \leq |F'| + \frac{\Delta - 3}{2}|F'| = \frac{\Delta - 1}{2}|F'|. \tag{4}$$

Recall that F' is the α-approximate solution for the original problem, and hence it holds that $|F'| \leq \alpha \cdot \mathsf{OPT}_{\mathsf{FVS}}(G)$. In addition, since every independent feedback vertex set of G is also a feedback vertex set of G, we have $\mathsf{OPT}_{\mathsf{FVS}}(G) \leq \mathsf{OPT}(G)$. Thus, together with (4), we can conclude that

$$|F| \leq \frac{\Delta - 1}{2} \cdot \alpha \cdot \mathsf{OPT}_{\mathsf{FVS}}(G) \leq \frac{\Delta - 1}{2} \cdot \alpha \cdot \mathsf{OPT}(G),$$

as claimed in (2).

Finally, we estimate the running time of Algorithm 1. Let $n = |V(G)|$. By assumption, the line 1 takes $t(G)$ time. The whole loop of lines 3–9 can be executed in $O(\Delta n)$ time in total, by using an appropriate data structure such as a queue. Therefore, the total running time of Algorithm 1 is $O(t(G) + \Delta n)$. This completes the proof of Theorem 2.   □

We now obtain two approximation algorithms for the independence variant on bipartite graphs, by applying Theorem 2 to two known approximation algorithms for the original problem. First, Bafna et al. [1] gave an algorithm whose approximation factor is $2 - 2/(3\Delta - 2)$ and running time is $O(n^2)$. Then, we have the following corollary.

Corollary 1. *The independent feedback vertex set problem for bipartite graphs of n vertices and maximum degree $\Delta \geq 3$ admits an algorithm whose approximation factor is $3(\Delta - 1)^2/(3\Delta - 2)$ and running time is $O(n^2)$.*

Second, Becker and Geiger [4] gave a 2-approximation algorithm for the original problem which runs in $O(m + n \log n)$ time, where m is the number of edges in a graph. Using this algorithm, we can obtain a faster and easily implementable approximation algorithm for the independence variant, although its approximation factor is slightly worse than that of Corollary 1.

Corollary 2. *The independent feedback vertex set problem for bipartite graphs of n vertices and maximum degree $\Delta \geq 3$ admits a $(\Delta - 1)$-approximation algorithm which runs in $O(n \log n + \Delta n)$ time.*

5   Conclusion

In this paper, we have shown that the independent feedback vertex set problem for bipartite planar graphs of n vertices admits no polynomial-time approximation algorithm within a factor $n^{1-\varepsilon}$, for any fixed $\varepsilon > 0$, unless P = NP. This gives a contrast to the fact that the original problem admits a polynomial-time 2-approximation algorithm for general graphs. We also have developed an $\alpha(\Delta - 1)/2$-approximation algorithm for the independence variant on bipartite graphs of maximum degree Δ, where α is the approximation factor of an algorithm for the original problem on bipartite graphs.

References

1. Bafna, V., Berman, P., Fujito, T.: Constant ratio approximations of the weighted feedback vertex set problem for undirected graphs. In: Staples, J., Eades, P., Katoh, N., Moffat, A. (eds.) ISAAC 1995. LNCS, vol. 1004, pp. 142–151. Springer, Heidelberg (1995). https://doi.org/10.1007/BFb0015417
2. Bafna, V., Berman, P., Fujito, T.: A 2-approximation algorithm for the undirected feedback vertex set problem. SIAM J. Discrete Math. **12**, 289–297 (1999)
3. Baker, B.S.: Approximation algorithms for NP-complete problems on planar graphs. J. ACM **41**(1), 153–180 (1994)
4. Becker, A., Geiger, D.: Optimization of Pearl's method of conditioning and greedy-like approximation algorithms for the vertex feedback set problem. Artif. Intell. **83**(1), 167–188 (1996)
5. Bonamy, M., Dabrowski, K.K., Feghali, C., Johnson, M., Paulusma, D.: Recognizing graphs close to bipartite graphs. In: Proceedings of the 42nd International Symposium on Mathematical Foundations of Computer Science (MFCS 2017), Leibniz International Proceedings in Informatics, vol. 83, pp. 70:1–70:14 (2017)
6. Bonamy, M., Dabrowski, K.K., Feghali, C., Johnson, M., Paulusma, D.: Independent feedback vertex sets for graphs of bounded diameter. Inf. Process. Lett. **131**, 26–32 (2018)
7. Bonamy, M., Dabrowski, K.K., Feghali, C., Johnson, M., Paulusma, D.: Independent feedback vertex set for P_5-free graphs. Algorithmica **81**(4), 1342–1369 (2019)
8. Festa, P., Pardalos, P.M., Resende, M.G.C.: Feedback Set Problems. Handbook of Combinatorial Optimization, pp. 209–258. Springer, Boston (1999). https://doi.org/10.1007/978-1-4757-3023-4_4
9. Kloks, T., Lee, C.M., Liu, J.: New algorithms for k-face cover, k-feedback vertex set, and k-disjoint cycles on plane and planar graphs. In: Goos, G., Hartmanis, J., van Leeuwen, J., Kučera, L. (eds.) WG 2002. LNCS, vol. 2573, pp. 282–295. Springer, Heidelberg (2002). https://doi.org/10.1007/3-540-36379-3_25
10. Kociumaka, T., Pilipczuk, M.: Faster deterministic feedback vertex set. Inf. Process. Lett. **114**(10), 556–560 (2014)
11. Li, S., Pilipczuk, M.: An improved FPT algorithm for independent feedback vertex set. In: Brandstädt, A., Köhler, E., Meer, K. (eds.) WG 2018. LNCS, vol. 11159, pp. 344–355. Springer, Cham (2018). https://doi.org/10.1007/978-3-030-00256-5_28
12. Lichtenstein, D.: Planar formulae and their uses. SIAM J. Comput. **11**(2), 329–343 (1982)
13. Misra, N., Philip, G., Raman, V., Saurabh, S.: On parameterized independent feedback vertex set. Theoret. Comput. Sci. **461**, 65–75 (2012)
14. Speckenmeyer, E.: On feedback vertex sets and nonseparating independent sets in cubic graphs. J. Graph Theory **12**, 405–412 (1988)
15. Takaoka, A., Tayu, S., Ueno, S.: On minimum feedback vertex sets in bipartite graphs and degree-constraint graphs. IEICE Trans. Inf. Syst. **E96–D**(11), 2327–2332 (2013)
16. Tamura, Y., Ito, T., Zhou, X.: Algorithms for the independent feedback vertex set problem. IEICE Trans. Fundam. Electron. Commun. Comput. Sci. **E98–A**(6), 1179–1188 (2015)
17. Yang, A., Yuan, J.: Partition the vertices of a graph into one independent set and one acyclic set. Discrete Math. **306**(12), 1207–1216 (2006)
18. Zuckerman, D.: Linear degree extractors and the inapproximability of max clique and chromatic number. Theory Comput. **3**(1), 103–128 (2007)

Efficient Enumeration of Non-isomorphic Ptolemaic Graphs

Dat Hoang Tran and Ryuhei Uehara[✉]

School of Information Science, Japan Advanced Institute of Science
and Technology (JAIST), Nomi, Ishikawa, Japan
{tran.hoangdat,uehara}@jaist.ac.jp

Abstract. Recently, a general framework for enumerating every element in a graph class was given. The main feature of this framework was that it was designed to enumerate only non-isomorphic graphs in a graph class. Applying this framework to the classes of interval graphs and permutation graphs, we gave efficient enumeration algorithms for these graph classes such that each element in the class was output in polynomial time delay. In this paper, we investigate the class of Ptolemaic graphs that consists of graphs that satisfy Ptolemy inequality for the distance. From the viewpoint of graph classes, it is an intersection of the class of chordal graphs and the class of distance-hereditary graphs. To enumerate Ptolemaic graphs, we need more tricks for applying the general framework. We develop an efficient enumeration algorithm for non-isomorphic Ptolemaic graphs.

1 Introduction

Recently we have to process huge amounts of data in the area of data mining, bioinformatics, etc. When we solve some problems on them, we sometimes use some certain structure common in data since we cannot solve the problems efficiently otherwise. When we deal with such a structure, essentially different instances have to be enumerated efficiently. From the viewpoint of graphs, it is natural to consider that two graphs are different when they are non-isomorphic. However, in general, it is unknown whether graph isomorphism can be solved efficiently, even on quite restricted graph classes (see [12]). On the other hand, even if we restrict ourselves to the graph classes that allow us to solve graph isomorphism efficiently, such graph classes still have certain structures with many applications.

We investigate the enumeration of all graphs that belong to a graph class from this viewpoint in this paper. In this context, there are two early results [9,10]. In these papers, the authors gave efficient enumeration algorithms for proper interval graphs and bipartite permutation graphs. Later, these algorithms have been implemented [3,5], and the lists of these graphs are published on a website[1]. For example, proper interval graphs are enumerated up to $n = 23$ so far, where

---

[1] http://www.jaist.ac.jp/~uehara/graphs.

M. S. Rahman et al. (Eds.): WALCOM 2020, LNCS 12049, pp. 296–307, 2020.
https://doi.org/10.1007/978-3-030-39881-1_25

n is the number of vertices. Since these graph classes have simple structures, the amount of hard disk drive is the bottleneck to publish on the website for them.

On the other hand, a general framework that gives us to enumerate all non-isomorphic graphs in some graph classes such that the graph isomorphism can be solved efficiently on the class is investigated recently [15]. In this context, enumeration algorithms for interval graphs and permutation graphs have been developed, and the data sets of these graph classes are on public on the above website. The class of permutation graphs is enumerated up to $n = 8$, and the class of interval graphs is enumerated up to $n = 12$ on the website. Recently, the result is updated to $n = 15$ for interval graphs [8]. For these graph classes, since they have richer structure, the computation time is the bottleneck to enumerate.

From the viewpoint of graph theory, the graph classes above have natural geometric representations, and these properties help to design the enumeration algorithms (see, e.g., [13] for further details). In this paper, we focus on Ptolemaic graphs. A graph is Ptolemaic if and only if any four vertices satisfy Ptolemaic inequality, which is originally defined in Euclidean geometry. This geometric property, however, does not give us any geometric representations. In the context of the graph classes, the class of Ptolemaic graphs is an intersection of the classes of chordal graphs and distance-hereditary graphs. The graph isomorphism problem is intractable for chordal graphs, while it is tractable for distance-hereditary graphs. From these graph classes, Ptolemaic graphs inherit useful properties, and based on these properties, a laminar structure of maximal cliques has been proved [14], and it admits us to solve the graph isomorphism efficiently. In this paper, we give an enumeration algorithm for Ptolemaic graphs. More precisely, our enumeration algorithm outputs all non-isomorphic Ptolemaic graphs with n vertices for a given integer n, and the time complexity of the interval between two consecutive Ptolemaic graphs is bounded above by a polynomial time.

2   Preliminaries

We only consider simple graphs $G = (V, E)$ with no self-loop and multi edges. We assume $V = \{v_1, v_2, \ldots, v_n\}$ for some n and $|E| = m$. Let K_n denote the complete graph of n vertices and P_n denote the path of n vertices of length $n - 1$. We define a *graph isomorphism* between two graphs $G_1 = (V_1, E_1)$ and $G_2 = (V_2, E_2)$ as follows. The graph G_1 is isomorphic to G_2 when there is a one-to-one mapping $\phi : V_1 \to V_2$ such that for any pair of vertices $u, v \in V_1$, $\{u, v\} \in E_1$ if and only if $\{\phi(u), \phi(v)\} \in E_2$. We denote by $G_1 \sim G_2$ for two isomorphic graphs G_1 and G_2.

The *neighborhood* of a vertex v in a graph $G = (V, E)$ is the set $N_G(v) = \{u \in V \mid \{u, v\} \in E\}$, and the *degree* of a vertex v is $|N_G(v)|$ and it is denoted by $\deg_G(v)$. For a subset U of V, we denote by $N_G(U)$ the set $\{v \in V \mid v \in N(u) \text{ for some } u \in U\}$. If no confusion can arise, we will omit the index G. We denote the closed neighborhood $N(v) \cup \{v\}$ by $N[v]$. Given a graph $G = (V, E)$ and a subset U of V, the *induced subgraph* by U, denoted by $G[U]$, is the graph (U, E'), where $E' = \{\{u, v\} \mid u, v \in U \text{ and } \{u, v\} \in E\}$. Given a graph

$G = (V, E)$, its *complement* is defined by $\bar{E} = \{\{u, v\} \mid \{u, v\} \notin E\}$, which is denoted by $\bar{G} = (V, \bar{E})$. A vertex set I is an *independent set* if $G[I]$ contains no edges, and then the graph $\bar{G}[I]$ is said to be a *clique*. Two vertices u and v are said to be a pair of *twins* if $N(u) \setminus \{v\} = N(v) \setminus \{u\}$. For a pair of twins u and v, we say that they are *strong twins* if $\{u, v\} \in E$, and *weak twins* if $\{u, v\} \notin E$.

Given a graph $G = (V, E)$, a sequence of the distinct vertices $v_1, v_2, \ldots, v_l$ is a *path*, denoted by $(v_1, v_2, \ldots, v_l)$, if $\{v_j, v_{j+1}\} \in E$ for each $1 \le j < l$. The *length* of a path is the number of edges on the path. For two vertices u and v, the *distance* of the vertices, denoted by $d(u, v)$, is the minimum length of the paths joining u and v. A *cycle* is a path beginning and ending at the same vertex.

An edge which joins two vertices of a cycle but it is not an edge of the cycle is a *chord* of the cycle. A graph is *chordal* if each cycle of length at least 4 has a chord. Given a graph $G = (V, E)$, a vertex $v \in V$ is *simplicial* in G if $G[N(v)]$ is a clique in G.

It is also known that a graph $G = (V, E)$ is chordal if and only if it is the intersection graph of subtrees of a tree T (see [2, Section 1.2] for further details). Let T_v denote the subtree of T corresponding to the vertex v in G. Then we can assume that each node c in T corresponds to a maximal clique C of G such that C contains v on G if and only if T_v contains c on T^2. Such a tree T is called a *clique tree* of G. From a perfect elimination ordering of a chordal graph G, we can construct a clique tree of G in linear time [11]. We sometimes identify a node c of a clique tree T with a maximal clique (or a vertex set) C of G.

Given a graph $G = (V, E)$ and a subset U of V, an induced connected subgraph $G[U]$ is *isometric* if the distances in $G[U]$ are the same as in G. A graph G is *distance-hereditary* if G is connected and every induced path in G is isometric. In other words, a connected graph G is distance-hereditary if and only if all induced paths are shortest paths.

A connected graph G is *Ptolemaic* if for any four vertices u, v, w, x of G, $d(u, v)d(w, x) \le d(u, w)d(v, x) + d(u, x)d(v, w)$. The following characterization of Ptolemaic graphs is due to Howorka [4]:

Theorem 1. *The following conditions are equivalent: (1) G is Ptolemaic; (2) G is distance-hereditary and chordal; (3) for all distinct non-disjoint maximal cliques P, Q of G, $P \cap Q$ separates $P \setminus Q$ and $Q \setminus P$.*

By Theorem 1(2), we can obtain the following:

Corollary 1 (Folklore). *Any Ptolemaic graph G can be obtained from K_1 by repeating the following two operations: (1) adding a leaf (of degree 1), and (2) split a vertex v into two strong twins v and v' with $N[v] = N[v']$.*

Let V be a set of n vertices. Two sets X and Y are said to be *overlapping* if $X \cap Y \ne \emptyset$, $X \setminus Y \ne \emptyset$, and $Y \setminus X \ne \emptyset$. A family $\mathcal{F} \subseteq 2^V \setminus \{\emptyset\}\}$ is said to be *laminar* if $\mathcal{F}$ contains no overlapping sets; that is, for any pair of two distinct

---

[2] We use two terms "node" and "vertex" to indicate an element in a graph. When we use "vertex," it indicates a vertex in the original chordal graph G. On the other hand, when we use "node," it indicates a node of the clique tree.

sets X and Y in $\mathcal{F}$ satisfy either $X \cap Y = \emptyset$, $X \subset Y$, or $Y \subset X$. Given a laminar family $\mathcal{F}$, we define *laminar digraph* $\overrightarrow{T}(\mathcal{F}) = (\mathcal{F}, \overrightarrow{E}_\mathcal{F})$ as follows; $\overrightarrow{E}_\mathcal{F}$ contains an arc (X, Y) if and only if $X \subset Y$ and there are no other subset Z such that $X \subset Z \subset Y$, for any sets X and Y. We denote the underlying graph of $\overrightarrow{T}(\mathcal{F})$ by $T(\mathcal{F}) = (\mathcal{F}, E_\mathcal{F})$. The following two lemmas for the laminar digraph are known (see, e.g., [7, Chapter 2.2]);

Lemma 1. *Let $\mathcal{F}$ be a laminar family over V. Then, (1) $T(\mathcal{F})$ is a forest, and (2) $|\mathcal{F}| \leq 2|V| - 1$.*

Hence, hereafter, we call $T(\mathcal{F})$ ($\overrightarrow{T}(\mathcal{F})$) a (directed) laminar forest. We regard each maximal (directed) tree in the laminar forest $T(\mathcal{F})$ ($\overrightarrow{T}(\mathcal{F})$) as a (directed) tree rooted at the maximal set, whose outdegree is 0 in $\overrightarrow{T}(\mathcal{F})$.

We define a *label* of each node S_0 in $\overrightarrow{T}(\mathcal{F})$, denoted by $\ell(S_0)$, as follows: If S_0 is a leaf, $\ell(S_0) = S_0$. If S_0 is not a leaf and has children $S_1, S_2, \ldots, S_h$, $\ell(S_0) = S_0 \setminus (S_1 \cup S_2 \cup \cdots \cup S_h)$. That is, each vertex v in V appears in $\ell(S)$ where S is the minimal set containing v. Since $\mathcal{F}$ is laminar, each vertex in V appears exactly once in $\ell(S)$ for some $S \subseteq V$, and its corresponding node is uniquely determined. We note that internal nodes in $\overrightarrow{T}(\mathcal{F})$ have a label $\emptyset$ when it is partitioned completely by its subsets in $\mathcal{F}$. (For example, for $V = \{a, b\}$ and $\mathcal{F} = \{X = \{a, b\}, Y = \{a\}, Z = \{b\}\}$, we have $\ell(X) = \{\emptyset\}, \ell(Y) = \{a\}$, and $\ell(Z) = \{b\}$.)

2.1 Tree Structure for a Ptolemaic Graph

Uehara and Uno characterized the class of Ptolemaic graphs by the laminar structure of maximal cliques [14]. We here describe the detail of the notion of *clique laminar tree*, *CL-tree* for short, and some properties which are not shown in [14] and that will be used in our algorithm. For a Ptolemaic graph $G = (V, E)$, let $\mathcal{M}(G)$ be the set of all maximal cliques, i.e.,

$$\mathcal{M}(G) := \{M \mid M \text{ is a maximal clique in } G\},$$

and $\mathcal{C}(G)$ be the set of nonempty vertex sets defined below:

$$\mathcal{C}(G) := \bigcup_{S \subseteq \mathcal{M}(G)} \{C \mid C = \cap_{M \in S} M, C \neq \emptyset\}.$$

That is, each vertex set $C \in \mathcal{C}(G)$ is a nonempty intersection of some maximal cliques. Hence, $\mathcal{C}(G)$ contains all maximal cliques, and each C in $\mathcal{C}(G)$ induces a clique. We also denote by $\mathcal{L}(G)$ the set $\mathcal{C}(G) \setminus \mathcal{M}(G)$. That is, each vertex set $L \in \mathcal{L}(G)$ is an intersection of two or more maximal cliques, and hence L is a non-maximal clique.

Theorem 2. *[14] Let $G = (V, E)$ be a Ptolemaic graph. Let $\mathcal{F}$ be a family of sets in $\mathcal{L}(G)$ such that $\cup_{L \in \mathcal{F}} L \subset M$ for some maximal clique $M \in \mathcal{M}(G)$. Then $\mathcal{F}$ is laminar.*

For a given Ptolemaic graph $G = (V, E)$, we generalize the notion of the (directed) laminar forest of a laminar family to a directed graph $\overrightarrow{T}(\mathcal{G}) = (\mathcal{C}(G), A(G))$ as follows; two nodes $C_1, C_2 \in \mathcal{C}(G)$ are joined by an arc (C_1, C_2) if and only if $C_1 \subset C_2$ and there is no other C in $\mathcal{C}(G)$ such that $C_1 \subset C \subset C_2$. We denote by $T(\mathcal{G})$ the underlying graph of $\overrightarrow{T}(\mathcal{G})$. Then a Ptolemaic graph can be characterized as follows:

Theorem 3. *[14] A graph $G = (V, E)$ is Ptolemaic if and only if the graph $T(\mathcal{G})$ is a tree. Moreover, $\overrightarrow{T}(\mathcal{G})$ is canonical; that is, $\overrightarrow{T}(\mathcal{G})$ is uniquely determined by G up to isomorphism.*

Given a Ptolemaic graph $G = (V, E)$, we call $T(\mathcal{G})$ ($\overrightarrow{T}(\mathcal{G})$) a *(directed) CL-tree* of G. We note that $\overrightarrow{T}(\mathcal{G})$ can contain two or more nodes of indegree 0. The nodes of indegree 0 in $\overrightarrow{T}(\mathcal{G})$ are called *roots*. If an arc (C, C') is in $\overrightarrow{T}(\mathcal{G})$, we say C is a *parent* of C', and C' is a *child* of C. We note that a node may have two or more parents. *Ancestors* and *descendants* are defined as in a normal way.

Hereafter, to simplify, we assume that a Ptolemaic graph G is not K_n. In this special case, its CL-tree contains one node. In the other cases below, we assume that any CL-tree contains at least two nodes without loss of generality.

We extend the label of a laminar forest to the directed CL-tree in a natural way: Each node C_0 in $\mathcal{C}(G)$ has a label $\ell(C_0) := C_0 \setminus (C_1 \cup C_2 \cup \cdots \cup C_h)$, where (C_i, C_0) is an arc on $\overrightarrow{T}(\mathcal{C}(G))$ for $1 \le i \le h$. Intuitively, we additionally define the label of a maximal clique as follows; the label of a maximal clique is the set of vertices which are not contained in any other maximal cliques. For the $\overrightarrow{T}(\mathcal{G})$, each node of degree 1 in $T(G)$ is said to be a *leaf*, and the other nodes are called *inner nodes*.

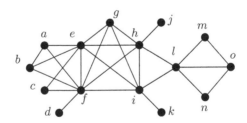

Fig. 1. An example of a Ptolemaic graph G.

For a Ptolemaic graph $G = (V, E)$ given in Fig. 1, its CL-tree $\overrightarrow{T}(\mathcal{G})$ is illustrated in Fig. 2. In Fig. 2, a black square represents a maximal clique, while a white square represents a non-maximal clique. Each upper label describes the clique C, and lower label describes its label $\ell(C)$. We note that the number of nodes of the CL-tree is at most $2|V|$, and each vertex in V appears exactly once in a label $\ell(C)$ for some node C which corresponds to a clique in G.

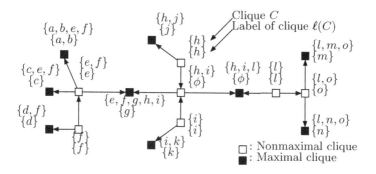

Fig. 2. The CL-tree of G.

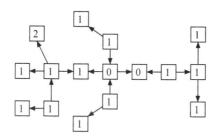

Fig. 3. The directed labeled tree induced by the CL-tree.

Here we show a lemma which is useful from the algorithmic point of view:

Lemma 2. *Let $G = (V, E)$ be a Ptolemaic graph, and $\overrightarrow{T}(\mathcal{G})$ the CL-tree of G. Then we have the following;*

(1) each node in $\mathcal{L}(G)$ has outdegree at least 2 in $\overrightarrow{T}(\mathcal{G})$,

(2) each node in $\mathcal{M}(G)$ has outdegree 0 in $\overrightarrow{T}(\mathcal{G})$, and

(3) each leaf in $\overrightarrow{T}(\mathcal{G})$ corresponds to a maximal clique in $\mathcal{M}(G)$, and hence the node of degree 1 in $T(\mathcal{G})$ has indegree 1 in $\overrightarrow{T}(\mathcal{G})$.

Proof. We first remind that $\mathcal{M}$ contains all maximal cliques, and $\mathcal{L}$ contains all cliques that are intersections of 2 or more cliques in $\mathcal{M}$. Since there is no clique that properly contains a maximal clique M, the node M in $\mathcal{M}(G)$ has outdegree 0 in $\overrightarrow{T}(\mathcal{G})$, which states (2).

Now we show that there is no clique in $\mathcal{C} = \mathcal{M} \cup \mathcal{L}$ with outdegree 1 in $\overrightarrow{T}(\mathcal{G})$. This fact implies (1) and (3). To derive contradictions, we assume that there is a node C in $\mathcal{C}$ with outdegree 1 in $\overrightarrow{T}(\mathcal{G})$. Let C' be the unique node with an arc (C, C') in $\overrightarrow{T}(\mathcal{G})$. Then C' properly contains C. Hence C cannot be a maximal clique (and hence we have (3)). Thus, from the definition, C can be represented by $C_1 \cap C_2 \cap \cdots \cap C_k$ for some maximal cliques C_i with $1 \leq i \leq k$. If $C' \not\subseteq C_i$ for some i, letting $C'' := C' \cap C_i$, we have $C \subset C'' \subset C'$, which contradicts that (C, C') is an arc in $\overrightarrow{T}(\mathcal{G})$. Hence we have $C' \subset C_i$ for all i, which implies that $C' \subset C_1 \cap C_2 \cap \cdots \cap C_k$. Hence we have $C_1 \cap C_2 \cap \cdots \cap C_k = C \subset C' \subset C_1 \cap C_2 \cap \cdots \cap C_k$, which is a contradiction. Hence we have (2).   □

We note that some maximal cliques in $\mathcal{M}(G)$ are not leaves (in Fig. 2, the maximal cliques $\{e, f, g, h, i\}$ and $\{h, i, l\}$ are in the case); they properly contain some independent cliques, have indegree greater than 1, and are inner nodes in $\overrightarrow{T}(\mathcal{G})$.

Let $\overrightarrow{T}(\mathcal{G})$ be the CL-tree for a Ptolemaic graph G. Then, by Lemma 2(1) and (2), we can determine if each node is maximal clique or not; maximal clique has outdegree 0 and non-maximal clique has outdegree at least 2. Hence we have the following observation:

Observation 1. *A CL-tree contains the following four kinds of nodes; (1) nodes C_1 of indegree 0 and of outdegree ≥ 2, which are roots of tree, (2) nodes C_2 of indegree ≥ 1 and of outdegree > 2, (3) nodes C_3 of indegree > 1 and of outdegree 0, which are maximal cliques that contain two or more distinct cliques in $\mathcal{L}$, and (4) nodes C_4 of indegree 1 and of outdegree 0, which are maximal cliques and leaves in $\overrightarrow{T}(\mathcal{G})$.*

We here define *size* of each node C by the number of vertices in the clique C, and it is denoted by $|C|$. Let C be any node and $P_1, P_2, \ldots, P_k$ the parents of C. We say C is *consistent* if $|C| \geq \sum_{i=1}^{k} |P_i|$ when $k \geq 2$ or $|C| > |P_1|$ when $k = 1$. The following fact is easy but important:

Fact 1. *Let $\overrightarrow{T}$ be a directed graph such that its underlying graph T is a tree. We also assume that some integer size is assigned to each node C in T, which is denoted by $|C|$. Then $\overrightarrow{T}$ is the CL-tree of some Ptolemaic graph G if and only if $\overrightarrow{T}$ contains four kinds of nodes stated in Observation 1 and each node C in $\overrightarrow{T}$ is consistent.*

Proof. The only-if part is easy. The if part can be shown by a simple induction for the number of nodes in $\overrightarrow{T}$.   □

We define a directed labeled tree $\overrightarrow{LT}(\mathcal{G})$ induced by $\overrightarrow{T}(\mathcal{G})$ as follows (Fig. 3): The tree itself is the same as $\overrightarrow{T}(\mathcal{G})$, and each label of a node C in $\overrightarrow{LT}(\mathcal{G})$ is the number $\ell(C)$. By Theorem 3 and Fact 1, we can observe that two Ptolemaic graphs G_1 and G_2 are isomorphic if and only if $\overrightarrow{LT}(\mathcal{G}_1)$ is isomorphic to $\overrightarrow{LT}(\mathcal{G}_2)$.

3   Enumeration Algorithms

As shown in Sect. 2.1, a Ptolemaic graph $G = (V, E)$ is identified by a directed labeled tree $\overrightarrow{LT}(\mathcal{G})$ induced by $\overrightarrow{T}(\mathcal{G})$ up to isomorphism. Therefore, basically, enumeration of Ptolemaic graphs is done by enumeration of non-isomorphic directed labeled trees. For the labels of nodes of a directed labeled tree, we have the following lemma (when G is not K_n):

Lemma 3. *A CL-tree contains the following four kinds of nodes; (1) nodes C_1 of indegree ≤ 1 and of outdegree ≥ 2 with $\ell(C_1) \geq 1$, (2) nodes C_2 of indegree ≥ 2 and of outdegree ≥ 2 with $\ell(C_2) \geq 0$, (3) nodes C_3 of indegree ≥ 2 and of outdegree 0 $\ell(C_3) \geq 0$, and (4) nodes C_4 of indegree ≤ 1 and of outdegree 0 with $\ell(C_4) \geq 1$. We note that C_1 and C_2 are non-maximal cliques, and C_3 and C_4 are maximal cliques. Especially, C_4 is a leaf. (The indegree of C_4 is 0 if and only if the graph is K_n; otherwise, it is 1.)*

Proof. Combining Lemma 2, Observation 1, and Fact 1, the claims follow. □

We call a Ptolemaic graph G *minimal* if each node satisfies $\ell(C_1) = 1$, $\ell(C_2) = 0$, $\ell(C_3) = 0$, or $\ell(C_4) = 1$ according to its type in Lemma 3 in its directed labeled tree $\overrightarrow{LT}(\mathcal{G})$ induced by the CL-tree $\overrightarrow{T}(\mathcal{G})$. Intuitively, a minimal Ptolemaic graph G has the minimum number of vertices in the Ptolemaic graphs having the isomorphic directed labeled tree regardless of labels. For example, we can obtain a minimal Ptolemaic graph from the graph in Fig. 1 by removing the vertices b and g. For a minimal Ptolemaic graph, we also call its CL-tree and directed labeled tree of it *minimal*.

Now we are ready to give an outline of our enumeration algorithm for Ptolemaic graphs:

1. Enumerate all minimal CL-trees $\overrightarrow{T}$ that contain at most n vertices.
2. For each $\overrightarrow{T}$, we add extra vertices into cliques to make it have n vertices.

Since these two steps can be considered separately, we first show the enumeration algorithm for minimal CL-trees, and we next show how we can distribute extra vertices into each CL-tree having less than n vertices. Due to space limitation, we only give the outline of each algorithm without proofs.

3.1   Enumeration of Minimal CL-trees

In this section, we enumerate minimal CL-trees. In order to do that efficiently, we maintain each CL-tree $\overrightarrow{T}$ by a directed tree (without labels) with the total number of vertices in it. Remind that we can distinguish maximal cliques from non-maximal cliques in a CL-tree by Lemma 2 by their degrees. We use *reverse search* technique to enumerate minimal CL-trees (see [1] for the details about reverse search). In reverse search, we define a family tree $\hat{T}$ over the CL-trees $\overrightarrow{T}(\mathcal{G})$ by introducing a parent-child relationship between two CL-trees $\overrightarrow{T}(\mathcal{G})$ and $\overrightarrow{T}(\mathcal{G}')$. In the class of minimal CL-trees, we first fix the root vertex[3] G_R in the family tree. The root node G_R is defined by $G_R = K_1$, which is a vertex. In this case, the minimal CL-tree also consists of one node with label 1. In order to define a parent-child relationship, we need two notions.

---

[3] In this context, when we use terms "node" in a minimal CL-tree $\overrightarrow{T}(\mathcal{G})$ and "vertex" in a family tree $\hat{T}$ to distinguish with them.

Canonical Representation: The first notion is a *canonical representation* of a CL-tree $\overrightarrow{T}(\mathcal{G})$, which is a directed tree. In this paper, we define a canonical representation of a CL-tree $\overrightarrow{T}(\mathcal{G})$ as follows. We first consider the underlying (undirected) tree $T(\mathcal{G})$. Then we pick up *central* nodes of $T(\mathcal{G})$ which are given by the farthest nodes from leaves. It is well known that any undirected tree has at most two central nodes. Thus we have two cases; one central node and two central nodes.

We first consider the case that the tree $T(\mathcal{G})$ has one central node c. From this root node c, we perform the depth first search (DFS) on $T(\mathcal{G})$. Then we draw $T(\mathcal{G})$ in the *left-heavy* manner (see [6] for the notion of left-heavy). We here note that it is easy to extend the notion of the left-heavy defined on undirected trees to directed trees. On this tree, by the DFS of depths (or integers) of nodes, we can obtain a canonical string representation of $T(\mathcal{G})$.

Next we consider the case that the tree $T(\mathcal{G})$ has two central nodes c_1 and c_2. In this case, c_1 and c_2 are neighbors on $T(\mathcal{G})$. We then define the parent by the arc on $\overrightarrow{T}(\mathcal{G})$. That is, if (c_1, c_2) is an arc in $\overrightarrow{T}(\mathcal{G})$, c_1 is the parent of c_2. Otherwise, c_2 is the parent of c_1.

Table 1. Three operations to generate minimal CL-trees

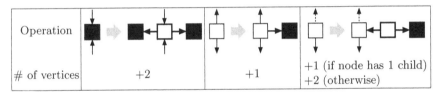

Operation			
# of vertices	+2	+1	+1 (if node has 1 child) +2 (otherwise)

Rules for Generation of CL-trees: Now we consider generation rules for CL-trees:

Lemma 4. *From G_R, applying one of the three operations given in Table 1, we can generate any minimal CL-tree $\overrightarrow{T}(\mathcal{G})$.*

Proof. (Sketch) Intuitively, these operations correspond to Corollary 1(1), and they add a leaf to the current graph. However, to update the laminar property, we sometimes need to add an extra vertex. Therefore, the number of vertices in the Ptolemaic graph increases 1 or 2 according to the case. In the third operation, we add two nodes into the CL-tree and add two vertices to the Ptolemaic graph if the clique has 0, 2 or more children. If it has only one child, we add two nodes into the CL-tree and add two vertices to the Ptolemaic graph, but we have to remove one vertex from the original clique. Therefore, the number of vertices is increased by +1(= +2 − 1) in total.   □

We note that we can apply the first operation on $G = G_R$. That is, G_R has the unique child by applying the first operation, and we obtain a path of length 2. Precisely, from a clique $\{v_0\}$, we obtain two maximal cliques $\{v_0, v_1\}$ and $\{v_0, v_2\}$.

Now we are ready to define the parent-child relationship in the class of CL-trees. For each graph $G \in \mathcal{C} \setminus \{G_R\}$, we have the corresponding CL-tree $\overrightarrow{T}(\mathcal{G})$. Then, since G is not G_R, the corresponding undirected labeled tree $T(\mathcal{G})$ has at least two leaves. Let $L_1, L_2, \ldots$ be these leaves. For each leaf, we can remove it by applying the reverse of the generating operation shown in Table 1. Now, in the canonical string representation of $T(\mathcal{G})$, each leaf L_i appears exactly once since the string is generated in the DFS manner. We apply the reverse operation on the last leaf L_i in the string. Let G' be the new graph obtained by this reverse operation. Intuitively, G' is obtained by removing one maximal clique from G. (Precisely, we remove one pendant vertex or one pendant vertex with its neighbor from G.) We define the unique *parent* of G by this G'.

We now show the main theorem in this section:

Theorem 4. *All minimal Ptolemaic graphs with at most n vertices can be enumerated in polynomial time delay.*

Proof. By the definitions and discussion above, for any Ptolemaic graph $G \in \mathcal{C} \setminus \{G_R\}$, we can compute its unique parent G' in polynomial time. Therefore, using the same framework in [15], we can enumerate all non-isomorphic children of any minimal Ptolemaic graph G in polynomial time delay. Thus we have the theorem. $\qquad\square$

3.2 Distribution of Extra Vertices

In this section, we consider how we can add extra $n - n'$ vertices into a minimal CL-tree $\overrightarrow{T}$ having n' vertices. That is, we are given a minimal CL-tree $\overrightarrow{T}$, and its corresponding Ptolemaic graph G has n' vertices with $n' < n$ for given n. Let m be the number of nodes in $\overrightarrow{T}$. Since each clique C satisfies $0 \le \ell(C) \le 1$ in a minimal CL-tree, we have $m \ge n'$.

The basic idea is simple: we distribute $n - n'$ extra vertices of G into m nodes in the minimal CL-tree $\overrightarrow{T}$. When a vertex v is added into a clique C in $\overrightarrow{T}$, the clique has one more vertex. If $\ell(C)$ contains another vertex u, we make v as a strong twin of u. That is, we have $N[u] = N[v]$ after adding. When $\ell(C)$ is $\emptyset$, each vertex in C is shared by some other maximal cliques. However, it is not difficult to see that $N(v)$ is given by $\cup_{u \in C} N[u]$.

The considerable point is the inner automorphism of the minimal CL-tree $\overrightarrow{T}$. When $n - n' = 1$ and $\overrightarrow{T}$ has two isomorphic directed subtrees $\overrightarrow{T_1}$ and $\overrightarrow{T_2}$, adding one extra vertex into T_1 and adding one extra vertex into T_2, the algorithm may enumerate one Ptolemaic graph twice. However, this happens only when T_1 and T_2 share the same parent. Therefore, the algorithm avoids the duplicates as follows:

1. We fix the central vertex of $\overrightarrow{T}$.
2. For each inner node, check if it has isomorphic subtrees.
3. When it has two or more subtrees $\overrightarrow{T_1}, \overrightarrow{T_2}, \ldots$, we assume that $\#\overrightarrow{T_1} < \#\overrightarrow{T_2} < \cdots$, where $\#\overrightarrow{T_i}$ is the number of extra vertices put into this subtree.

After checking the inner automorphism, the distribution of the extra vertices is not difficult to do that in polynomial time for $n - n'$ and m. Since $m = O(n')$, we have the following theorem:

Theorem 5. *Let $\overrightarrow{T}$ be a minimal CL-tree, and n', m be the number of total vertices in $\overrightarrow{T}$ and the number of nodes of $\overrightarrow{T}$, respectively. For a given integer n with $n > n'$, we can enumerate all CL-trees that contain exactly n vertices in total in polynomial time delay.*

In fact, the delay can be linear of n' and n, which is omitted here.

By Theorems 4 and 5, we can enumerate all Ptolemaic graphs with n vertices in polynomial time delay.

4   Concluding Remarks

In this paper, we show that we can enumerate non-isomorphic Ptolemaic graphs with n vertices in polynomial time delay. As described in Introduction, we have already published some catalogs of some graph classes. The implementation of our algorithm and publishing the catalog of Ptolemaic graphs are future work. We also mention that the graph isomorphism is tractable on distance-hereditary graphs. Therefore, enumeration of distance-hereditary graphs is the next reasonable target.

Acknowledgements. This work is partially supported by KAKENHI grant numbers 17H06287 and 18H04091.

References

1. Avis, D., Fukuda, K.: Reverse search for enumeration. Discrete Appl. Math. **65**, 21–46 (1996)
2. Brandstädt, A., Le, V.B., Spinrad, J.P.: Graph Classes: A Survey. SIAM, Philadelphia (1999)
3. Harasawa, S., Uehara, R.: Efficient enumeration of connected proper interval graphs. IEICE Technical report COMP2018-44, IEICE, pp. 9–16, March 2019. (in Japanese)
4. Howorka, E.: A characterization of ptolemaic graphs. J. Graph Theory **5**, 323–331 (1981)

5. Ikeda, S., Uehara, R.: Implementation of enumeration algorithm for connected bipartite permutation graphs. IEICE Technical report COMP2018-45, IEICE, pp. 17–23, March 2019. (in Japanese)
6. Nakano, S., Uno, T.: Constant time generation of trees with specified diameter. In: Hromkovič, J., Nagl, M., Westfechtel, B. (eds.) WG 2004. LNCS, vol. 3353, pp. 33–45. Springer, Heidelberg (2004). https://doi.org/10.1007/978-3-540-30559-0_3
7. Korte, B., Vygen, J.: Combinatorial Optimization. AC, vol. 21. Springer, Heidelberg (2018). https://doi.org/10.1007/978-3-662-56039-6
8. Mikos, P.: Efficient enumeration of non-isomorphic interval graphs. arXiv:1906.04094, June 2019
9. Saitoh, T., Otachi, Y., Yamanaka, K., Uehara, R.: Random generation and enumeration of bipartite permutation graphs. J. Discrete Algorithms **10**, 84–97 (2012). https://doi.org/10.1016/j.jda.2011.11.001
10. Saitoh, T., Yamanaka, K., Kiyomi, M., Uehara, R.: Random generation and enumeration of proper interval graphs. IEICE Trans. Inf. Syst. **E93**(D(7)), 1816–1823 (2010)
11. Spinrad, J.P.: Efficient Graph Representations. American Mathematical Society, Providence (2003)
12. Uehara, R., Toda, S., Nagoya, T.: Graph isomorphism completeness for chordal bipartite graphs and strongly chordal graphs. Discrete Appl. Math. **145**(3), 479–482 (2004)
13. Uehara, R.: The graph isomorphism problem on geometric graphs. Discrete Math. Theoret. Comput. Sci. **16**(2), 87–96 (2014)
14. Uehara, R., Uno, Y.: Laminar structure of ptolemaic graphs with applications. Discrete Appl. Math. **157**(7), 1533–1543 (2009)
15. Yamazaki, K., Saitoh, T., Kiyomi, M., Uehara, R.: Enumeration of nonisomorphic interval graphs and nonisomorphic permutation graphs. Theoret. Comput. Sci. (2019, accepted). https://doi.org/10.1016/j.tcs.2019.04.017

Faster Privacy-Preserving Computation
of Edit Distance with Moves

Yohei Yoshimoto[1], Masaharu Kataoka[1], Yoshimasa Takabatake[1], Tomohiro I[1],
Kilho Shin[2], and Hiroshi Sakamoto[1(✉)]

[1] Kyushu Institute of Technology, 680-4 Kawazu, Iizuka, Fukuoka 820-8502, Japan
y_yoshimoto@donald.ai.kyutech.ac.jp, m_kataoka@donald.ai.kyutech.ac.jp,
takabatake@ai.kyutech.ac.jp, tomohiro@ai.kyutech.ac.jp,
hiroshi@ai.kyutech.ac.jp
[2] Gakushuin University, 1-5-1 Mejiro, Toshimaku, Tokyo 171-8588, Japan
kilhoshin314@gmail.com

Abstract. We consider an efficient two-party protocol for securely computing the similarity of strings w.r.t. an extended edit distance measure. Here, two parties possessing strings x and y, respectively, want to jointly compute an approximate value for $\mathrm{EDM}(x,y)$, the minimum number of edit operations including substring moves needed to transform x into y, without revealing any private information. Recently, the first secure two-party protocol for this was proposed, based on homomorphic encryption, but this approach is not suitable for long strings due to its high communication and round complexities. In this paper, we propose an improved algorithm that significantly reduces the round complexity without sacrificing its cryptographic strength. We examine the performance of our algorithm for DNA sequences compared to previous one.

1 Introduction

1.1 Motivation

As the number of strings containing personal information has increased, privacy-preserving computation has become more and more important. Secure computation based on public key encryption is one of the great achievements of modern cryptography, as it enables untrusted parties to compute a function based on their private inputs while revealing nothing but the result.

In addition, edit distance is a well-established metric for measuring the similarity or dissimilarity of two strings. The rapid progress of gene sequencing technology has expanded the range of edit distance applications to include personalized genomic medicine, disease diagnosis, and preventive treatment (for example, see [1]). A person's genome is, however, ultimately individual information that uniquely identifies its owner, so the parties involved should not share their personal genomic data as plaintext.

Thus, we consider a secure multi-party edit distance computation based on the public key encryption model. Here, untrusted two parties generating their

© Springer Nature Switzerland AG 2020
M. S. Rahman et al. (Eds.): WALCOM 2020, LNCS 12049, pp. 308–320, 2020.
https://doi.org/10.1007/978-3-030-39881-1_26

own public and private keys have strings x and y, respectively, and want to jointly compute $f(x, y)$ for a given metric f without revealing anything about their individual strings.

1.2   Related Work

Homomorphic encryption (HE) based on the public key encryption model is an emerging technique that is being used for secure multi-party computation. The Paillier encryption system [20] possesses additive homomorphism, enabling us to perform additive operations on two encrypted integers *without decryption*. This means that parties can jointly compute the encrypted value $E(x + y)$ directly based only on two encrypted integers $E(x)$ and $E(y)$. HE is also *probabilistic*, i.e., an adversary can hardly predicts x given $E(x)$, even if they can observe some number of $(x', E(x'))$ pairs for any x'.

By taking advantage of these characteristics, researchers have proposed several HE-based privacy-preserving protocols for computing the Levenshtein distance $d(x, y)$. For example, Inan et al. [16] designed a three-party protocol where two parties securely compute $d(x, y)$ by enlisting the help of a reliable third party. Rane and Sun [21] then improved this three-party protocol to develop the first two-party one.

In this paper, we focus on an interesting metric called the *edit distance with moves* (EDM), where we allow any substring to be moved with unit cost in addition to the standard Levenshtein distance operations. Based on the EDM, we can find a set of approximately maximal common substrings appearing in two strings, which can be used to detect plagiarism in documents or long repeated segments in DNA sequences. As an example, consider two unambiguously similar strings $x = a^N b^N$ and $y = b^N a^N$, which can be transformed into each other by a single move. Whereas the exact EDM is simply $\text{EDM}(x, y) = 1$, the Levenshtein distance has the undesirable value $d(x, y) = 2N$. The n-gram distance is preferable to the Levenshtein's in this case, but it requires huge time/space complexity depending on N.

Although computing $\text{EDM}(x, y)$ is NP-hard [22], Cormode and Muthukrishnan [8] were able to find an almost linear-time approximation algorithm for it. Many techniques have been proposed for computing the EDM; for example, Ganczorz et al. [12] proposed a lightweight probabilistic algorithm. In these algorithms, each string x is transformed into a characteristic vector v_x consisting of nonnegative integers representing the frequencies of particular substrings of x. For two strings x and y, we then have the approximate distance guaranteeing $L_1(v_x, v_y) = O(\lg^* N \lg N)\text{EDM}(x, y)$ for $N = |x| + |y|$.[1] Since $\lg^* N$ increases extremely slowly[2], we employ $L_1(v_x, v_y)$ as a reasonable approximation to $\text{EDM}(x, y)$.

---

[1] In Appendix A of [10], the authors point out that there is a subtle flaw in the ESP algorithm [8] that achieves this $O(\lg^* N \lg N)$ bound. However, this flaw can be remedied by an alternative algorithm called HSP [10].

[2] $\lg^* N$ is the number of times the logarithm function $\lg$ must be iteratively applied to N until the result is at most 1.

Recently, Nakagawa et al. proposed the first secure two-party protocol for EDM (sEDM) [19] based on HE, but, their algorithm suffers from a bottleneck during the step where the parties construct a shared labeling scheme. This motivated us to improve the previous algorithm to make it easier to use in practice.

Table 1. Comparison of the communication and round complexities of secure EDM computation models. Here, N is the total length of both parties' input strings, n is the number of characteristic substrings determining the approximate EDM, and m is the range of the rolling hash $H(\cdot)$ for the substrings satisfying $m > n$. "Naive" is the baseline method that uses $H(\cdot)$ as the labeling function for the characteristic substrings. In this table, we omit the security parameter or the unit cost of encryption and decryption because the models use a same key length (e.g., 256-bit).

Method	Communication	Round
Ours	$O(n \lg n + m)$	$O(1)$
Naive	$O(m \lg m)$	$O(1)$
sEDM [19]	$O(n \lg n)$	$O(\lg N)$

1.3 Our Contribution

The complexities of our algorithm and related ones are summarized in Table 1. Computing the approximate EDM involves two phases: the shared labeling of characteristic substrings (Phase 1) and the L_1-distance computation of characteristic vectors (Phase 2). First, we outline those phases below. Let the parties have strings x and y, respectively. In the offline case (i.e., there is no need for privacy-preserving communication), they construct the respective parsing trees T_x and T_y by the bottom-up parsing called ESP [8] where the node labels must be *consistent*, i.e., two labels are equal if they correspond to the same substring. In such an ESP tree, a substring derived by an internal node is called a characteristic substring. In a privacy-preserving model, the two parties should jointly compute such consistent labels without revealing whether or not a characteristic substring is common to both of them (Phase 1). After computing all the labels in T_x and T_y, they jointly compute the L_1-distance of two characteristic vectors consisting the frequencies of all labels in T_x and T_y (Phase 2).

As reported in [19], in terms of usefulness, a bottleneck exists in Phase 1. The task is to design a bijection $f : X \cup Y \to \{1, 2, \dots, n\}$ where X and Y ($|X \cup Y| = n$) are the sets of characteristic substrings for the parties, respectively. Since X and Y are computable without communication, the goal is to jointly compute $f(w)$ for any $w \in X$ without revealing whether or not $w \in Y$. Here, this problem is closely related to the private set operation (PSO) where parties possessing their private sets want to obtain the results for several set operations, e.g., intersection or union. Applying the Bloom filter [4] and HE techniques, various protocols for PSO have been proposed [3,9,18]. However, these protocols cannot be directly applicable to our problem because these protocols require at least three parties for the security constrained. Thus, we propose a novel secure two-party protocol for Phase 1.

As shown in Table 1, we eliminate the $O(\lg N)$ round complexity using the proposed method that can achieve $O(1)$ round complexity while maintaining the efficiency of communication complexity. Furthermore, we examine the practical performance of our algorithm for real DNA sequences.

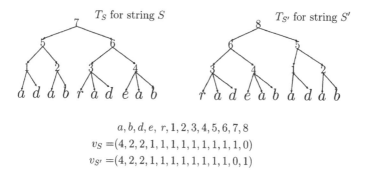

$$a, b, d, e, \, r, 1, 2, 3, 4, 5, 6, 7, 8$$
$$v_S = (4, 2, 2, 1, 1, 1, 1, 1, 1, 1, 1, 1, 0)$$
$$v_{S'} = (4, 2, 2, 1, 1, 1, 1, 1, 1, 1, 1, 0, 1)$$

Fig. 1. Example of approximate EDM computation for strings S and S'. Here, the ESP trees T_S and $T_{S'}$ are constructed by applying a shared labeling scheme for all internal nodes. After constructing T_S and $T_{S'}$, the corresponding characteristic vectors v_S and $v_{S'}$ are computed offline. Finally, the exact $\mathrm{EDM}(S, S')$ is approximated by $L_1(v_S, v_{S'}) = 2$.

2   Preliminaries

2.1   EDM

Let Σ be a finite set of alphabet symbols and Σ^* be its closure. Denote the set of all strings of the length N by Σ^N and the length of a string S by $|S|$. For simplicity, we also denote the cardinality of a set U by $|U|$. In addition, $S[i]$ denotes the i-th symbol of S and $S[i..j]$ denotes the substring $S[i]S[i+1]\cdots S[j]$.

Next, we define $\mathrm{EDM}(S, S')$ as the length of the shortest sequence of edit operations that transforms S into S', where the permitted operations (each with unit cost) are inserting, deleting, or renaming one symbol at any position and moving an arbitrary substring. Unfortunately, as Theorem 1 shows, computing $\mathrm{EDM}(S, S')$ is NP-hard even if renaming operations are not allowed [22], so we focus on an approximation algorithm for EDM, called ESP (Edit-Sensitive Parsing) [8].

Theorem 1 (Shapira and Storer [22]). *Determining* $\mathrm{EDM}(x, y)$ *is NP-hard even if only three unit-cost operations namely inserting or deleting a character and moving a substring are allowed.*

To illustrate the ESP algorithm, we consider a string $S \in \Sigma^*$. First, S is deterministically partitioned into blocks as $S = s_1 s_2 \cdots s_k$ such that $2 \leq |s_i| \leq 3$. Here, we omit the details since partitions are determined based on S alone, without communication. Next, a *consistent* label $\ell(s_i)$ is assigned to each block s_i, where $\ell(x) = \ell(y)$ if $x = y$. The resulting string $L = \ell(s_1) \cdots \ell(s_k)$ is then processed recursively until $|L| = 1$. Finally, a parsing tree T_S is obtained for S, which can be used to approximate $EDM(S, S')$ by applying the following result.

Theorem 2 (Cormode and Muthukrishnan [8]). *Let T_S and $T_{S'}$ be consistently labeled ESP trees for $S, S' \in \Sigma^*$, and let v_S be the characteristic vector for S, where $v_S[k]$ is the frequency of label k in T_S. Then,*

$$\frac{1}{2} EDM(S, S') \leq L_1(v_S, v_{S'}) = O(\lg^* N \lg N) EDM(S, S')$$

for $L_1(v_S, v_{S'}) = \sum_{i=1}^{k} |v_S[i] - v_{S'}[i]|$.

Figure 1 shows an example of applying consistent labeling to the trees T_S and $T_{S'}$, together with the resulting characteristic vectors. The strings S and S' are partitioned offline, so the problem of preserving privacy reduces to designing secure protocol for creating consistent labels and computing the L_1-distance between the trees. In this study, we propose a novel HE-based algorithm for the former problem.

2.2 Homomorphic Encryption

Here, we briefly review the framework of homomorphic encryption. Let (pk, sk) be a key pair for a public-key encryption scheme, and let $E_{pk}(x)$ be the encrypted values of message x and $D_{sk}(C)$ be the decrypted value of ciphertext C, respectively. We say that the encryption scheme is *additively homomorphic* if we have the properties: (1) There is an operation $h_+(\cdot, \cdot)$ for $E_{pk}(x)$ and $E_{pk}(y)$ such that $D_{sk}(h_+(E_{pk}(x), E_{pk}(y))) = x + y$. (2). For any r, we can compute the scalar multiplication such that $D_{sk}(r \cdot E_{pk}(x)) = r \cdot x$.

An additive homomorphic encryption scheme allowing sufficient number of these operations is called an additive HE[3]. Paillier's encryption scheme [20] is the first secure additive HE, but we cannot evaluate many functions by only the additive homomorphism and scalar multiplication.

On the other hand, the multiplication $D_{sk}(h_\times(E_{pk}(x), E_{pk}(y))) = x \cdot y$ is another important homomorphism. If we allow both additive and multiplicative homomorphism as well as scalar multiplication (called a fully homomorphic encryption, FHE [13] for short), it follows that we can perform any arithmetic operation on ciphertexts. For example, if we can use sufficiently number of additive operations and a single multiplicative operation over ciphertexts, we obtain the inner-product of two encrypted vectors.

---

[3] In general, the number of applicable operations over ciphertexts is bounded by the size of (pk, sk).

However, there is a tradeoff between the available homomorphic operations and their computational cost. To avoid this difficulty, we focus on the Leveled HE (LHE) where the number of homomorphic multiplications is restricted beforehand. In particular, *two-level* HE (Additive HE that allows a single homomorphic multiplication) has attracted a great deal of attention. BGN encryption system is the first two-level HE invented by Boneh et al. [5] assuming a single multiplication and sufficient numbers of additions. Using the BGN, we can securely evaluate formulas in disjunctive normal form (DNF). After this pioneering study, many practical two-level HE protocols have been proposed [2,7,11,15].

For the EDM computation, Nakagawa et al. [19] introduced an algorithm for computing the EDM based on two-level HE, but their algorithm is very slow for large strings. So, we propose another novel secure computation of EDM for large strings based on the faster two-level HE proposed by Attrapadung et al. [2]. As far as we know, there are no secure two-party protocols for the EDM computation that only use additive homomorphic property. Whether we can compute EDM on a two-party protocol based on additive HE only is an interesting question.

3 Two-Party Secure Consistent Labeling

3.1 Hash Function

In our protocol, two parties, $\mathcal{A}$ (Alice) and $\mathcal{B}$ (Bob), agree to use a shared hash function to assign tentative labels to their ESP trees. First, we consider the conditions that a hash function should satisfy. One desirable property of any hash function used for our algorithm is that its hash value should be uniformly distributed, and we assume this is true in this paper. In addition, the function involves a parameter m that represents the number of possible hash values, and this affects our algorithm's computational complexity, as well as the hash function's conflict resistance and one-wayness.

Computational Complexity. In our algorithm, $\mathcal{A}$ encrypts m individual bits and sends the resulting m ciphertexts to $\mathcal{B}$, who then adds or multiplies pairs of ciphertexts. Thus, our first requirement is that m be small enough for these computation to be performed efficiently.

Conflict Resistance. The hash function's conflict resistance affects the accuracy of the edit distance estimated by Algorithm 1. We say that a conflict occurs when two distinct texts happen to be hashed to the same value. Very roughly, if conflicts occur with probability p, the average proportional error in the edit distances is also $O(p)$. That said, we can create conditions where the probability of conflict is below some threshold p, as follows. Let n denote the number of labels (hash values) computed by the algorithm. To avoid conflicts, n must be sufficiently small relative to m. After computing n hash values at random, we can estimate the probability of at least one conflict having occurred as follows.

$$\Pr[\text{Conflicts}] = 1 - \Pr[\text{No conflict}]$$

$$= 1 - \left(1 - \frac{1}{m}\right)\left(1 - \frac{2}{m}\right) \cdots \left(1 - \frac{n-1}{m}\right)$$

$$\approx 1 - \exp\left(-\frac{n^2}{2m}\right).$$

Thus, to ensure this probability is below a given (small) threshold p, we require

$$n \leq -\ln(1 - p)\sqrt{2m}. \tag{1}$$

One-wayness. One-wayness is important for security. For example, if $\mathcal{A}$ happens to have two texts with hash values h and $h+2$, respectively, but does not have a text with hash value $h+1$, then $\mathcal{A}$ would know that $\mathcal{B}$ has a text with hash value $h+1$. If the hash function is not one-way, $\mathcal{A}$ could then guess the next. A function is theoretically one-way if it is computationally difficult to find an x such that $H(x) = y$ given y with non-negligible probability. Given an ideal hash oracle that selects $H(x)$ uniformly at random from $\{1, \ldots, m\}$ for any x, the probability of any guess x' being correct for an unknown x is exactly $\frac{1}{m}$. Thus, to ensure the function is effectively one-way, m must be sufficiently large.

It is known that, if the problem of finding a pair of distinct inputs that hash to the same value is computationally intractable (strong conflict-resistance), the hash function is also one-way. For cryptographic hash algorithms, such as, MD5 and SHA-1, strong conflict-resistance is required and the conflict probability must be negligibly small, such as less than $\frac{1}{2^{100}}$. This indeed requires that m be very large (2^{128} and 2^{160} for MD5 and SHA-1, respectively), so it would be computationally unfeasible to use a cryptographic hash function in our algorithm.

Nevertheless, if we relax the requirements somewhat, it is not difficult in practice to select an m that meets our needs. For example, if $n = 100$ and $p = 0.05$, then $m = 1,900,416$ satisfies inequality (1). Generating and transmitting so many ciphertexts would be time-consuming but still feasible, and would reduce the probability of breaking one-wayness to a very low value.

We should, however, note that these requirements on m are merely necessary conditions for conflict resistance and one-wayness. Even under these conditions, using a well-designed hash algorithm is still crucial.

The rolling hash algorithm [17], defined as follows based on two parameters m and b, is expected to be sufficiently conflict-resistant and one-way. For a given input $x = (s_1, \ldots, s_\ell) \in [0, b)^\ell$, the hash function is given by $H(x) = \sum_{i=1}^{\ell} s_i \cdot b^{\ell-i} \bmod m$. This algorithm has the useful advantage that we can compute $H(xy)$ from $H(x)$ and $H(y)$ in constant time, independent of the lengths of x and y.

3.2   Algorithm

Two parties $\mathcal{A}$ and $\mathcal{B}$ have strings $S_\mathcal{A}$ and $S_\mathcal{B}$, respectively. First, they compute the corresponding ESP trees $T_\mathcal{A}$ and $T_\mathcal{B}$ offline, using the rolling hash function to generate (tentative) consistent labels, thereby defining a set $X \subseteq \{0, 1, \ldots, m\}$ of n different labels in T_A and T_B with a fixed m. The algorithm's goal is to securely relabel X using by a bijection: $X \to \{1, 2, \ldots, n\}$, as described in Algorithm 1, where A and B have their own public and private keys.

In our algorithm, we assume a FHE (LHE) system supporting both additive and multiplicative operations. Since these operations are usually implemented by AND ($\cdot$) and XOR ($\oplus$) logic gates (e.g. [6]), we introduce several notations using such gates as follows. First, $E_A(x)$ denotes the ciphertext generated by encrypting plaintext x with A's public key, and $E_A(x, y, z)$ is an abbreviation for the vector $(E_A(x), E_A(y), E_A(z))$. Here, $E_A(x, y, z) \cdot E_A(a, b, c)$ denotes $(E_A(x \cdot a), E_A(y \cdot b), E_A(z \cdot c))$ and $E_A(x, y, z) \oplus E_A(a, b, c)$ denotes $(E_A(x \oplus a), E_A(y \oplus b), E_A(z \oplus c))$ for each bits $x, y, z, a, b, c \in \{0, 1\}$. Using these notations, we describe the proposed protocol in Algorithm 1.

Algorithm 1 for consistently labeling T_A and T_B

Preprocessing (tentative labeling): Parties A and B agree to use a shared hash function H with a range $\{0, \ldots, m\}$, where m is chosen so as to meet the requirements given in Section 3.1. Both parties compute the ESP trees T_A and T_B corresponding to their respective strings offline, then assign labels $H(w)$ to all the nodes in their trees based on their computed blocks w. Now, parties A and B have tentative label sets $[T_A], [T_B] \subseteq \{0, \ldots, m\}$, respectively.

Goal: Change all the labels using a bijection: $[T_A] \cup [T_B] \to \{1, \ldots, n\}$ without either party having to reveal anything about their private strings.

Notations: $E_A(x)$ denotes the ciphertext of a message x encrypted by a two-level HE with A's public key.

Sharing a dictionary:
Step 1: Party A computes the bit vector $\mathbf{X}[1..m]$ such that $\mathbf{X}[\ell] = 1$ iff $\ell \in [T_A]$. Similarly, party B computes $\mathbf{Y}[1..m]$ such that $\mathbf{Y}[\ell] = 1$ iff $\ell \in [T_B]$.
Step 2: A sends $E_A(\mathbf{X})$ to B and B sends $E_B(\mathbf{Y})$ to A.
Step 3: B computes $(E_A(\mathbf{X}) \oplus E_A(\mathbf{Y})) \oplus (E_A(\mathbf{X}) \cdot E_A(\mathbf{Y})) = E_A(\mathbf{X} \cup \mathbf{Y})$ and A computes $(E_B(\mathbf{X}) \oplus E_B(\mathbf{Y})) \oplus (E_B(\mathbf{X}) \cdot E_B(\mathbf{Y})) = E_B(\mathbf{X} \cup \mathbf{Y})$.

Relabeling $[T_A]$ using $E_A(\mathbf{X} \cup \mathbf{Y})$ ($[T_B]$ is relabeling in the symmetrical way)

Step 4: A computes $E_B(L_\ell) = E_B \left(\sum_{i=1}^{\ell} (\mathbf{X} \cup \mathbf{Y})[i] \right)$ for all $\ell \in [T_A]$.
Step 5: A sends all $E_B(L_\ell + r_\ell)$ to B choosing r_ℓ uniformly at random from $\mathbb{N}$.
Step 6: B decrypts all $L_\ell + r_\ell$ and sends them back to A.
Step 7: A recreates $L_\ell \in \{1, \ldots, n\}$ for all $\ell \in [T_A]$ by subtracting r_ℓ.

Next, we define our protocol's security based on a model where we assume that both parties are *semi-honest*, i.e., corrupted parties merely cooperate to gather information out of the protocol, but do not deviate from the protocol specification. The security is defined as follows.

Definition 1 (Semi-honest security [14]). *A protocol is secure against semi-honest adversaries if each party's observation of the protocol can be simulated using only the input they hold and the output that they receive from the protocol.*

Intuitively, this definition tells us that a corrupted party is unable to learn any extra information that cannot be derived from the input and output explicitly (For details, see [14]). Under this assumption, since the algorithm is symmetric with respect to $\mathcal{A}$ and $\mathcal{B}$, the following theorem proves our algorithm's security against semi-honest adversaries.

Theorem 3. Let $[T_{\mathcal{A}}]$ be the set of labels appearing in $T_{\mathcal{A}}$. The only knowledge that a semi-honest $\mathcal{A}$ can gain by executing Algorithm 1 is the distribution of the labels $\{L_\ell \mid \ell \in [T_{\mathcal{A}}]\}$ over $[1, \ldots, n]$.

Proof. First, the preprocessing phase gives $\mathcal{A}$ no new information, since it is conducted offline. Second, the dictionary sharing phase does not provide any new knowledge either, since all the information that $\mathcal{A}$ receives from $\mathcal{B}$ is encrypted using $\mathcal{B}$'s public key. Third, when $\mathcal{A}$ is relabeling $[T_{\mathcal{A}}]$, they only receive L_ℓ for $\ell \in [T_{\mathcal{A}}]$. Finally, when $\mathcal{B}$ is relabeling $[T_{\mathcal{B}}]$, $\mathcal{A}$ knows $L_\ell + r_\ell$ for $\ell \in [T_{\mathcal{B}}]$, but the r_ℓ are secret random numbers that $\mathcal{B}$ has generated uniformly at random, and $\mathcal{A}$ cannot know their values. Hence, the $L_\ell + r_\ell$ are distributed uniformly at random from $\mathcal{A}$'s perspective.   □

Although $\mathcal{A}$ can guess n as being either $\max\{L_\ell \mid \ell \in [T_{\mathcal{A}}]\}$ or a value just above this, and can obtain knowledge about $\mathcal{B}$'s labels by investigating $\{1, \ldots, n\} \setminus \{L_\ell \mid \ell \in [T_{\mathcal{A}}]\}$, since we have assumed that the hash function is (probabilistically) one-way, this does not give $\mathcal{A}$ any knowledge about $\mathcal{B}$'s text.

Theorem 4. Algorithm 1 assigns consistent labels using the injection: $[T_{\mathcal{A}}] \cup [T_{\mathcal{B}}] \to \{1, 2, \ldots, n\}$ without revealing the parties' private information. Its round and communication complexities are $O(1)$ and $O(\alpha(n \lg n + m + rn))$, respectively, where $n = |[T_{\mathcal{A}}] \cup [T_{\mathcal{B}}]|$, m is the modulus of the rolling hash used for preprocessing, $r = \max\{r_1, \ldots, r_n\}$ is the security parameter, and α is the cost of executing a single encryption, decryption, or homomorphic operation.

Proof. A two-level HE scheme allows sufficient number of additions and a single multiplication of encrypted integers. Thus, we can securely represent the set $[T_{\mathcal{A}}] \cup [T_{\mathcal{B}}]$ by $E_{\mathcal{A}}(\mathbf{X} \cup \mathbf{Y}) = (E_{\mathcal{A}}(\mathbf{X}) \oplus E_{\mathcal{A}}(\mathbf{Y})) \oplus (E_{\mathcal{A}}(\mathbf{X}) \cdot E_{\mathcal{A}}(\mathbf{Y}))$, and assign consistent labels $L_\ell = \mathbf{rank}_1(\ell, \mathbf{X} \cup \mathbf{Y})$ for all $\ell \in [T_{\mathcal{A}}] \cup [T_{\mathcal{B}}]$. Thus, the parties can securely obtain consistent labels, with the security level depending on the encryption strength. Regarding the communication complexity, the plaintexts sent have size of m bits (Step2) and $n \lg n$ bits (Step5), from which we can immediately derive the complexity using the other parameters. The round complexity is evident.   □

Table 2. Execution time (seconds) comparison for Phase 1, showing the preprocessing and relabeling time per label for the number n of characteristic substrings to be relabeled. Here, "Preprocessing" denotes the time required to construct the shared dictionary and "Relabeling (per label)" denotes the time needed to change a single label using the dictionary.

	n	sEDM [19]	Ours
Preprocessing	100	9.772	3.147
	1000	76.996	31.150
	10000	725.463	304.314
	100000	7264.354	3030.031
Relabeling (per label)	100	13.977	0.010
	1000	160.995	0.047
	10000	1066.259	0.319
	100000	NA (>10000)	2.124

Table 3. Execution time (seconds) of approximated EDM computation for Escherichia coli (100MB). Here, n is the number of characteristic substrings used for EDM and the same rolling hash in Table 2 is used for each n. L_1-distance is computed by the sEDM [19].

Detail of EDM computation	n	time
Relabeling by our algorithm (Phase 1)	100	4.055
	1000	63.597
	10000	2506.431
L_1-distance computation by sEDM [19] (Phase 2)	100	4.097
	1000	4.135
	10000	4.689

4 Experimental Results

Finally, we compared the practical performance of our algorithm with that of sEDM [19]. Both algorithms were implemented in C++ based on the two-level HE [2] and library available from GitHub[4], and compiled using Apple LLVM version 8.0.0 (clang-800.0.42.1) under MacOS Mojave 10.14.5. The algorithm's performance was evaluated on a system with a 2.7 GHz Intel Core i5 CPU and 8 GB 1867 MHz DDR3 RAM.

Table 2 shows the results for generating $n \in \{100, 1000, 10000, 100000\}$ different labels shared by the parties. The key length of encryption is fixed to 256 bits. Our algorithm uses the rolling hash modulo $p \in \{1031, 10313, 103123, 1031347\}$ for each n, respectively. This shows the running times for each n, where "Preprocessing" gives the time t_1 required to construct the shared dictionary and

---

[4] https://github.com/herumi/mcl.

"Relabeling (per label)" gives the response time t_2 needed to change a single label. Thus, the total time for each algorithm is $t_1 + nt_2$ for each n. Note that the total time is mainly occupied by the relabeling time for both algorithms. Therefore, these results confirm that our algorithm's computation was significantly lower than that of sEDM in all cases.

Table 3 shows the total time of approximate EDM computation for real DNA sequence available from Pizza&Chili Corpus[5]. From this corpus, we use Escherichia coli, known as *highly repetitive string* where 110MB original string is compressed to 5 MB by 7-zip[6]. This means that the number of characteristic substrings is relatively smaller, so the restriction of the examined n up to 10000 is reasonable. However, in reality, we cannot execute the sEDM [19] even for these repetitive strings due to the cost of relabeling shown in Table 2. By the results in Tables 2 and 3, we confirm the efficiency of our algorithm for large-scale data.

5   Conclusion

In this paper, we have presented an improvement to a previously proposed HE-based secure two-party protocol for computing approximate EDM. The problem we tackled is reduced to jointly assigning minimum consistent labels from $X \cup Y \subseteq \{1, 2, \ldots, m\}$ to $\{1, 2, \ldots, n\}$. The fact that recent two-level HE systems allow sufficient number of additions and a single multiplication over ciphertexts enabled us to significantly improve the execution time. From a cryptographic point of view, m should be sufficiently large (i.e., we assume that $X \cup Y$ is sparse) so that it is difficult for Alice to learn about Bob's labels from the distribution of ones in his label vector. In contrast, m should be smaller for saving the communication cost. We plan to investigate this problem further in future work.

To the best of our knowledge, existing two-party protocols for consistent labeling need both additive and multiplicative homomorphic operations over the ciphertexts. Since an HE system that only involves additive operations is computationally less taxing, whether or not we can solve the relabeling problem by only exploiting additive homomorphism is an important practical question.

Acknowledgments. This work was supported by JST CREST (JPMJCR1402), KAKENHI (16K16009, 17H01791, 17H00762 and 18K18111) and Fujitsu Laboratories Ltd. The authors thank anonymous reviewers for their helpful comments.

---

[5] http://pizzachili.dcc.uchile.cl.

[6] https://www.7-zip.org/.

References

1. Akgün, M., Bayrak, A.O., Ozer, B., Sağiroğlu, M.S.: Privacy preserving processing of genomic data: a survey. J. Biomed. Inform. **56**, 103–111 (2015)
2. Attrapadung, N., Hanaoka, G., Mitsunari, S., Sakai, Y., Shimizu, K., Teruya, T.: Efficient two-level homomorphic encryption in prime-order bilinear groups and a fast implementation in webassembly. In: ASIACCS, pp. 685–697 (2018)
3. Blanton, M., Aguiar, E.: Private and oblivious set and multiset operations. In: ASIACCS, pp. 40–41 (2012)
4. Bloom, B.H.: Space/time trade-offs in hash coding with allowable errors. Commun. ACM **13**(7), 422–426 (1970)
5. Boneh, D., Goh, E.-J., Nissim, K.: Evaluating 2-DNF formulas on ciphertexts. In: Kilian, J. (ed.) TCC 2005. LNCS, vol. 3378, pp. 325–341. Springer, Heidelberg (2005). https://doi.org/10.1007/978-3-540-30576-7_18
6. Brakerski, Z., Gentry, C., Vaikuntanathan, V.: (Leveled) fully homomorphic encryption without bootstrapping. In: ITCS, pp. 309–325 (2012)
7. Catalano, D., Fiore, D.: Using linearly-homomorphic encryption to evaluate degree-2 functions on encrypted data. In: CCS, pp. 1518–1529 (2015)
8. Cormode, G., Muthukrishnan, S.: The string edit distance matching problem with moves. ACM Trans. Algor. **3**(1), 1–19 (2007). Article 2
9. Davidson, A., Cid, C.: An efficient toolkit for computing private set operations. In: Pieprzyk, J., Suriadi, S. (eds.) ACISP 2017. LNCS, Part II, vol. 10343, pp. 261–278. Springer, Cham (2017). https://doi.org/10.1007/978-3-319-59870-3_15
10. Fischer, J., I, T., Köppl, D.: Deterministic sparse suffix sorting on rewritable texts. In: Kranakis, E., Navarro, G., Chávez, E. (eds.) LATIN 2016. LNCS, vol. 9644, pp. 483–496. Springer, Heidelberg (2016). https://doi.org/10.1007/978-3-662-49529-2_36
11. Freeman, D.M.: Converting pairing-based cryptosystems from composite-order groups to prime-order groups. In: Gilbert, H. (ed.) EUROCRYPT 2010. LNCS, vol. 6110, pp. 44–61. Springer, Heidelberg (2010). https://doi.org/10.1007/978-3-642-13190-5_3
12. Ganczorz, M., Gawrychowski, P., Jez, A., Kociumaka, T.: Edit distance with block operations. In: ESA, pp. 33:1–33:14 (2018)
13. Gentry, C.: Fully homomorphic encryption using ideal lattices. In: STOC, pp. 169–178 (2009)
14. Goldreich, O.: Foundations of Cryptography, vol. II. Cambridge University Press, New York (2004)
15. Herold, G., Hesse, J., Hofheinz, D., Ràfols, C., Rupp, A.: Polynomial spaces: a new framework for composite-to-prime-order transformations. In: Garay, J.A., Gennaro, R. (eds.) CRYPTO 2014, Part I. LNCS, vol. 8616, pp. 261–279. Springer, Heidelberg (2014). https://doi.org/10.1007/978-3-662-44371-2_15
16. Inan, A., Kaya, S., Saygin, Y., Savas, E., Hintoglu, A., Levi, A.: Privacy preserving clustering on horizontally partitioned data. Data Knowl. Eng. **63**(3), 646–666 (2007)
17. Karp, R.M., Rabin, M.O.: Efficient randomized pattern-matching algorithms. IBM J. Res. Dev. **31**(2), 249–260 (1987)
18. Kissner, L., Song, D.: Privacy-preserving set operations. In: Shoup, V. (ed.) CRYPTO 2005. LNCS, vol. 3621, pp. 241–257. Springer, Heidelberg (2005). https://doi.org/10.1007/11535218_15

19. Nakagawa, S., Sakamoto, T., Takabatake, Y., I, T., Shin, K., Sakamoto, H.: Privacy-preserving string edit distance with moves. In: Marchand-Maillet, S., Silva, Y.N., Chávez, E. (eds.) SISAP 2018. LNCS, vol. 11223, pp. 226–240. Springer, Cham (2018). https://doi.org/10.1007/978-3-030-02224-2_18
20. Paillier, P.: Public-key cryptosystems based on composite degree residuosity classes. In: Stern, J. (ed.) EUROCRYPT 1999. LNCS, vol. 1592, pp. 223–238. Springer, Heidelberg (1999). https://doi.org/10.1007/3-540-48910-X_16
21. Rane, S., Sun, W.: Privacy preserving string comparisons based on Levenshtein distance. In: WIFS, pp. 1–6 (2010)
22. Shapira, D., Storer, J.A.: Edit distance with move operations. J. Discrete Algorithms **5**(2), 380–392 (2007)

Short Papers

Parameterized Algorithms for the Happy Set Problem

Yuichi Asahiro[1], Hiroshi Eto[2], Tesshu Hanaka[3], Guohui Lin[4], Eiji Miyano[5(✉)],
and Ippei Terabaru[5]

[1] Kyushu Sangyo University, Fukuoka, Japan
asahiro@is.kyusan-u.ac.jp
[2] Kyushu University, Fukuoka, Japan
h-eto@econ.kyushu-u.ac.jp
[3] Chuo University, Tokyo, Japan
hanaka.91t@g.chuo-u.ac.jp
[4] University of Alberta, Edmonton, Canada
guohui@ualberta.ca
[5] Kyushu Institute of Technology, Iizuka, Japan
miyano@ces.kyutech.ac.jp, terabaru.ippei704@mail.kyutech.jp

Abstract. In this paper we introduce the MAXIMUM HAPPY SET problem (MaxHS) and study its parameterized complexity: For an undirected graph $G = (V, E)$ and a subset $S \subseteq V$ of vertices, a vertex v is *happy* if v and all its neighbors are in S; and otherwise *unhappy*. Given an undirected graph $G = (V, E)$ and an integer k, the goal of MaxHS is to find a subset $S \subseteq V$ of k vertices such that the number of happy vertices is maximized. In this paper we first show that MaxHS is W[1]-hard when parameterized by k. Then, we prove the fixed-parameter tractability of MaxHS when parameterized by the tree-width, the clique-width and k, the neighborhood diversity, or the twin-cover number.

1 Introduction

A social network is represented by a graph, where a vertex corresponds to a person in the network, and an edge between two vertices denotes that corresponding persons are connected with the network. In [15], Easley and Kleinberg mentioned that one of the most basic laws governing the structure of social networks is *homophily*, that is, the principle that in social networks people are more likely to connect with people sharing similar interests with them. Motivated by a study of algorithmic aspects of the homophily laws in social networks, Zhang and Li [25] introduced an optimization problem in terms of *graph coloring*, called the MAXIMUM HAPPY VERTICES problem (MaxHV): We are given an unweighted, undirected graph $G = (V, E)$, a color set $C = \{1, 2, \ldots, \ell\}$, and a partial vertex coloring function $c : V \rightarrow C$. That is, c assigns colors only to a part of the vertices in V. A vertex is *happy* if it shares the same color with all its neighbors. The goal of MaxHV is to color all the uncolored vertices such that the number of happy vertices is maximized.

M. S. Rahman et al. (Eds.): WALCOM 2020, LNCS 12049, pp. 323–328, 2020.
https://doi.org/10.1007/978-3-030-39881-1_27

1.1 Our Problem and Contributions

In this paper we introduce a new variant of MaxHV; we express the happy problem motivated by the homophily laws as a "vertex-subset" problem like the MAXIMUM INDEPENDENT SET and the MINIMUM DOMINATING SET problems [18]: For an undirected graph $G = (V, E)$ and a subset $S \subseteq V$ of vertices, a vertex v is *happy* if v and all its neighbors are in S. Our new problem, called the MAXIMUM HAPPY SET problem (MaxHS), is formally defined as follows: Given an undirected graph $G = (V, E)$ and an integer k, the goal of MaxHS is to find a subset $S \subseteq V$ of k vertices such that the number of happy vertices is maximized.

In this paper we mainly focus on the parameterized complexity and the fixed-parameter tractability (FPT) for MaxHS on several graph parameters such as the tree-width, the clique-width, the neighborhood diversity, and the twin-cover number of an input graph G. Let n and m be the number of vertices and the number of edges of an input graph, respectively. In this paper we show the following results (but all the proofs are omitted, due to page limitation):

1. The problem MaxHS is W[1]-hard with respect to k even if an input graph is restricted to split graphs. Furthermore, MaxHS on split graphs cannot be solved in time $f(k)n^{o(\sqrt{k})}$ for any computable function f unless the *Exponential-Time Hypothesis* (ETH) fails.
2. Given a tree decomposition of width tw for an input graph, MaxHS can be solved in time $O(3^{\text{tw}} \cdot \text{tw} \cdot k^2 \cdot n)$.
3. Suppose that the clique-width of an input graph is cw. Given the cw-expression tree of the graph, MaxHS can be solved in time $O(8^{\text{cw}} \cdot k^{3(\text{cw}+1)} \cdot n)$, i.e., MaxHS is FPT when parameterized by the clique-width and k; on the other hand, MaxHS is in XP when parameterized only by the clique-width.
4. The problem MaxHS can be solved in time $O(2^{\text{nd}} \cdot \text{nd}^3 + n + m)$ for an input graph of neighborhood diversity nd.
5. The problem MaxHS can be solved in time $O(3^{\text{tc}}(\text{tc} + k^2)n + m)$ for an input graph of twin-cover number tc.

1.2 Related Work

In [25], Zhang and Li also introduced an "edge-variant" of MaxHV as one of the vertex coloring problems, called the MAXIMUM HAPPY EDGES problem (MaxHE). In a vertex-colored graph, an edge is *happy* if its two endpoints have the same color. Then, the goal of MaxHE is to color all the uncolored vertices in a partially vertex-colored input graph such that the number of happy edges is maximized. By straightforwardly following this formulation, we can define an edge-variant of our new problem MaxHS: Given an (uncolored) graph $G = (V, E)$ and an integer k, the goal of MAXIMUM EDGE HAPPY SET problem (MaxEHS) is to find a subset $S \subseteq V$ of k vertices such that the number of happy edges is maximized. Actually, however, MaxEHS is identical to the DENSEST k-SUBGRAPH problem (DkS), which is defined as a problem of finding a subgraph of the given graph with exactly k vertices such that the number of edges in the subgraph is maximized. The problem DkS is well known in the literature under various names, such as the k-CLUSTER problem [12], the HEAVIEST UNWEIGHTED SUB-GRAPH problem [20], and the k-CARDINALITY SUBGRAPH problem [9]. The problem MaxEHS is generally NP-hard since it is a generalization of the MAXIMUM CLIQUE problem [18]. Moreover, it is NP-complete even to decide if there exists a solution with at least $k^{1+\varepsilon}$ happy edges for any positive constant ε [4]. In [16] it was shown that MaxEHS is NP-hard for graphs whose maximum degree is equal to three. The problem MaxEHS is NP-hard even for very restricted classes of graphs, such as bipartite and chordal graphs [12], or planar graphs [19]. Fortunately, however, MaxEHS is solvable in polynomial time on graphs whose maximum degree is equal to two, cographs, split graphs, and k-trees [12]. As for the parameterized complexity, Cai proved [10] that MaxEHS is not FPT, i.e., it is W[1]-hard, with respect to k even for regular graphs. This result implies that MaxEHS is W[1]-hard with respect to the size of the solution since any solution cannot contain more than $k(k-1)/2$ edges. Bourgeois, Giannakos, Lucarelli, Milis, and Paschos proposed [7] two FPT algorithms; one algorithm is parameterized by the tree-width $\mathtt{tw}(G)$ of the input graph G and uses exponential space and the other is parameterized by the size $\mathtt{vc}(G)$ of the minimum vertex cover and uses polynomial space. In [8], Broersma, Golovach, and Patel proved that for an n-vertex graph G, MaxEHS can be solved in time $k^{O(\mathtt{cw}(G))} \times n$, but it cannot be solved in time $2^{o(\mathtt{cw}(G)\log k)} \times n^{O(1)}$ unless the ETH fails. As surveyed previous results above, there are a huge number of previous results for MaxEHS. To the best of our knowledge, however, there are no previously known results for the vertex-variant MaxHS.

Recently, the original, coloring variants MaxHV and MaxHE have attracted growing attention and thus there is a large literature [1–3,11,22,24–26]: First of all, one sees that if an input graph G is an uncolored graph, then MaxHV and MaxHE become trivial, i.e., it must be the optimal solution to assign an unique color to all the vertices in $V(G)$. Zhang and Li [25] proved that MaxHV and MaxHE can be solved in polynomial time if the color number ℓ is (at most) two [25]. Unfortunately, however, MaxHV and MaxHE are NP-hard for general graphs if $\ell \geq 3$ [25]. Misra and Reddy [23] proved that MaxHV and MaxHE remain NP-hard even if an input graph is restricted to split graphs or bipartite graphs. On the other hand, Aravind, Kalyanasundaram, and Kare [2] showed that MaxHV and MaxHE can be solved in $O(nk\log k)$ and in $O(nk)$ time, respectively, if an input graph is a tree. Also, a lot of results on the parameterized complexity and FPT algorithms of MaxHV and MaxHE were shown in [1–3,5,11,23]. For example, MaxHV is FPT when parameterized by the solution size, by the tree-width $\mathtt{tw}(G)$ and the number ℓ of colors, or the neighborhood diversity $\mathtt{nd}$; but, MaxHV is W[1]-hard with respect to only $\mathtt{tw}(G)$. On the other hand, the complexity parameterized by the clique-width $\mathtt{cw}(G)$ is still unknown. As for the approximability, Zhang et al. [26] showed that MaxHV can be approximated within $\frac{1}{\Delta+1}$, where Δ is the maximum degree of the input graph, and MaxHE can be approximated within $\frac{1}{2} + \frac{\sqrt{2}}{4}f(k)$, where $f(k) = \frac{(1-1/k)\sqrt{k(k-1)}+1/\sqrt{2}}{k-1+1/2k} \leq 1$.

2   Preliminaries

In this paper, we consider several graph parameters including *tree-width, clique-width, neighborhood diversity, twin-cover number,* and *vertex cover number* [13, 14, 17]. Let $\mathtt{tw}(G)$, $\mathtt{cw}(G)$, $\mathtt{nd}(G)$, $\mathtt{tc}(G)$ and $\mathtt{vc}(G)$ be the tree-width, the clique-width, the neighborhood diversity, the twin-cover number, and the vertex cover number of an input graph G, respectively. For simplicity, however, we omit "(G)" of $\mathtt{cw}(G)$, $\mathtt{tw}(G)$, $\mathtt{nd}(G)$, $\mathtt{tc}(G)$, and $\mathtt{vc}(G)$ since the graph G is clear in the following.

For those parameters, the following relations are known [6, 13, 17, 21]:

Proposition 1 ([6,13,17,21]). *Let* $\mathtt{tw}$, $\mathtt{cw}$, $\mathtt{nd}$, $\mathtt{tc}$, $\mathtt{vc}$ *be the tree-width, the clique-width, the neighborhood diversity, the twin-cover number, and the vertex cover number of a graph* G, *respectively. Then the following inequalities hold: (i)* $\mathtt{cw} \leq 2^{\mathtt{tw}+1} + 1$; *(ii)* $\mathtt{tw} \leq \mathtt{vc}$; *(iii)* $\mathtt{cw} \leq \mathtt{nd} + 1$; *(iv)* $\mathtt{nd} \leq 2^{\mathtt{vc}} + \mathtt{vc}$; *(v)* $\mathtt{cw} \leq 2^{\mathtt{tc}} + \mathtt{tc}$; *and (vi)* $\mathtt{tc} \leq \mathtt{vc}$.

3   Parameterized Complexity of MaxHS

In this section we study the parameterized complexity and FPT algorithms for the vertex variant MaxHS when parameterized by the solution size, tree-width, clique-width, neighborhood diversity, and twin-cover number.

(Solution size). Recall that if an input graph is restricted to split graphs, then the edge variant MaxEHS can be solved in polynomial time as shown in [12]. On the other hand, we can show that MaxHS on split graphs is more difficult even when an input integer k is a small constant:

Theorem 1. *The problem* MaxHS *is* W[1]-*hard with respect to* k *even for split graphs. Furthermore,* MaxHS *on split graphs cannot be solved in time* $f(k)n^{o(\sqrt{k})}$ *for any computable function* f *unless the ETH fails.*

(Tree-width). One sees that the edge variant MaxEHS is very trivial on trees since the optimal solution-size of MaxEHS must be $k-1$ and thus any connected subtree of k vertices is an optimal solution. In contrast, MaxHS is not so trivial, but, we can design a polynomial-time algorithm for trees and also an FPT algorithm parameterized by the tree-width:

Theorem 2. *Given a tree decomposition of width* $\mathtt{tw}$ *for an input graph,* MaxHS *can be solved in time* $O(3^{\mathtt{tw}} \cdot \mathtt{tw} \cdot k^2 \cdot n)$.

(Clique-width). We design a parameterized algorithm for MaxHS by the clique-width. Our algorithm runs in time $O(8^{\mathtt{cw}} \cdot k^{O(\mathtt{cw})} \cdot n)$ when a $\mathtt{cw}$-expression tree of an input graph G of clique-width $\mathtt{cw}$ is given. This means that MaxHS is FPT when parameterized by the clique-width $\mathtt{cw}$ and an input integer k; on the other hand MaxHS is in XP when parameterized only by the clique-width $\mathtt{cw}$.

Theorem 3. *Given a cw-expression tree of an input graph of clique-width cw, MaxHS can be solved in time* $O(8^{cw} \cdot k^{3(cw+1)} \cdot n)$.

(Neighborhood diversity). We consider the neighborhood diversity as a graph parameter in the following:

Theorem 4. *The problem MaxHS can be solved in time* $O(2^{nd} \cdot nd^3 + n + m)$ *for an input graph of neighborhood diversity* nd.

(Twin-cover number). We can design the following FPT algorithm:

Theorem 5. *The problem MaxHS can be solved in time* $O(3^{tc}(tc + k^2)n + m)$ *for an input graph of twin-cover number* tc.

Acknowledgments. This work was partially supported by the Natural Sciences and Engineering Research Council of Canada, the Grants-in-Aid for Scientific Research of Japan (KAKENHI) Grant Numbers JP17K00016 and JP17K00024, JP19K21537, and JST CREST JPMJR1402.

References

1. Agrawal, A.: On the parameterized complexity of happy vertex coloring. IWOCA **2017**, 103–115 (2017)
2. Aravind, N., Kalyanasundaram, S., Kare, A.: Linear time algorithms for happy vertex coloring problems for trees. IWOCA **2016**, 281–292 (2016)
3. Aravind, N., Kalyanasundaram, S., Kare, A., Lauri, J.: Algorithms and hardness results for happy coloring problems. arXiv preprint arXiv:1705.08282 (2017)
4. Asahiro, Y., Hassin, R., Iwama, K.: Complexity of finding dense subgraphs. Discrete Appl. Math. **121**, 15–26 (2002)
5. Bliznets, I., Sagunov, S.: On happy colorings, cuts, and structual parameterizations. arXiv preprint arXiv:1907.06172 (2019)
6. Bodlaender, H.L., Gilbert, J.R., Hafsteinsson, H., Kloks, T.: Approximating treewidth, pathwidth, frontsize, and shortest elimination tree. J. Algorithms **18**(2), 238–255 (1995)
7. Bourgeois, N., Giannakos, A., Lucarelli, G., Milis, I., Paschos, V.: Exact and approximation algorithms for densest k-subgraph. WALCOM **2013**, 114–125 (2013)
8. Broersma, H., Golovach, P., Patel, V.: Tight complexity bounds for FPT subgraph problems parameterized by the clique-width. Theor. Comput. Sci. **485**, 69–84 (2013)
9. Bruglieri, M., Ehrgott, M., Hamacher, H., Maffioli, F.: An annotated bibliography of combinatorial optimization problems with fixed cardinality constraints. Discrete Appl. Math. **154**(9), 1344–1357 (2006)
10. Cai, L.: Parameterized complexity of cardinality constrained optimization problems. Comput. J. **51**, 102–121 (2007)
11. Choudhari, J., Reddy, I.: On structual parameterizations of happy coloring, empire coloring and boxicity. WALCOM **2018**, 228–239 (2018)
12. Corneil, D., Perl, Y.: Clustering and domination in perfect graphs. Discrete Appl. Math. **9**(1), 27–39 (1984)

13. Courcelle, B., Olariu, S.: Upper bounds to the clique width of graphs. Discrete Appl. Math. **101**(1), 77–114 (2000)
14. Cygan, M., et al.: Parameterized Algorithms. Springer, Cham (2015). https://doi.org/10.1007/978-3-319-21275-3
15. Easley, D., Kleinberg, J.: Networks Crowds and Markets: Reasoning about a Highly Connected World. Cambridge University Press, Cambridge (2010)
16. Feige, U., Seltser, M.: On the densest k-subgraph problem. Technical report CS97-16, Weizmann Institute, Rehovot (1997).http://www.wisdom.weizmann.ac.il
17. Ganian, R.: Improving vertex cover as a graph parameter. Discrete Math. Theor. Comput. Sci. **17**(2), 77–100 (2015)
18. Garey, M., Johnson, D.: Computers and Intractability: A Guide to the Theory of NP-Completeness. W. H. Freeman & Co., New York (1979)
19. Keil, J., Brecht, T.: The complexity of clustering in planar graphs. J. Comb. Math. Comb. Comput. **9**, 155–159 (1991)
20. Kortsarz, G., Peleg, D.: On choosing a dense subgraph. FOCS **1993**, 692–701 (1993)
21. Lampis, M.: Algorithmic meta-theorems for restrictions of treewidth. Algorithmica **64**(1), 19–37 (2012)
22. Lewis, R., Thiruvady, D., Morgan, K.: Finding happiness: an analysis of the maximum happy vertices problem. Comput. Oper. Res. **103**, 265–276 (2019)
23. Misra, N., Reddy, I.: The parameterized complexity of happy colorings. IWOCA **2017**, 142–153 (2017)
24. Zhang, P., Jiang, T., Li, A.: Improved approximation algorithms for the maximum happy vertices and edges problems. COCOON **2015**, 159–170 (2015)
25. Zhang, P., Li, A.: Algorithmic aspects of homophyly of networks. Theor. Comput. Sci. **593**, 117–131 (2015)
26. Zhang, P., Xu, Y., Jiang, T., Li, A., Lin, G., Miyano, E.: Improved approximation algorithms for the maximum happy vertices and edges problems. Algorithmica **80**(5), 1412–1438 (2018)

An Experimental Study of a 1-Planarity Testing and Embedding Algorithm

Carla Binucci$^{(\boxtimes)}$ ⓘ, Walter Didimo ⓘ, and Fabrizio Montecchiani ⓘ

Università degli Studi di Perugia, Perugia, Italy
{carla.binucci,walter.didimo,fabrizio.montecchiani}@unipg.it

Abstract. A graph is 1-planar if it can be drawn in the plane with at most one crossing per edge. The 1-planarity testing problem is NP-complete, even for restricted classes of graphs. We present the first general 1-planarity testing and embedding algorithm, and we experimentally investigate its feasibility in practice. The results suggest that our approach can be successfully applied to graphs with up to 30 vertices, while more sophisticated techniques are needed to attack larger graphs.

1 Introduction

One of the most studied families of sparse nonplanar graphs, whose definition naturally extends that of planar graphs, is the family of 1-*planar graphs*, i.e., graphs that can be drawn in the plane with at most one crossing per edge. An n-vertex 1-planar graph has $O(n)$ edges, $O(\sqrt{n})$ separators and treewidth, and $O(1)$ stack and queue number. Refer to [14,18] for recent surveys on 1-planar graphs and related families. Despite these similarities with planar graphs, testing whether a graph is 1-planar is NP-complete [15,19], even for restricted classes of graphs [4,9] and for graphs with a fixed rotation system [3]. The problem is fixed-parameter tractable in the vertex-cover number, the cyclomatic number, or the tree-depth [4]. Polynomial-time testing algorithms are known only for subfamilies of 1-planar graphs (e.g., [2,7,8,17]). In contrast to this rich set of theoretical results, there is a lack of general 1-planarity testing algorithms that can be effectively implemented and adopted in applications. Our research goes in the direction of filling this gap by investigating practical approaches and by providing indications for further advances. Namely:

(i) We describe an easy to implement 1-planarity testing strategy based on a backtracking approach. If the test is positive, the algorithm returns a 1-planar embedding of the graph.

(ii) We report the results of an experimental study on two well-established real-world graph benchmarks, the ROME and NORTH graphs [1,12]. They suggest that our approach can be successfully applied to the majority of instances with up to 30 vertices, while more sophisticated techniques are needed for larger graphs.

Work partially supported by MIUR, under Grant 20174LF3T8 AHeAD: efficient Algorithms for HArnessing networked Data.

ⓒ Springer Nature Switzerland AG 2020
M. S. Rahman et al. (Eds.): WALCOM 2020, LNCS 12049, pp. 329–335, 2020.
https://doi.org/10.1007/978-3-030-39881-1_28

(*iii*) We make publicly available the solved instances (http://mozart.diei.unipg. it/montecchiani/1planarity/labels.xlsx), with a labeling that specifies whether each instance is 1-planar or not. An interesting finding is that most of the solved instances in the ROME and NORTH sets are 1-planar.

For space restrictions, several details are omitted and can be found in [5].

Preliminaries. We assume familiarity with basic concepts of graph drawing and planarity [11]. We only consider *simple* drawings, where adjacent edges do not cross and two independent edges cross at most in one of their interior points. For a graph G, we denote by $V(G)$ and $E(G)$ the sets of vertices and edges of G, respectively. A *plane* graph is a graph with a given planar embedding. If G is not planar, the *planarization* of a drawing of G is a plane graph obtained by replacing each crossing point with a *dummy vertex*. An *embedding* of G is an equivalence class of drawings of G whose planarizations yield the same planar embedding. A 1-*planar drawing* is a drawing where each edge is crossed at most once. A 1-*planar embedding* is the embedding induced by a 1-planar drawing. A 1-*plane* graph is a graph with a given 1-planar embedding. A *kite* is a 1-plane graph isomorphic to K_4, in which the outer face is bounded by a cycle composed of four vertices and four crossing-free edges, called *kite edges*, while the remaining two edges cross (see, e.g., [6,13]). As for planar graphs, a graph G is 1-planar if and only if all its subgraphs are 1-planar; also, the following property holds.

Property 1. G is 1-planar if and only if every block (biconnected component) of G is 1-planar. If G is 1-planar, a 1-planar embedding of G can be obtained in linear time by suitably merging the 1-planar embeddings of its blocks.

2   Algorithm Design and Experiments

Our algorithm, called `1PlanarTester`, takes as input a connected graph G and, based on Property 1, it processes each block of G independently. For each block C, with n_C vertices and m_C edges, it executes a quick preliminary test to check if C can be immediately labeled as 1-planar (when C is planar or has less than 7 vertices) or as not 1-planar (when $m_C > 4n_C - 8$ [20]). In the former case, `1PlanarTester` processes the next block, while in the latter case it halts and returns that G is not 1-planar. If none of these conditions applies, `1PlanarTester` runs a backtracking procedure on C; it returns either a negative answer, which means that C (and thus G) is not 1-planar, or a 1-planar embedding of C if it exists. At the end, `1PlanarTester` either outputs an embedding for each block of G or it returns a block that is not 1-planar.

Backtraking Procedure. Let C be a block of G such that $m_C \leq 4n_C - 8$ edges. We define as *candidate solution* for the 1-planarity testing problem on C a set of pairs of crossing edges. Namely, let $\overline{E}$ be the set of all (unordered) pairs of edges that can cross in some embedding of C, i.e., the pairs $\{e_1, e_2\}$ such that $e_1 \in E(C)$ and $e_2 \in E(C)$ are independent. Let $k = |\overline{E}|$ and observe that $k = O(n_C^2)$ because $m_C \leq 4n_C - 8$. Let σ be any ordering of $\overline{E}$, and let $\sigma(i)$

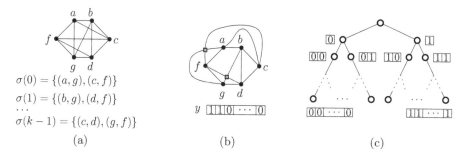

Fig. 1. (a) A block C and an ordering σ of $\overline{E}$. (b) A planarization C^* of C (dummy vertices are gray squares) with the corresponding TRUE solution. (c) The search tree T.

denote the i-th pair of edges in such an ordering. We encode a candidate solution by a binary array y of length k such that $y[i] = 0$ (resp. $y[i] = 1$) means that the two edges $\sigma(i)$ do not cross (resp. cross) in a 1-planar embedding of C (if it exists). Refer to Fig. 1(a) for an illustration. We say that y is a TRUE solution of C if: (1) each edge is crossed at most once, and (2) by replacing each crossing with a dummy vertex, the resulting graph C^* is planar (see also Fig. 1(b)). Else, we say that y is a FALSE solution. Note that Condition (2) is well defined because each edge is crossed at most once and hence we do not need to know the order of the crossings along the edges. It is easy to see that C is 1-planar if and only if the set of candidate solutions contains a TRUE solution.

The backtracking procedure generates the set of candidate solutions incrementally, by computing a binary search tree T as follows, see also Fig. 1(c). Each node ν of T is equipped with an index $i_\nu < k$ and with an array y_ν of length i_ν that represents a partial candidate solution. Node ν has two children, ν_0 and ν_1, such that $i_{\nu_0} = i_{\nu_1} = i_\nu + 1$, $y_{\nu_0}[i] = y_{\nu_1}[i] = y_\nu[i]$, for $i < i_\nu$, $y_{\nu_0}[i_\nu] = 0$, and $y_{\nu_1}[i_\nu] = 1$. We say that y_μ *extends* y_ν if μ is a descendant of ν in T. Tree T is visited top-down starting from the root. When visiting a node ν of T, the backtracking procedure runs a routine called VerifyNode(y_ν,C), which returns one of three possible values: (i) SOL, if y_ν is a TRUE solution or can be extended to a TRUE one; in this case the algorithm returns a 1-planar embedding of C; (ii) CUT, if y_ν is either a FALSE solution or cannot be extended to a TRUE one; in this case the algorithm will not visit the children of ν (i.e., the whole subtree of T rooted at ν is pruned); (iii) CNT, if none of the previous condition applies; in this case the routine adds the two children of ν to the set of nodes to be visited.

To describe how the VerifyNode(y_ν,C) routine works, we need some definitions. An edge is *crossed in* y_ν if it is in a pair $\sigma(j)$, with $j < i_\nu$, such that $y[j] = 1$. An edge e is *saturated in* y_ν if at least one of the following conditions applies: (a) e is crossed in y_ν; (b) the greatest index j such that $\sigma(j)$ contains e is smaller than i_ν; (c) every pair of edges of $\overline{E}$ that contains e is such that the other edge of the pair is crossed in y_ν.

Lemma 1. *Let e be a saturated edge in y_ν. Then e is either crossed in y_ν or it is not crossed in every array y_μ that extends y_ν.*

First of all, if VerifyNode(y_ν,C) finds an edge crossed twice in y_ν, it returns CUT. Else, it computes the graph C_ν induced by the edges saturated in y_ν, and the graph C_ν^* obtained from C_ν by replacing each crossing with a dummy vertex.

Lemma 2. *If y_μ extends y_ν and is a TRUE solution, then C_ν^* is planar.*

The routine verifies whether C_ν coincides with C. Note that this happens if $|y_\nu| = k$, or if $|y_\nu| < k$ and all edges of C are part of C_ν because are all saturated in y_ν. If $C_\nu \equiv C$, based on Lemma 2, the routine tests if C_ν^* is planar. In the positive case, it returns SOL and a 1-planar embedding of C is obtained from a planar embedding of C_ν^* by replacing the dummy vertices with crossing points. In the negative case, it returns CUT because y_ν cannot be extended to a TRUE solution. If $C_\nu \not\equiv C$, the routine computes an arbitrary extension of y_ν and verifies whether it is a TRUE solution. This corresponds to exploring a leaf of the subtree of T rooted at ν. If such a leaf is a TRUE solution, a 1-planar embedding is computed and returned. Else CNT is returned and the traversal of T continues.

Further Optimizations. To speed-up the backtracking algorithm in some cases, we consider some additional issues. An edge is a *kite edge in y_ν* if it is a kite edge with respect to a pair of edges that cross in y_ν. In the VerifyNode routine, we extend the definition of a saturated edge e by adding a fourth condition: (d) e is a kite edge in y_ν. In fact, an array y_μ that extends y_ν and in which a kite edge is crossed can be discarded by the routine, because kite edges can always be redrawn without crossings. This allows us to increase the number of edges in C_ν and hence to increase the probability that C_ν coincides with C or that C_ν^* is not planar. Further optimization criteria are described in [5].

Experimental Analysis. We implemented 1PlanarTester in the C# language and we exploit some planarity testing subroutines of the OGDF library [10] (written in C++). In our implementation, the search tree T is traversed by a depth-first search, which guarantees that the space complexity of the algorithm is $O(n^2)$ (with a breadth-first-search the space requirement may be $2^{\Omega(n^2)}$).

Our experiments have two main objectives: **G1 - Performance Analysis**: Evaluating the largest instances that our algorithm can handle in a reasonable time. For those instances that are 1-planar, we also want to compare the total number of crossings produced by 1PlanarTester with respect to a state-of-the-art planarizer that is allowed to cross an edge more than once. **G2 - Labeling of Popular Graph Benchmarks**: Labeling as 1-planar or not 1-planar the instances of well-established benchmarks in graph drawing. Our ultimate goal is to stimulate further practical research on beyond-planar graphs [14].

We ran the experiments on a set of computers, each having 16 GB of RAM, an Intel i7 CPU, and the Windows 10 operating system. We used the ROME and the NORTH graphs as benchmarks for answering both G1 and G2. We preliminary removed the planar instances from these two sets of graphs, as they are of no interest for us. Table 1 reports some information about the considered nonplanar

Table 1. Instances and Labeling.

	# INSTANCES	AVG DENSITY	AVG # BLOCKS	SOLVED Number	%	LABEL 1-PLANAR	NOT 1-PLANAR
ROME 10-20	91	1.49	5.8	83	91.2%	100.0%	0.0%
ROME 21-30	164	1.37	12.4	114	69.5%	100.0%	0.0%
ROME 31-40	388	1.31	15.7	170	43.8%	100.0%	0.0%
ROME 41-50	119	1.29	25.1	45	37.8%	100.0%	0.0%
NORTH 10-20	121	2.05	4.6	89	73.6%	88.8%	11.2%
NORTH 21-30	69	2.07	12.4	27	39.1%	77.8%	22.2%
NORTH 31-40	55	2.00	14.0	21	38.2%	57.1%	42.9%
NORTH 41-50	32	1.79	18.2	6	18.8%	83.3%	16.7%

instances grouped by size. The first columns contain the number of instances in each sample, the average density (i.e., the average ratio between number of edges and number of vertices), and the average number of blocks (recall that the algorithm processes each block independently). We halted the computations that took more than 3 h. As state-of-the-art planarizer, we used an implementation available in the OGDF library and discussed in [16].

Concerning G1, Table 1 summarizes the performance of `1PlanarTester` in terms of number of solved instances. Our algorithm could solve a high percentage of instances up to 20 vertices (91.2% of ROME and 73.6% of NORTH). For the larger instances up to 40 vertices, the NORTH graphs become more challenging, with a percentage of solved instances that is stable around 39%. The percentage of solved instances on the ROME graphs is higher, namely above 69% for the instances with 21–30 vertices, and above 43% for the instances with 31–40 vertices. The results are much worse for larger graphs. The better performance of our algorithm on the ROME graphs is partly justified by their lower density with respect to the NORTH graphs. On average, a computation took less than 4 min, while the slowest solved instance took about 115 min. The instances for which the algorithm needed to run the backtracking procedure, i.e., those for which the preliminary tests did not suffice, are mainly in the subsets ROME 21–30, ROME 31–40, and NORTH 10–20. Concerning the positive instances, we report that most frequently they were found by running the completion subroutine on y_ν and by checking that C_ν^* was planar. Also, during the backtracking procedure, the most frequent type of cuts was an edge crossed twice. About the number of crossings, both `1PlanarTester` and OGDF produced drawings with very few crossings on average. The average over all ratios between the crossings made by `1PlanarTester` and those made by the OGDF planarizer is below 1.78.

Concerning G2, Table 1 reports the percentage of solved instances divided by 1-planar and not. All solved instances in the ROME set are 1-planar, while the NORTH graphs contain a percentage of non-1-planar instances varying between 11% and 43% over the different samples. The fact that in both sets the percentage of 1-planar graphs is greater than the percentage of not 1-planar graphs may also suggest that the unsolved instances are more likely to be not 1-planar.

3   Open Problems

The main bottleneck of our algorithm is its $O(n^2)$ encoding scheme. Can we reduce this size by using a balanced separating curves approach like that in [4]? Also, more sophisticated rules can be studied to prune a subtree or to complete a partial solution during a backtracking execution. For example, can we use SPQR-trees to test different triconnected components independently?

References

1. http://www.graphdrawing.org/data.html. Accessed June 2019
2. Auer, C., et al.: Outer 1-planar graphs. Algorithmica **74**(4), 1293–1320 (2016)
3. Auer, C., Brandenburg, F.J., Gleißner, A., Reislhuber, J.: 1-planarity of graphs with a rotation system. J. Graph Algorithms Appl. **19**(1), 67–86 (2015)
4. Bannister, M.J., Cabello, S., Eppstein, D.: Parameterized complexity of 1-planarity. J. Graph Algorithms Appl. **22**(1), 23–49 (2018)
5. Binucci, C., Didimo, W., Montecchiani, F.: An experimental study of a 1-planarity testing and embedding algorithm. CoRR arXiv:1911.00573 (2019)
6. Brandenburg, F.J.: 1-visibility representations of 1-planar graphs. J. Graph Algorithms Appl. **18**(3), 421–438 (2014)
7. Brandenburg, F.J.: Recognizing optimal 1-planar graphs in linear time. Algorithmica **80**(1), 1–28 (2018)
8. Brandenburg, F.J.: Characterizing and recognizing 4-map graphs. Algorithmica **81**(5), 1818–1843 (2019)
9. Cabello, S., Mohar, B.: Adding one edge to planar graphs makes crossing number and 1-planarity hard. SIAM J. Comput. **42**(5), 1803–1829 (2013)
10. Chimani, M., Gutwenger, C., Jünger, M., Klau, G.W., Klein, K., Mutzel, P.: The open graph drawing framework (OGDF). In: Handbook of Graph Drawing and Visualization, pp. 543–569. Chapman and Hall/CRC (2013)
11. Di Battista, G., Eades, P., Tamassia, R., Tollis, I.G.: Graph Drawing: Algorithms for the Visualization of Graphs. Prentice-Hall, Upper Saddle River (1999)
12. Di Battista, G., Garg, A., Liotta, G., Tamassia, R., Tassinari, E., Vargiu, F.: An experimental comparison of four graph drawing algorithms. Comput. Geom. **7**, 303–325 (1997)
13. Di Giacomo, E., et al.: Ortho-polygon visibility representations of embedded graphs. Algorithmica **80**(8), 2345–2383 (2018)
14. Didimo, W., Liotta, G., Montecchiani, F.: A survey on graph drawing beyond planarity. ACM Comput. Surv. **52**(1), 4:1–4:37 (2019)
15. Grigoriev, A., Bodlaender, H.L.: Algorithms for graphs embeddable with few crossings per edge. Algorithmica **49**(1), 1–11 (2007)
16. Gutwenger, C., Mutzel, P.: An experimental study of crossing minimization heuristics. In: Liotta, G. (ed.) GD 2003. LNCS, vol. 2912, pp. 13–24. Springer, Heidelberg (2004). https://doi.org/10.1007/978-3-540-24595-7_2
17. Hong, S., Eades, P., Katoh, N., Liotta, G., Schweitzer, P., Suzuki, Y.: A linear-time algorithm for testing outer-1-planarity. Algorithmica **72**(4), 1033–1054 (2015)

18. Kobourov, S.G., Liotta, G., Montecchiani, F.: An annotated bibliography on 1-planarity. Comput. Sci. Rev. **25**, 49–67 (2017)
19. Korzhik, V.P., Mohar, B.: Minimal obstructions for 1-immersions and hardness of 1-planarity testing. J. Graph Theory **72**(1), 30–71 (2013)
20. Pach, J., Tóth, G.: Graphs drawn with few crossings per edge. Combinatorica **17**(3), 427–439 (1997)

Trichotomy for the Reconfiguration Problem of Integer Linear Systems

Kei Kimura[1](✉) and Akira Suzuki[2]

[1] Saitama University, Saitama-city, Saitama 338-8570, Japan
kkimura@mail.saitama-u.ac.jp
[2] Tohoku University, Aoba-ku, Sendai 980-8579, Japan
a.suzuki@ecei.tohoku.ac.jp

Abstract. In this paper, we consider the reconfiguration problem of integer linear systems. In this problem, we are given an integer linear system I and two feasible solutions s and t of I, and then asked to transform s to t by changing a value of only one variable at a time, while maintaining a feasible solution of I throughout. $Z(I)$ for I is the complexity index introduced by Kimura and Makino (Discrete Applied Mathematics 200:67–78, 2016), which is defined by the sign pattern of the input matrix. We analyze the complexity of the reconfiguration problem of integer linear systems based on the complexity index $Z(I)$ of given I. We show that the problem is (i) solvable in constant time if $Z(I)$ is less than one, (ii) weakly coNP-complete and pseudo-polynomially solvable if $Z(I)$ is exactly one, and (iii) PSPACE-complete if $Z(I)$ is greater than one. Since the complexity indices of Horn and two-variable-par-inequality integer linear systems are at most one, our results imply that the reconfiguration of these systems are in coNP and pseudo-polynomially solvable. Moreover, this is the first result that reveals coNP-completeness for a reconfiguration problem, to the best of our knowledge.

Keywords: Combinatorial reconfiguration · Integer linear systems · Complexity index

1 Introduction

In *reconfiguration problem* we are asked to transform the current configuration into a desired one by step-by-step operations. Formally, in this problem, we are given two feasible solutions of a combinatorial problem, then we find the transformation between them, such that all intermediate results are also feasible, and each step conforms to an adjacency relation defined on feasible solutions. The

The first author is partially supported by JSPS KAKENHI Grant Number JP17K12636. The second author is partially supported by JST CREST Grant Number JPMJCR1402, and JSPS KAKENHI Grant Numbers JP17K12636 and JP18H04091, Japan.

M. S. Rahman et al. (Eds.): WALCOM 2020, LNCS 12049, pp. 336–341, 2020.
https://doi.org/10.1007/978-3-030-39881-1_29

reconfiguration problem investigates the properties of solution spaces of combinatorial problems, and has a deep relationship to the optimization variants of them. After Ito et al. [9] introduced this reconfiguration framework, many researchers applied this framework to a variety of combinatorial problems, including not only graph problems such as independent set, vertex cover, and coloring, but also set cover, knapsack problem, and general integer programming problem. For recent surveys, see [7,15].

In this paper, we investigate the reconfiguration problem of integer linear system (ILS) through the complexity index for ILS introduced in [11]. In an ILS, we are given a matrix $A = (A_{ij}) \in \mathbb{Q}^{m \times n}$, a vector $\boldsymbol{b} \in \mathbb{Q}^m$, and a positive integer d, where m and n denote positive integers and $\mathbb{Q}$ denotes the set of rational numbers. We denote an ILS by $I = (A, \boldsymbol{b}, d)$. For an ILS $I = (A, \boldsymbol{b}, d)$, the complexity index $Z(I)$ of I is the optimal value of the following linear programming problem (LP) with variables $Z, \alpha_1, \ldots, \alpha_n$.

$$\text{minimize } Z$$
$$\text{subject to } \sum_{j : \text{sgn}(A_{ij}) = +} \alpha_j + \sum_{j : \text{sgn}(A_{ij}) = -} (1 - \alpha_j) \leq Z \ (i = 1, \ldots, m) \tag{1}$$
$$0 \leq \alpha_j \leq 1 \qquad (j = 1, \ldots, n),$$

where for a real number a, $\text{sgn}(a) = +, 0, -$ if $a > 0, = 0, < 0$, respectively. The complexity index extends a complexity index for SAT introduced in [3], and classifies the complexity of the feasibility problem of ILS in terms of the sign structure of the input matrix; see Table 1 for the results. We note that LP (1) depends *only* on the sign pattern of A, and captures the sign structure of ILSes. In particular, if the index is at most one, an ILS can be decomposed to Horn and two-variable-per-inequality (TVPI) ILSes in a certain way [11]. Here, a Horn ILS is an ILS where each row of the input matrix contains at most one positive element, and a TVPI ILS is an ILS where each row of the input matrix contains at most two nonzero elements. These subclasses arise in, e.g., program verification and scheduling, respectively, and many algorithms have been devised to solve the feasibility problems of these subclasses [1,5,8,14]. For $\gamma \geq 0$, we denote by ILS(γ) the family of ILSes I with $Z(I) \leq \gamma$.

In this paper, we consider the reconfiguration problem of ILS. Namely, we are given an ILS I and two feasible solutions $\boldsymbol{s}$ and $\boldsymbol{t}$ of I, and then asked to transform $\boldsymbol{s}$ to $\boldsymbol{t}$ by changing a value of only one variable at a time, while maintaining a feasible solution of I throughout. We analyze the complexity of this problem using the complexity index defined above and show the following three results: the reconfiguration problem of ILS is (i) always yes if the complexity index is less than one, (ii) weakly coNP-complete and pseudo-polynomially solvable, and (iii) PSPACE-complete if the complexity index is greater than one. Thus, we obtain a complexity trichotomy for the reconfiguration problem of ILS. See also Table 1. In Table 1, "pseudo-P" stands for "pseudo-polynomially solvable", i.e., solvable in polynomial time in the numeric value of the input. Moreover, a problem is weakly NP-complete (resp., weakly coNP-complete) if it is NP-complete (resp., coNP-complete) in the usual sense, and strongly NP-complete

Table 1. Results for integer linear systems (results of this paper are in bold). Here, n is the number of variables and d is the upper bound of the values of the variables.

ILS(γ)	Feasibility [11]	Reconfiguration	Diameter
$\gamma < 1$	P (linear time)	**P (always yes)**	$\Theta(n)$
$\gamma = 1$	weakly NP-complete pseudo-P	**weakly coNP-complete pseudo-P**	$\Theta(dn)$
$\gamma > 1$	strongly NP-complete	**PSPACE-complete**	$\Omega(d \cdot 3^{\sqrt{\frac{(\gamma-1)n}{8}}})$

if it is NP-complete even when all of its numerical parameters are bounded by a polynomial in the size of the input.

We also analyze how far two feasible solutions can be, namely, the maximum of the minimum number of value changes between any two feasible solutions. This also can be cast as the analysis of diameter of solution graph of ILS, which will be defined in next section. See Table 1 for the results.

Finally, we obtain some positive results complementing the hardness results for ILS(1). An ILS $I = (A, \boldsymbol{b}, d)$ is called *unit* if $A \in \{0, \pm 1\}^{m \times n}$ for positive integers m and n. We show that the reconfiguration problem of unit ILS(1) is solvable in polynomial time. We also show that unit ILS(γ) is PSPACE-complete for $\gamma > 1$. Therefore, we obtain a dichotomy result for unit ILS. Interestingly, the diameter of the solution graph of an ILS in unit ILS(1) is still $\Theta(dn)$ and thus the minimum number of value changes between two feasible solutions can be exponential in the input size, where we note that d is a part of the input and its input size is $\log d$.

To the best of our knowledge, our result for unit ILS(1) provides the first example that the reconfiguration problem is in P even if the diameter of the solution graph can be exponential in the input size and the reconfiguration problem is not always yes. Furthermore, we obtain the first coNP-completeness result for reconfiguration problems as far as we know. We hope that our results give a new insight to the complexity of reconfiguration problems.

The rest of the paper is organized as follows. Section 2 formally defines the reconfiguration problem of ILS. Sections 3 and 4 consider the case of $Z(I) = 1$. Section 5 presents our results for $Z(I) < 1$ and $Z(I) > 1$.

Due to the space limitation, we omit proofs of most results.

2   Preliminaries

We assume that the reader is familiar with the standard graph theoretic terminology as contained, e.g., in [2].

For an ILS $I = (A, \boldsymbol{b}, d)$, a *feasible solution* of I is an integer vector $\boldsymbol{x} \in D^n$ satisfying $A\boldsymbol{x} \geq \boldsymbol{b}$, where $D = \{0, 1, \ldots, d\}$. Note that the bounds on variables, i.e., the domain D, allow us to analyze the problem in more details, and also ensure that the solution graph defined below is finite.

For an integer n, a subset $R \subseteq D^n$ is called an *(n-ary) relation* on D.

Definition 1 (Solution graph). *For a relation $R \subseteq D^n$, we define the solution graph $G(R) = (V(R), E(R))$ as follows: $V(R) := R$ and $E(R) := \{\{x, y\} \mid x, y \in V(R), \mathrm{dist}(x, y) = 1\}$, where $\mathrm{dist}(x, y) := |\{j \mid x_j \neq y_j\}|$ is the Hamming distance of x and y. For an ILS $I = (A, b, d)$, we denote by $G(I)$ the solution graph of the set of the feasible solutions of I, that is, $G(R)$ with $R := \{x \in D^n \mid Ax \geq b\}$.*

We call a path from s to t an s-t path. Using this definition, we can treat the reconfiguration problem of ILS as following: in the reconfiguration problem of ILS, we are given an ILS I and two feasible solutions s and t of I, and then we are asked whether there exists an s-t path in $G(I)$ or not.

For an ILS, a feasible solution x^* is called a *unique minimal solution* of the ILS if it satisfies $x^* \leq x$ for all the feasible solutions x of the ILS. Here, for two vectors x and y, $x \geq y$ holds if $x_j \geq y_j$ for all j.

3   The General Case of $Z(I) = 1$

Our pseudo-polynomial solvability is based on the decomposition of any ILS in ILS(1) into Horn and TVPI ILSes introduced in [11]. In fact, we first show that the reconfiguration problems of these two ILSes are pseudo-polynomially solvable. For Horn ILS, this is done by extending the greedy algorithm for Horn SAT [6], using the fact that the set of solutions of any Horn ILS is closed under a minimum operation. On the other hand, for TVPI ILS, extending the algorithm for 2-SAT in [6] is not straightforward. We reveal that the solution sets of any TVPI ILS are closed under a median operation. Using this closedness property, we can induce a partial order on the set of solutions and devise an algorithm that changes values of variables according to this partial order. Finally, using the decomposition and the algorithms for Horn and TVPI ILSes, we obtain the following result.

Theorem 1. *The reconfiguration problem of ILS(1) is pseudo-polynomially solvable.*

Since each solution of a Horn ILS is connected to the unique minimal solution in the same component by a monotone path of length at most dn, the diameter of the solution graph of a Horn ILS is at most $2dn$. By a similar argument, we can also have that the diameter of $G(I)$ with $Z(I) = 1$ is at most $2dn$. On the other hand, we can construct a family of instances of ILS(1) with the diameter of $G(I)$ at least dn. Thus, we obtain the following.

Theorem 2. *The diameter of each component of $G(I)$ is $\Theta(dn)$ for the case where I is in ILS(1).*

For our coNP-completeness result, we use the reduction by Lagarias [12] that shows the weak NP-hardness of the feasibility problem of monotone quadratic ILS, showing the following theorem.

Theorem 3. *The reconfiguration problem of ILS(1) is weakly coNP-complete.*

4   Tractable Subclass of $Z(I) = 1$

In this section, we show that the reconfiguration problem of unit ILS(1) is solvable in polynomial time. Recall that an ILS $I = (A, \boldsymbol{b}, d)$ is called *unit* if $A \in \{0, \pm 1\}^{m \times n}$ holds for positive integers m and n. For the feasibility problem, it is known that unit ILS(1) is polynomially solvable [11]. In this subsection, we consider the reconfiguration problem of unit ILS(1). We note that unit ILS(1) includes a well-studied subclass of ILS such as unit Horn ILS (e.g., [4,16]) and unit TVPI (UTVPI) ILS (e.g., [10,13]).

We can show that the reconfiguration problem of unit Horn and UTVPI ILSes are polynomially solvable. Using the decomposition of any instance unit ILS(1) into unit Horn and UTVPI ILSes, we obtain the following.

Theorem 4. *The reconfiguration problem of unit ILS(1) is polynomially solvable.*

For the diameter of ILSes in unit ILS(1), we can show the following result.

Theorem 5. *The diameter of each component of $G(I)$ is $\Theta(dn)$ for the case where I is in ILS(1).*

5   The Cases of $Z(I) < 1$ and $Z(I) > 1$

To show the results for the case of $Z(I) < 1$, we use the following lemma.

Lemma 1. *Let I be an ILS which has at least one feasible solution, and with $Z(I) < 1$. Let $\boldsymbol{x}^*$ be a unique minimal solution of I and $\boldsymbol{s}$ be any feasible solution of I. Then, there exists a path from $\boldsymbol{s}$ to $\boldsymbol{x}^*$ on $G(I)$. Consequently, $G(I)$ is a connected graph.*

Using Lemma 1, we show the following theorems.

Theorem 6. *The reconfiguration problem of ILS(γ) is always yes for any $\gamma < 1$.*

Theorem 7. *The diameter of $G(I)$ is $\Theta(n)$ if $Z(I) < 1$.*

For the case of $Z(I) > 1$, we show that the reconfiguration problem is PSPACE-complete even for SAT. Note that ILS can formulate SAT by representing each clause $(\bigvee_{j \in L^+} x_j \vee \bigvee_{j \in L^-} \overline{x}_j)$ as $\sum_{j \in L^+} x_j + \sum_{j \in L^-} (1 - x_j) \geq 1$ and setting $d = 1$. Through this formulation, we can also define a complexity index for SAT, and this index actually coincides with the complexity index for SAT introduced by Boros et al. [3]. Using the structural expression introduced in [6], we show the following results.

Theorem 8. *For any $\gamma > 1$, the reconfiguration problems of SAT(γ) and ILS(γ) are respectively PSPACE-complete.*

Theorem 9. *For infinitely many n, $d \geq 2$, and $\gamma > 1$, there exists an ILS I_n in ILS(γ) with n variables such that $G(I_n)$ has diameter $2(d+2) \cdot 3^{\sqrt{\frac{(\gamma-1)n}{8}} - 1} - 2$.*

6   Conclusion

This paper investigates the complexity of the reconfiguration problem of ILS based on the complexity index introduced in [11] and obtains a complexity trichotomy. On the way of showing this result, we also reveal the complexity of the reconfiguration problems of Horn and TVPI ILSes, ones of the most studied subclasses of ILSes. We also obtain a complexity dichotomy for the reconfiguration problem of unit integer linear systems and Boolean satisfiability problem.

References

1. Bar-Yehuda, R., Rawitz, D.: Efficient algorithms for integer programs with two variables per constraint. Algorithmica **29**(4), 595–609 (2001)
2. Bondy, J.A., Murty, U.S.R.: Graph Theory. Springer, London (2008)
3. Boros, E., Crama, Y., Hammer, P.L., Saks, M.: A complexity index for satisfiability problems. SIAM J. Comput. **23**(1), 45–49 (1994)
4. Chandrasekaran, R.: Integer programming problems for which a simple rounding type algorithm works. Prog. Comb. Optim. **8**, 101–106 (1984)
5. Glover, F.: A bound escalation method for the solution of integer linear programs. Cahiers du Centre d'Etudes de Recherche Operationelle **6**(3), 131–168 (1964)
6. Gopalan, P., Kolaitis, P.G., Maneva, E., Papadimitriou, C.H.: The connectivity of Boolean satisfiability: computational and structural dichotomies. SIAM J. Comput. **38**, 2330–2355 (2009)
7. van den Heuvel, J.: The complexity of change. In: Surveys in Combinatorics 2013, London Mathematical Society Lecture Note Series, vol. 409, pp. 127–160. Cambridge University Press (2013)
8. Hochbaum, D.S., Megiddo, N., Naor, J.S., Tamir, A.: Tight bounds and 2-approximation algorithms for integer programs with two variables per inequality. Math. Program. **62**, 69–83 (1993)
9. Ito, T., Demaine, E.D., Harvey, N.J., Papadimitriou, C.H., Sideri, M., Uehara, R., Uno, Y.: On the complexity of reconfiguration problems. Theor. Comput. Sci. **412**(12–14), 1054–1065 (2011)
10. Jaffar, J., Maher, M.J., Stuckey, P.J., Yap, R.H.C.: Beyond finite domains. In: Borning, A. (ed.) PPCP 1994. LNCS, vol. 874, pp. 86–94. Springer, Heidelberg (1994). https://doi.org/10.1007/3-540-58601-6_92
11. Kimura, K., Makino, K.: Trichotomy for integer linear systems based on their sign patterns. Disc. Appl. Math. **200**, 67–78 (2016)
12. Lagarias, J.C.: The computational complexity of simultaneous diophantine approximation problems. SIAM J. Comput. **14**(1), 196–209 (1985)
13. Lahiri, S.K., Musuvathi, M.: An efficient decision procedure for UTVPI constraints. In: Proceedings of the 5th International Workshop on Frontiers of Combining Systems, pp. 168–183 (2005)
14. van Maaren, H., Dang, C.: Simplicial pivoting algorithms for a tractable class of integer programs. J. Comb. Optim. **6**(2), 133–142 (2002)
15. Nishimura, N.: Introduction to reconfiguration. Algorithms **11**(52), 1–25 (2018)
16. Subramani, K., Worthington, J.: Feasibility checking in Horn constraint systems through a reduction based approach. Theor. Comput. Sci. **576**, 1–17 (2015)

Train Scheduling: Hardness and Algorithms

Christian Scheffer[(✉)] [ID]

Department of Computer Science, TU Braunschweig, 38106 Braunschweig, Germany
`c.scheffer@tu-bs.de`

Abstract. We introduce the TRAIN SCHEDULING PROBLEM which can be described as follows: Given m trains via their tracks, i.e., curves in the plane, and the trains' lengths, we want to compute a schedule that moves collision-free and with limited speed the trains along their tracks such that the maximal travel time is minimized. We prove that there is no FPTAS for the TRAIN SCHEDULING PROBLEM unless P = NP. Furthermore, we provide near-optimal runtime algorithms extending existing schedules.

Keywords: Motion planning · Path coordination · Reparametrization

1 Introduction

In this paper, we introduce a new parallel motion planning problem, the TRAIN SCHEDULING PROBLEM defined as follows: Consider k given *trains* each one defined as a pair which is made up of a curve in the plane, called the *track* of the train and a value, called the *length* of the train. We want to compute a schedule moving collision-free and with bounded speed all trains along their tracks from their start points to their end points such that the maximal travel time called the *makespan* is minimized.

Furthermore, we consider the situation that there is a fixed schedule for m trains which has been established over several years. Such an established schedule causes that arrival and departure times of the existing trains at intermediate stops are fixed which means that we are given fixed reparametrizations of the k given trains. The problems EXTMINTIME and EXTMAXLENGTH ask for a collision-free schedule of a new train avoiding collisions with the other already existing trains and their fixed schedules. In particular, EXTMINTIME considers a fixed length of the new train and asks for its smallest possible travel time. Furthermore, EXTMAXLENGTH considers a fixed travel time of the new train and asks for its largest possible length.

1.1 Our Results

1. We prove that there is no FPTAS for the TRAIN SCHEDULING PROBLEM unless $P = NP$, see Theorem 1.
2. We provide algorithms with near-optimal runtime for both EXTMINTIME and EXTMAXLENGTH, see Theorems 2 and 3.

© Springer Nature Switzerland AG 2020
M. S. Rahman et al. (Eds.): WALCOM 2020, LNCS 12049, pp. 342–347, 2020.
https://doi.org/10.1007/978-3-030-39881-1_30

1.2   Related Work

Multi-robot coordination is one of the most famous and traditional interfaces between robotics and computational geometry. Due to the amazingly large landscape of parallel motion planning topics and corresponding results, we refer to surveys as [5–7] for detailed overviews.

In their pioneering work, Hopcroft, Schwartz, and Sharir [3] show that even the simple WAREHOUSEMAN'S PROBLEM which requires to coordinate a set of rectangles from a start configuration to a target configuration inside a rectangular box is PSPACE-hard.

O'Donnell and Lozano-Perez [8] consider a problem setting in which disks have to be moved along given tracks. They give a $\mathcal{O}(q^2 \log q)$ runtime algorithm for coordinating two robots at which only forward movements are allowed and q is the maximal number of segments on the considered curves. Akella and Hutchinson [1] consider a problem variant in which both the curves and the speed at which the disks traverse the curves are known. They showed that it is NP-complete to compute departure times for arbitrarily many disks such that a minimum-time collision-free coordination is achieved.

Reif and Sharir [9] consider *dynamic movers* problems in which a given polyhedral body B has to be moved collision-free within some 1D, 2D, or 3D space by translation and rotation from a start position to a target position amid a set of obstacles that rotate and move along known trajectories. They provide PSPACE-hardness of the 3D dynamic movement problem if the body B has to hold a velocity bound and NP-hardness if the body's velocity is unbounded. Furthermore, Reif and Sharir [9] consider *asteroid avoidance* problems as a special variant of dynamic movers problems. In particular, they require that neither the moving body B nor the obstacles may rotate and provide a near-linear time algorithm for the 1-dimensional asteroid avoidance problem in which each of the obstacles is a polyhedron traveling with fixed (possible distinct) translational velocity along a 1-dimensional line. The 1-dimensional asteroid avoidance problem assumes the obstacles and the body B moving along 1-dimensional lines where our TRAIN SCHEDULE EXTENSION PROBLEM considers all trains to move along curves in the two-dimensional plane which is an important difference. Reif and Sharir provide a polynomial time algorithm for the 2D asteroid avoidance problem if the number of the obstacles is a constant and for the three-dimensional asteroid avoidance problem a single exponential time and a polynomial space algorithm for a convex polyhedron B and arbitrary many obstacles.

Kant and Zucker [4] consider a problem that is equivalent to the TRAIN SCHEDULE EXTENSION PROBLEM arising in the context of computing shortest trajectories among a set of moving obstacles. Let w be the complexity of the parameter space. Their algorithm computes a shortest path in the whole visibility graph of the corresponding parameter space with complexity w which leads to running time of $\Omega(w^2)$ while our algorithm has a running time of $\mathcal{O}(w \log w)$.

2 Preliminaries

A train is a pair (H, L_h) where $L_h \in \mathbb{R}_{>0}$ is the *length* of the train and H is the *track* of the train which is defined as a polygonal curve $H : [0,1] \to \mathbb{R}^2$. We simultaneously denote by H, the function $H : [0,1] \to \mathbb{R}^2$ and its image $\{p \in \mathbb{R}^2 \mid$ there is a $t \in [0,1]$ with $p = H(t)\}$. The length $|T|$ of a track T : $[0,1] \to \mathbb{R}^2$ in the ambient space is defined as the total length of its segments w.r.t. the Euclidean norm. A *k-fleet* is a k-tuple of trains. Two trains (H, L_h) and (X, L_x) *collide* for the parameters λ_h and λ_x if the subcurves of H and X with midpoints $H(\lambda_h)$ and $X(\lambda_x)$ and lengths L_h and L_x are intersecting each other. A *reparametrization* of a train (H, L_h) is a continuous and piecewise linear function $\alpha : [0, M] \to [0,1]$ such that (1) $\alpha(0) = 0$, (2) there is a minimal value $\lambda \geq 0$ with $\alpha(\mu) = 1$ for all $\mu \geq \lambda$, and (3) the speed of the train is upper-bounded by 1, i.e., for each point in time $t \in [0, M]$, both left and right derivative of $H(\alpha(\cdot))$ have Euclidean length of at most 1. A schedule for a k-fleet $((T_1, L_1), \ldots, (T_k, L_k))$ is a tuple $(\alpha_1 : [0, M_1] \to [0,1], \ldots, \alpha_k : [0, M_k \to [0,1])$ such that (1) α_i is a reparametrization for the train (T_i, L_i) for all $i \in \{1, \ldots, k\}$ and (2) T_i and T_j do not collide for the parameters $\alpha_i(t)$ and $\alpha_j(t)$ for all $i \neq j \in \{1, \ldots, k\}$ and $t \geq 0$. The *makespan* of the schedule $(\alpha_1 : [0, M_1] \to [0,1], \ldots, \alpha_k : [0, M_k \to [0,1])$ is defined as the maximal $M_{\max}$ of $M_1, \ldots, M_k$. W.l.o.g., we assume that all travel times are equal to $M_{\max}$. If this is not the case, we extend α_i with $\alpha_i(t) = \alpha_i(M_i)$ for all $M_i < t < M_{\max}$. Given a k-fleet F, the TRAIN SCHEDULING PROBLEM asks for a schedule with minimal makespan.

The *parameter space* P of a k-fleet $((T_1, L_1), \ldots, (T_k, L_k))$ is defined as $P :=$ $[0, |T_1|] \times \cdots \times [0, |T_k|]$. The *forbidden* or *black* space $\mathcal{B}$ of P is the set of all parameter points $p = (\lambda_1, \ldots, \lambda_k) \in P$ such that there are two trains T_i and T_j that collide with the parameters λ_i and λ_j. The *allowed* or *white space* $\mathcal{W}$ is the closure of $P \setminus \mathcal{B}$. The *free-space diagram* is the partitioning of the parameter space into the allowed and the forbidden space.

A *path* is a curve $\pi : [0,1] \to P$ and the *length* $\ell(\pi)$ of π is the length of π w.r.t. the maximum metric, i.e., $\ell(\pi) := \int_0^1 ||\pi'(t)||_\infty \, dt$. An a-b-*path in the free space diagram* of $((T_1, L_1), \ldots, (T_k, L_k))$ is a path $\pi \subset \mathcal{W}$ between a and b. If not other stated, a path in the free space diagram is a path $\pi \subset \mathcal{W}$ connecting the points $(0, \ldots, 0)$ and $(|T_1|, \ldots, |T_k|)$.

3 Hardness of Scheduling Trains

In this section, we show that there is no hope for an FPTAS solving the TRAIN SCHEDULING PROBLEM for arbitrary many trains, TRAIN for short.

Theorem 1. *There is no FPTAS for* TRAIN *unless* $P = NP$.

In order to prove Theorem 1, we provide a polynomial time reduction of the 3-SAT PROBLEM to an $(1 + \varepsilon)$-approximation of TRAIN.

Let Φ be an arbitrarily chosen 3-SAT formula made up of m clauses $C_i :=$ $(\ell_{i,1} \vee \ell_{i,2} \vee \ell_{i,3})$ and $v_1, \ldots, v_n$ the variables appearing in Φ. We construct an

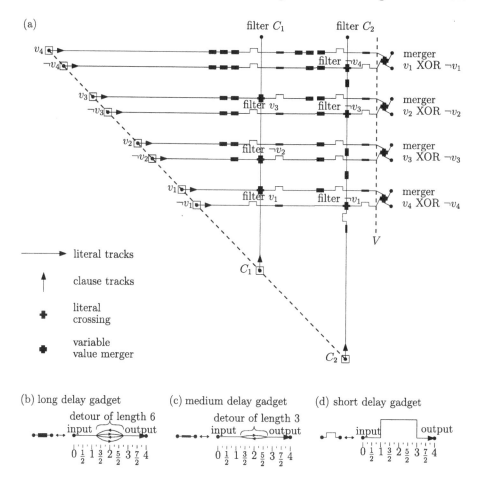

Fig. 1. (a) Reduction of Theorem 1 for a 3-SAT formula $\Phi = (\neg v_1 \vee v_2 \vee \neg v_3) \wedge (v_1 \vee v_3 \vee v_4)$. Each variable v_i is modeled by two tracks moving from left to right and representing the two literals v_i and $\neg v_i$. Each clause is modeled by a clause track moving from bottom to top. Literal crossings are crossings of a long delay gadget and a medium delay gadget, and variable value mergers are crossings of two long delay gadgets. (b) The gadgets used in the hardness construction of Theorem 1. The long delay gadget, the medium delay gadget, and the short delay gadget enforcing delays of 6, 3, and 2 by inserting corresponding detours into the addressed literal or clause track.

$(2n + m)$-fleet F such that there is schedule for F with makespan $M + 6$ if Φ is satisfiable where $M \in \Theta(poly(n + m))$, see Fig. 1. If Φ is not satisfiable, there will be at least one train (T_i, L_i) with travel time $M + 9$.

Next, we give the construction of the fleet $F := F(\Phi)$, see Fig. 1(a). We define the length of each train as 2. All tracks of F have their start points on a diagonal line ℓ with slope -1. Furthermore, we construct all tracks having the same length M. In particular, we construct the tracks in two steps. Step 1: Initially, we construct the tracks of F as segments. Step 2: We modify the segments of Step 1 by substituting subsegments by curves, called *delay gadgets*, see Fig. 1. Due to space constraints, technical details of the construction and the corresponding analysis are omitted.

4   Extending Schedules Within Near-Optimal Runtime

Let $(\alpha_1, \ldots, \alpha_k)$ be a given schedule of a given k-fleet $((T_1, L_1), \ldots, (T_k, L_k))$ and L_{k+1} an additional track. We consider the two following problem variants:

(1) Given a train length L_{k+1}, the EXTMINTIME problem asks for the smallest travel time M_{k+1} of the train (T_{k+1}, L_{k+1}) such that (T_{k+1}, L_{k+1}) does not collide with the existing k trains and their fixed schedule.
(2) Given a travel time M_{k+1}, the EXTMAXLENGTH problem asks for the maximal train length L_{k+1} such that there is a reparametrization α_{k+1} with travel time M_{k+1} such that collisions with the existing k trains and their fixed schedule are avoided.

In the remainder of this section, we give near-optimal runtime algorithms solving EXTMINTIME and EXTMAXLENGTH.

Let q and τ be the maximal numbers of segments of an input track and of an input reparametrization, respectively. We give an algorithm solving EXTMINTIME in $\mathcal{O}(\tau k q^2 \log(\tau k q))$ time. We observe that $\Omega(\tau k q^2)$ is a lower bound for the runtime of an algorithm solving EXTMINTIME or EXTMAXLENGTH implying that our runtimes are near-optimal.

Theorem 2. *There is an $\mathcal{O}\left(\tau k q^2 \log(\tau k q)\right)$ runtime algorithm for* EXTMINTIME.

Based on Theorem 2, we prove the following:

Theorem 3. EXTMAXLENGTH *can be solved in $\mathcal{O}\left(\tau k q^2 \log^2(\tau k q)\right)$ time.*

Next we give the proof for Theorem 3. In order to do this, we observe that we have to compute a maximal train length $L_{\max}$ such that there is a regular path $\pi \subset \mathcal{W}$ connecting $(0, 0)$ and $(|T_{k+1}|, M_{k+1})$. We first define *critical values* such that $L_{\max}$ has to be a critical value.

Definition 1. *We define types of critical values for* L_{k+1}:

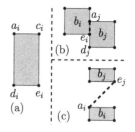

- *Type 1: Largest* L_{k+1} *with* $(0,0), (|T_{k+1}|, M_{k+1}) \in \mathcal{W}$.
- *Type 2: A maximal value for* L_{k+1} *such that* $c_i.x = a_j$ *for a pair* (b_i, b_j) *of blocks with* $e_i.y \in [a_j.y, d_j.y]$ *or* $c_i.y \in [a_j.y, d_j.y]$ *(a vertical passage closes between two neighboring blocks, i.e., that share a common y-coordinate), see Fig. 2(b).*
- *Type 3: A maximal value for* L_{k+1} *such that* $|a_i.x - e_j.x| = d_j.y - a_i.y$ *for a pair* (b_i, b_j) *of blocks (a diagonal passage closes between two blocks), see Fig. 2(c).*

Fig. 2. Critical values

While the number of Type 1 and Type 2 values are upper-bounded by 1 and $\mathcal{O}(\tau k q^2)$, there may be $\mathcal{O}(\tau^2 k^2 q^4)$ critical values of Type 3 where each value could be computed in $\mathcal{O}(1)$ time. It is obvious that $L_{\max}$ has to be a critical value. Finally, we apply Theorem 2 as a subroutine in the improved parametric search framework introduced by Cole [2] concluding the proof of Theorem 3.

References

1. Akella, S., Hutchinson, S.: Coordinating the motions of multiple robots with specified trajectories. In: Proceedings of the 2002 IEEE International Conference on Robotics and Automation, ICRA 2002, Washington, DC, USA, 11–15 May 2002, pp. 624–631 (2002)
2. Cole, R.: Slowing down sorting networks to obtain faster sorting algorithms. J. ACM **34**(1), 200–208 (1987)
3. Hopcroft, J.E., Schwartz, J.T., Sharir, M.: On the complexity of motion planning for multiple independent objects; PSPACE-hardness of the warehouseman's problem. Int. J. Rob. Res. **3**(4), 76–88 (1984)
4. Kant, K., Zucker, S.W.: Toward efficient trajectory planning: the path-velocity decomposition. Int. J. Rob. Res. **5**(3), 72–89 (1986)
5. Kavraki, L.E., LaValle, S.M.: Motion planning. In: Siciliano, B., Khatib, O. (eds.) Springer Handbook of Robotics, pp. 139–162. Springer, Cham (2016). https://doi.org/10.1007/978-3-319-32552-1_7
6. Latombe, J.C.: Robot Motion Planning. Kluwer Academic Publishers, Norwell (1991)
7. LaValle, S.M.: Planning Algorithms. Cambridge University Press, Cambridge (2006)
8. O'Donnell, P.A., Lozano-Pérez, T.: Deadlock-free and collision-free coordination of two robot manipulators. In: Proceedings of the 1989 IEEE International Conference on Robotics and Automation, Scottsdale, Arizona, USA, 14–19 May 1989, pp. 484–489 (1989)
9. Reif, J.H., Sharir, M.: Motion planning in the presence of moving obstacles. J. ACM **41**(4), 764–790 (1994)

Author Index

Printed in the United States
By Bookmasters